ECONOMICS OF AGRICULTURAL MARKETS

Ronald Schrimper

North Carolina State University

Upper Saddle River, New Jersey 07458

Library of Congress Cataloging-in-Publication Data

Schrimper, Ronald A. (Ronald Arthur)
Economics of agricultural markets / Ronald A. Schrimper.
p. cm.
Includes bibliographical references and index.
ISBN 0-13-775776-X
1. Agriculture–Economic aspects. 2. Food industry and trade. 3. Produce trade. I. Title.

HD1433 .S37 2001
380.1'41–dc21 00-033972

Publisher: Charles Stewart, Jr.
Executive Editor: Debbie Yarnell
Associate Editor: Kate Linsner
Assistant Editor: Kimberly Yehle
Production Editor: Mary Jo Graham, Carlisle Publishers Services
Production Liaison: Eileen O'Sullivan
Director of Manufacturing & Production: Bruce Johnson
Managing Editor: Mary Carnis
Manufacturing Manager: Ed O'Dougherty
Art Director: Marianne Frasco
Cover Design Coordinator: Miguel Ortiz
Cover Designer: Wanda España
Interior Design: Carlisle Communications, Ltd.
Composition: Carlisle Communications, Ltd.
Printing and Binding: R.R. Donnelley/Harrisonburg

Photo Credits: p. 186, Corbis Digital Stock; p. 123, courtesy of the Kroger Company; p. 1, David Young-Wolff/PhotoEdit; p. 264, Dernbinsky Photo Associated; p. 203, Mark Segal/Stone; p. 239, Mike Wieder, courtesy IFSTA; p. 149, Peter Dean/Stone; p. 67, Phil McGarton/PhotoEdit; pp. 1, 22, 91, 170, United States Department of Agriculture (USDA); p. 218, Win McNemage/Reuters/Corbis

Prentice-Hall International (UK) Limited, *London*
Prentice-Hall of Australia Pty. Limited, *Sydney*
Prentice-Hall Canada Inc., *Toronto*
Prentice-Hall Hispanoamericana, S.A., *Mexico*
Prentice-Hall of India Private Limited, *New Delhi*
Prentice-Hall of Japan, Inc., *Tokyo*
Pearson Education Asia Pte. Ltd., *Singapore*
Editora Prentice-Hall do Brasil, Ltda., *Rio de Janeiro*

10 9 8 7 6 5 4 3 2 1
ISBN 0-13-775776-X

DEDICATION

To Freddie, my supportive wife, and Kevin and Ronda, our loving children, who have provided abundant happiness.

CONTENTS

PART I

PART II

PREFACE

Economics of Agricultural Markets is about world agricultural and food markets. It includes useful principles and concepts of economic analysis that facilitate understanding how markets for agricultural and food products operate. Many of the analytical concepts and processes are equally applicable for understanding how markets for products other than agricultural and food products work, however the latter have some distinct characteristics worthy of special attention. For example, all of the world's 6 billion plus people are affected daily by how well markets provide a linkage between the production of agricultural products and food consumption. The use of agricultural and food products to illustrate economic principles and concepts as much as possible is appropriate because the kind of approach to market analysis contained in this book is most frequently offered in agricultural colleges.

This book has evolved over several years from a set of notes used for teaching an upper-level undergraduate course. The primary purpose of the course is to help students develop analytical tools for *thinking* about the overall business environment in which they will operate in some fashion for the rest of their lives. Even if students do not intend to make their livelihood from the sale of agricultural or food products, they will definitely be spending some of their income in these markets. Hopefully, the kind of skills and knowledge provided by the material in this book will equip students to better understand opportunities and challenges that those markets will provide. Economic history clearly indicates that markets are very dynamic institutions and are likely to continue to change.

Getting individuals to think about the environment in which they will be making future buying and selling decisions is a useful starting point for identifying all kinds of marketing issues. All too often the emphasis of marketing textbooks is on selling from an individual firm's perspective and adequate attention is not given to the overall environment in which transactions occur, including purchasing decisions. Thus the focus of this book is more on markets than on marketing; emphasizing the effects of group behavior more so than how individual transactions can be made. Consequently, this book is intended to complement other courses that concentrate primarily on an individual firm's marketing strategies and decisions.

One of the goals of the book is to help students understand the mysterious and marvelous nature of how agricultural markets influence the variety of food and other agricultural products produced and transformed into many alternative forms on a daily basis for consumers around the world. Also, hopefully, students will obtain an increased appreciation for the important role that agricultural markets play in the world economy.

The book has four major parts. Part I of the book contains three chapters. Chapter 1 describes what the subject matter of agricultural markets is all about. Chapter 2 describes several economic measures frequently encountered in discussions about agricultural and food markets. Finally, Chapter 3 contains a discussion of how different kinds of index numbers are computed and used to summarize and analyze changes in prices for different groups of products.

Part II contains five chapters designed to build a basic framework for thinking about how production and consumption activities are linked together. Chapter 4 describes key analytical concepts underlying the demand for products while Chapter 5 focuses on supply concepts. Chapter 6 combines the concepts of the previous two chapters to show how changes in prices and quantities of products at the farm and retail levels are linked together even though there is considerable modification in the form of products as well as spatial and temporal separation between the production and consumption activities. This framework is developed initially assuming a competitive market structure, but modifications associated with alternative market structures are considered in Chapter 7. Finally, Chapter 8 discusses the variety of ways that buyers and sellers discover a mutually satisfactory price at which they are willing to exchange ownership of commodities.

Part III consists of four chapters that examine the spatial dimensions of agricultural markets. Chapter 9 introduces the importance of the spatial dimension that arises from the geographical distribution of agricultural producers and consumers, and the distances that products are moved between initial production and consumption. Chapter 10 introduces a two-region trading model to illustrate how interregional trading affects producers and consumers. Chapters 11 and 12 extend the basic ideas of interregional trade to international trade. These two chapters include some of the additional complexities of international trade and show how the two-region interregional trade model can be adapted to analyze the effects of adjustments in exchange rates and governmental policies on the flow of commodities between countries.

Part IV concentrates on the temporal dimensions of agricultural markets. Chapter 13 considers explicit and implicit storage decisions associated with agricultural and food products because of the amount of time that occurs between production and consumption. The two-region model introduced in Part III is modified to illustrate how storage decisions depend on current and future market conditions. The final three chapters of the book are devoted to a discussion of futures and options markets that have become increasingly important to buyers and sellers of agricultural and food products. Chapters 14 and 15 are devoted to futures markets and Chapter 16 considers options markets. Chapter 14 introduces some specialized vocabulary and describes some of the basic mechanics of how futures markets operate. Chapter 15 shows how futures markets can be used for hedging purposes or to target price anticipated production of commodities. After students have an understanding of futures markets and how they can be used, the similarities and differences between futures and options markets and their uses are handled in Chapter 16.

In developing this book for teaching purposes, it has been assumed that students would either have had one basic economics course or at least be familiar with some basic economic vocabulary. Basic demand and supply concepts are introduced in Part II of the book assuming little carryover from previous coursework. Demand and supply concepts are introduced as abstract representations of producers' and consumers' behavior without developing all of the underlying theoretical principles of maximization behavior usually contained in an intermediate microeconomics course. The only mathematical tools required for certain parts of the book are a basic understanding of algebra and geometry, since no calculus is used. Diagrams are used to graphically represent economic relationships that can also be expressed algebraically, to determine the unknown values for quantities and prices of commodities.

I have found that following the chapters in the order presented in the book has been a good process for helping students develop an analytical framework before concentrating on spatial and temporal aspects of markets. Chapter 8 can be skipped without much loss in continuity if there is not enough time for all the material in the book. Some instructors also may decide to skip Chapter 3 if there are other courses at their institution that are specifically designed to expose students to index numbers. It is important, however, for students to be able to distinguish between nominal and real price changes in doing market analysis, and to feel comfortable using price indices.

In addition to reading the material in the various chapters, students need to be given ample homework assignments that provide opportunities to practice using the concepts and tools introduced in each of the chapters. In addition to a set of questions included at the end of each chapter, an Instructor's Manual contains some suggested homework exercises.

Finally, I want to express appreciation to many individuals who have contributed in numerous ways to the final product, but are in no way responsible for any shortcomings or remaining errors. First, a debt of gratitude is owed to the many students who have been guinea pigs for earlier versions of the chapters in this book and who made suggestions for improvements in exposition. Valuable suggestions have also been received from several reviewers including: James Beierlein, Penn State University; Michael Boland, Kansas State University; Donald Larson, Ohio State University; and Gary White, Dickinson State University. Thank you. Also, I want to express appreciation for the support of family, friends, and university administrators who have encouraged and supported the development of this book. I want to especially acknowledge the comments and suggestions received from N. Piggott and J. Russ on earlier drafts of several chapters. The hard work of Dawn Hartley, Pam Speight, and other support staff in the Department of Agricultural and Resource Economics at North Carolina State who assisted in this project is deeply appreciated. This book would never have occurred, however, were it not for the intellectual stimulation received from several of my colleagues and instructors over the years. A special gratitude is owed to George E. Brandow who first introduced me to some of the concepts contained in this book as well as to Richard A. King who reinforced the importance of thinking about time, space, and form dimensions of agricultural markets.

R. A. Schrimper

Part I

Overview and Quantitative Measures

Part I of the book consists of three chapters that provide an overview of the subject matter and introduces some quantitative measures frequently used for analyzing various aspects of agricultural and food marketing.

One of the objectives of Chapter 1 is to indicate the breadth and importance of the topics to be considered in the rest of the book. In particular, the variety of ways in which the words *market* and *marketing* are used to refer to different activities associated with coordinating production and consumption of food and other agricultural products is stressed. The different activities include exchanges of ownership, physical value-adding activities, and several facilitating functions that are involved in all the intermediate processes linking production and consumption of agricultural products. A framework for analyzing the changes in the form, spatial, and temporal characteristics of products is basically the content of the remaining three parts of this book. Although markets are the coordinating link between the production and consumption for all products, several characteristics of agricultural and food products contribute some unique aspects to the marketing of the products. At the end of Chapter 1, some of the important factors that have been affecting U.S. food markets in recent years are noted.

Chapter 2 describes some statistical measures frequently used in discussing the economic size or importance of markets for agricultural and food products in the United States. Some conceptualization issues associated with these kinds of measures are discussed initially. Various sources of data that can be used to calculate alternative measures are described. Differences in governmental measures of aggregate retail food expenditures and recent changes in some of its components are noted. Then attention is turned to how retail and farm level data can be used to calculate retail-farm price spreads and other aggregate measures of the value of economic resources used in transforming basic agricultural products into retail products and making them available where and when consumers ultimately purchase and use them.

Chapter 3, the final chapter in Part I, is about index numbers. In particular the chapter indicates alternative ways price information can be combined to obtain summary measures of changes in prices for groups of products and how the resulting index values can be used. It is especially important for students to understand how price indexes are calculated and used to differentiate between changes in nominal and real prices of products. The chapter begins with simple averages of price relatives and then shows that Laspeyres price indices are weighted averages of price relatives that can be calculated in a couple of different ways. Finally, some details about how the Consumer Price Index is calculated in the United States and some of the ways in which it is frequently used for adjusting expenditure, income, and other economic information are described.

CHAPTER

1

OVERVIEW OF MARKETS AND MARKETING

The purpose of this chapter is to provide an introduction and overview of the subject matter in this book.

The major points of the chapter are:

1. The important role of agricultural and food markets in coordinating production and consumption activities.

2. The different connotations associated with the way the words *markets* and *marketing* are commonly used.

3. The alternative ways of classifying economic activities that markets perform.

4. Some special characteristics of agricultural and food products that contribute to some unique marketing challenges.

5. Important factors causing changes in U.S. food markets in recent years.

INTRODUCTION

This book describes and analyzes many aspects about changes in ownership of agricultural and food products that occur under all kinds of circumstances around the world. Examples of exchanges of ownership are the sale of 12,000 bushels of corn by a farmer to a grain merchandiser, the purchase of 7,000 tons of bacon by a retail food chain from a pork processor, and a household's purchase of a week's supply of food at a retail outlet. Each of these examples is part of the process that links production of agricultural products with consumption of food. Subsequent chapters in this book are designed to provide a framework for considering key spatial, temporal, and form characteristics of market transactions of agricultural and food products. Many of these topics are also applicable to a broader set of markets than just agricultural and food products. Nevertheless, markets for these kind of commodities have several special characteristics that require some specialized knowledge to fully appreciate how they operate.

Coordinating Production and Consumption Activities

The challenge of trying to understand how producers, consumers, and **agribusiness** firms around the world make buying and selling decisions involving various agricultural and food products is a fascinating topic. One of the reasons that this subject matter can be fascinating is trying to figure out how all the activities required to provide daily nutrition for the 6.0 plus billion people in the world get coordinated. Another intriguing aspect is understanding and trying to figure out some of the changing nature and dynamics of agricultural markets. A cursory review of the history of agricultural production, distribution, and consumption activities indicates that the linkages between these activities have undergone tremendous changes in the past. Trying to anticipate the changes in these linkages that are likely to occur in the future is challenging. Agribusiness firms need to be aware of evolving changes in markets to capitalize on new opportunities.

One way to begin thinking about the tremendous role that agricultural markets perform in responding to the world's population daily **demand** for food is to think initially about a smaller hypothetical situation. For example, suppose you were responsible for planning and coordinating what needed to be done to provide one hot dog, a small order of french fries, and some kind of beverage to everyone attending a football game on a typical U.S. college or university campus on a particular Saturday afternoon. At first glance you might consider this task as primarily consisting of estimating how many fans (or customers) were likely to show up on a given day. Once you had forecast the number of potential customers, you could then determine the quantity of products you were going to need to purchase and how many workers would be required to ensure that the food was prepared and served in a timely manner. It would not be surprising for someone in this situation to be more concerned about the availability of enough workers than food products. This is because of the way consumers have become accustomed to assuming that they will be able to order or purchase whatever quantity of food is required at a reasonable price, whenever desired.

A much broader view of the task of providing food to a group of spectators in a stadium on any college campus on a Saturday afternoon would be to assume you also had the responsibility for planning the required production of the livestock, potatoes, grain (for the rolls), and the other basic ingredients required for the simple menu of a hot dog, an order of french fries, and a beverage. In addition to making decisions about who would produce the necessary quantities of potatoes and other items, a number of decisions about other kinds of economic activities would also have to be made. For example, several kinds of processing services would be required to transform agricultural products into edible forms. Transportation services would be necessary to move the products from where they are produced to where consumption was going to occur. Finally, arrangements would be necessary to make sure sufficient storage space or services were available at various locations and at critical times because production, consumption, and all the intermediate activities do not occur instantaneously. The preceding list of activities are essentially what modern complex agricultural marketing systems accomplish on a continuing basis without anyone overseeing or worrying about coordinating the entire process.

After thinking about what a coordinator associated with providing three food items to everyone in a college stadium for one meal would have to handle, a better appreciation of how complex the world's agricultural marketing system really is can be obtained by considering larger hypothetical situations. For example, suppose you were considering providing the same kind of menu to every individual in your state once a year. One way to begin thinking about the increase in effort that would be necessary for this larger task is to estimate how many stadiums like the one on your campus would be required to hold the total population of your state. If a state contained 6 million people, 100 stadiums each with 60,000 people would be required.

The above example can be extended to consider what would be involved in providing food on a continuing basis to the population in your state multiple times every day of the year. Additional extrapolation can be used to illustrate the magnitude of economic activities required to provide food on a daily basis for approximately 270 million people in the United States. The amount of effort required for the U.S. population would need to be multiplied by a factor of at least 20 to consider the magnitude of effort associated with providing food for the other 5.7 plus billion people in the world.

The above discussion illustrates that every person in the world is affected, on a daily basis, by the way agricultural markets provide alternative quantities and qualities of food for consumption. The efficiency with which all of these activities occur affects prices that consumers pay for the food they purchase as well as the prices that farmers receive for their products.[1] Less than 2% of the U.S.

population reside on farms and obtain income directly from producing agricultural products. However, a much larger fraction of the world's population depends on the production and marketing of agricultural products as a major source of their income. For example, more than half of the economically active populations in developing countries work in agriculture (International Food Policy Research Institute 1997). Also, many more individuals in the United States than just those living on farms earn income from adding value to products moving through food marketing channels from production to final consumption.

WHAT IS MEANT BY MARKETS OR MARKETING?

Multiple definitions and interpretations for the words *market* and *marketing* (without the agricultural or food modifiers) can be found in most dictionaries. The fact that the two words can be used as nouns as well as verbs contributes to versatility in how the words are used.

Exchange Processes

When the verb (or action) form of market or marketing is used, frequently it is in the context of having something to do with selling a particular item or service. For example, if one refers to a producer getting ready to market lettuce or strawberries, it may imply that the producer is trying to find someone interested in purchasing his crop or is evaluating alternative offers or outlets. The verb form of marketing is not used very often when describing activities that a prospective buyer might undertake to evaluate alternative sources of a product. The latter kind of activities is more often described by the words *procurement* or *procuring.* Market activity, however, involves buying as well as selling. The exchange process, involving the transfer of ownership, is a result of interaction between buyers and sellers and is one of the key functions performed by markets.

Historically, many business firms and organizations have viewed selling and buying activities as separate business activities. This is demonstrated by the existence of separate marketing and purchasing (or procurement) divisions in large organizations. Consequently, it is not surprising that marketing often is considered primarily from a selling standpoint and the dual importance of buyers and sellers in any market transaction is not always fully appreciated. In the case of small firms, the same individual may be responsible for coordinating the sale of merchandise as well as decisions about how, when, and where to purchase various items. Today, in some larger organizations, some common elements associated with selling and buying (or procurement) are handled by logistical support units that coordinate the flow of products into and out of production activities.

There are basically two ways in which the words market and marketing are used as a noun. One is in describing a particular location in conjunction with particular transactions. For example, one might refer to The Chicago Board of Trade or the Minneapolis Grain Exchange. The other way in which the words get used as nouns is to characterize all of the intermediate, value-adding activities associated with getting products into the hands of consumers—the ultimate users. Additional discussion of each of these ideas follows.

Geographical Context of a Market

A geographical connotation is certainly applicable to describe situations where buyers and sellers meet at a specific location to conduct business related to exchanging ownership of commodities. In this sense, the geographical connotation is similar to the way markets first began. In some cases, churches or governments designated certain locations and times for trading merchandise. In this way, prospective buyers and sellers knew when and where to assemble to arrange transactions. The designation of specific locations at which prospective buyers and sellers could get together was an efficient way for households to begin trading (perhaps by bartering) excess production in exchange for products from other households. The exchange process enabled households to take advantage of some specialization in production rather than trying to be self-sufficient in producing all the goods and services they desired to consume.

Geographical terminology is still frequently used when referring to different kinds of markets. Sometimes the geographical location describes a specific location where buying and selling actually occur, but more frequently it has a different and broader connotation, such as referring to the U.S. cotton market or corn market. An analytical interpretation of the latter concept will be developed more completely in Chapters 9 and 10. At this point, it is sufficient to note that a geographical adjective may be used to convey the general scope of demand and supply factors affecting the terms of trade (or price) at which transfer of ownership of particular commodities occurs. In some cases, a fairly narrow geographical context might be of interest and in other cases a much broader context may be appropriate. For example, referring to the U.S. or the North Carolina corn market differentiates among sets of participants involved and prices of particular exchanges of ownership more so than indicating specific locations at which negotiations or transfer of ownership actually occur. With continuing changes in the kind of communication capabilities now available, transactions can be negotiated without buyers and sellers of agricultural and food products actually meeting each other.

Even though a broad or narrow geographical characterization is at least implied when a given market is being discussed, it is appropriate to think of markets basically as sociopolitical institutions. A market is a sociopolitical institution in that it represents a mechanism by which people interact to make economic decisions about what goods and services get produced and consumed. The socio- component is appropriate to convey the sense that a market consists of people interacting to accomplish exchanges of ownership. The political suffix is appropriate in the sense of a market being a mechanism for making decisions that coordinate producer and consumer behavior and ultimately influence how resources of an economy are used. In this sense, markets perform the role of the invisible hand for coordinating economic activity as envisioned by Adam Smith, one of the founding fathers of modern-day economics.[2]

Physical, Value-Adding Activities

The other major use of the words market and marketing as nouns is in reference to all the activities that add value to products once they are produced. These activities involve transforming agricultural commodities into alternative products desired by consumers, transporting products from where they are produced to locations desired by consumers, and making necessary arrangements for intermediate storage to handle differences in the rate of production and desired rate of consumption. Thus the marketing system makes sure that processing, transporting, and storage is performed. The three activities of transformation, transportation, and storage are considered to be physical functions of a marketing system because all three involve some physical handling of products. Each of these activities requires using resources and consequently can be economically justified only if the expected change in value of goods is adequate to match the cost of the required resources that could be allocated to other uses.

The description of marketing activities associated with adding value to products after they leave the farm gate, characterizes a lot of the activities in the agribusiness world, but it omits or overlooks a similar set of activities related to getting nonagricultural and agricultural inputs to farmers. The activities associated with getting desired quantities and qualities of inputs into the hands of producers of agricultural commodities are a large component of agribusiness. As agricultural producers throughout the world have become increasingly dependent on inputs produced off the farm, it is not entirely appropriate to consider agricultural and food markets as just the linkage between production and consumption, or what happens to products after they leave the farm.

The focus of this book will be primarily on what happens to agricultural products as they get transformed, relocated, and stored as they move through market channels to final consumption. One of the reasons for this approach is that in practice it is not easy to totally classify items as inputs or outputs except in the context of a specific situation. An agricultural product that is the output of one enterprise may be an input for a subsequent stage of economic activity. For example, corn is clearly an output for one set of producers, but as ownership of this product changes hands, some of the products become inputs for livestock producers or food manufacturers. Similarly, the production of products such as fuels, chemicals, and similar nonagricultural products can be considered the initial steps of food production since these inputs are combined with soil, water, and energy (including human) through various biological processes to produce a variety of agricultural products.

OTHER FUNCTIONS OF MARKETS

There are several other important functions that are an integral part of markets besides exchange and physical functions. This additional set of functions is generally referred to as facilitating functions. The name is appropriate in that the activities do not directly add value to products and are not related to transferring ownership, but they help other aspects of the marketing process operate. Some of these additional functions are performed by private businesses, but often governments play an important role in providing infrastructure for markets to operate more efficiently. The facilitating functions of a marketing system consist of maintaining an environment conducive for business transactions, standardization of weights and measures, provision of market information, and support for a sound and stable financial sector.

Business Climate

In order for exchange processes to operate smoothly there needs to be a climate conducive to conducting business transactions. This includes a clear understanding of the rights and privileges of ownership. The idea of establishing and maintaining this kind of environment involves establishing and maintaining processes for the resolution of disputes about business transactions that might arise. A formal or informal code of behavior for conducting business transactions is useful for individuals to be able to negotiate and execute business transactions efficiently. The lack of a code of legal conduct or process for resolving business disputes in a quick and easy manner has been a handicap in some countries that have tried to make quick transitions to more market-oriented economies.[3]

Standardization of Measures

The standardization of weights and measures or some set of agreed-upon measurement process for the units of products being purchased and sold is a second kind of activity that facilitates the operation of markets. Governments often take a major role in establishing and ensuring proper measurements. Sometimes this activity may also extend to inspecting and determining the quality of merchandise being traded. Significant differences exist around the world in terms of the extent of governmental involvement in these activities. An often-debated issue concerns when it is appropriate for user fees to be assessed for particular services and which ones should governments provide free of charge.

Economic Information

The provision of certain kinds of economic information about current and prospective market situations is a service that many governments have historically provided free of charge. Often the government's role in this activity has been justified because of the vital importance of agricultural and food markets; assuming improved decisions by market participants would occur with access to economic information.[4] These days many specialized private firms gather, interpret, and deliver information about market conditions to clients who are willing to pay for this kind of service. Obviously, these firms must be providing a useful service in the eyes of their clients or they would not continue to exist. The private sector has also assisted this facilitating function through the establishment and operation of futures markets that provide additional information about prospective prices or values of various agricultural and food commodities.

Financial Services

A sound banking and financial sector is another important component that facilitates the operation of a marketing system. Financial institutions can help reduce the cost of transferring payments for goods and services as well as provide financing for businesses engaged in various marketing activities. Also a considerable amount of capital is always tied up with goods somewhere in the marketing process as a result of the time required for value-adding activities to occur between production and consumption. A great deal of capital is required for facilities and equipment required for the transforming, transporting, and storing of commodities. All of these marketing activities are risky and it is important that firms have access to capital markets to be able to undertake business ventures.

Responsibility for provision of the preceding types of facilitating functions is often shared jointly by government and private enterprise in varying proportions around the world. In response to changes in market characteristics there have been changes in the kinds of facilitating functions over time. For example, the growth of international trade has created new opportunities for services to facilitate the exchange process.

SPECIAL CHARACTERISTICS OF AGRICULTURAL AND FOOD PRODUCTS

Several product characteristics contribute to some unique aspects about agricultural and food markets and marketing activities. Some of these characteristics are the biological nature of products, the global and diffuse distribution of producers and consumers, bulky and generally low value products per unit of weight, and product homogeneity.

Biological Characteristics

Three biological characteristics of agricultural and food products are especially relevant in terms of how they affect the way in which markets for these products operate. One characteristic is that basic biological processes essentially determine how long it takes between initial production decisions and when the results of those decisions are realized. A second characteristic of the biological nature of the agricultural production processes is a certain degree of perishability between production and consumption. Finally the biological nature of the production process means that the output of many agricultural products cannot be controlled or regulated as easily as many nonagricultural products.

The amount of time it takes to obtain a change in output from a planting or breeding decision for most agricultural products is not very flexible. For example, gestation periods are biologically constrained. This means that little can be done to speed up or slow down the production process once it is underway. Some progress in developing crops that require slightly shorter growing seasons and improvements in feed efficiency that affect the rate of growth of animals have occurred, but at a given point in time there is not much that a producer can do to speed up production. Consequently, considerable lead time is required before production at one end of the agricultural-food-marketing spectrum can respond to changes in consumer desires at the other end of the spectrum. The lag between decisions to alter production and the availability of a product may be several months or longer. For some fruits, nuts, and forestry-related products the lag between new plantings and harvest can be several years.

Even though it is not possible to instantaneously change output of most agricultural products in response to an alteration in market conditions, a more rapid adjustment in the flow of agricultural products through the market channels may occur by varying the intended rate of expansion or depletion of existing stock. Additional discussion of the different ways in which producers can respond to market conditions is included in Chapter 5.

Not only do most agricultural products have a fixed production cycle, but also many production activities including harvesting must occur at very specific times. Most crop and tree products have an annual or semiannual growing and harvest schedule, but livestock products tend to enter market channels on a more continuing basis. Differing degrees of perishability of agricultural products affect the rate at which agricultural products must be consumed or transformed into alternative forms to decrease their vulnerability to deterioration. The transformation of agricultural commodities into alternative products by food-processing firms can extend the shelf life of food items by decreasing their vulnerability to deterioration. The development of processing and preservation technologies over time has been important in changing the marketing of many agricultural products.

The susceptibility of agricultural production processes to random changes in weather and other uncontrollable factors that affect the total output of many crops can have major implications on prospective market conditions. The availability of irrigation technology to offset extreme drought conditions has decreased some of the variability in output associated with weather conditions. The availability of improved varieties and chemical products also has helped decrease some of the variability associated with diseases and pests. Nevertheless, there are still a number of things about nature that man cannot control that influence the annual quantities of agricultural products available for consumption. Effects of an unexpected change in production may be observed over an extended period of time because of the annual nature of production of many major crops used for human consumption and feed for livestock production. These factors can cause considerable fluctuations in agricultural and food markets.

Distribution of Production and Consumption

Global patterns of production and consumption of agricultural and food products contribute some interesting dimensions to the marketing process. In general, the production of many agricultural products is dispersed quite widely around the world and often at considerable distances from where consumption occurs. This means that extensive transportation services are required to transport products to consumers within countries. Also, many markets for agricultural and food products are international in dimension. This means a change in weather conditions influencing the production of a commodity in one part of the world may result in changes in market conditions around the globe in a very short period of time.

The international nature of agriculture and food production and consumption is a very decentralized activity and the entire process is not subject to the decisions of just a few individuals. The size of many agricultural production firms has increased over time in the United States and other parts of the world, but the production of many products is still distributed among many million producers spread over vast geographical areas around the world. Generally, there are fewer decision-makers associated with some of the marketing activities than production or consumption. One reason for this is the economies of scale associated with certain market activities and the fact that some of the marketing activities can be more geographically concentrated than production or consumption. Differences in the number of decision-makers and geographical concentrations at different stages of the marketing system for agricultural and food products suggest that the entire food system can be depicted graphically as an hourglass on its side.[5] The extreme left-hand side of the hourglass represents the millions of producers around the world engaged in agricultural production and the extreme right-hand side represents the 6.0 plus billion consumers of food products throughout the world. The narrower intermediate connecting link between the two ends of the hourglass indicates that there are fewer decision-makers or firms involved with the marketing activities linking production with consumption.

Bulkiness and Value Per Unit of Weight

A third characteristic of most agricultural products is that many are bulky or at least have a low value per unit of weight at an initial stage of the marketing process relative to many nonagricultural products. Some of the bulkiness and/or unnecessary weight often can be eliminated by some initial transformation or processing. These activities may also eliminate some perishability of products and improve storability. There is an incentive to conduct some of these activities relatively close to where initial production occurs to the extent that the product can be converted into a more highly valued product with less weight. Often the output from several producers will be collected before doing some of the initial processing because of **economies of scale.**

Other value-adding activities may be performed more efficiently closer to the time or places where individuals will purchase and/or consume products. If these activities involve adding weight to the product there certainly would be an incentive to do it as close as possible to the final destination. An example is the addition of water to concentrated orange juice products within households or somewhere close to the retail level. Even with considerable initial processing fairly close to production locations, a large volume of products must be transported long distances in the most economical manner because of the distance between locations of production and consumption.

Product Homogeneity

Agricultural products tend to be generally more **homogeneous** than many nonagricultural products at least at initial stages of production. For example, wheat grown by farmers in Montana can be very similar to wheat grown by producers in Canada, India, or other places in the world. As long as potential buyers cannot distinguish any difference in the characteristics of who produced the item or where it came from, it is a homogeneous product as far as potential buyers are concerned. Also as long as producers receive a given amount of remuneration they do not care who buys their products. This means that competitive market forces can operate over a wide range of producers and potential buyers. As items move through the marketing system and acquire more specialized characteristics, they may lose some of their homogeneity. Also as biotechnology developments alter genetic characteristics of plants and animals, some of the homogeneous characteristics of basic agricultural products may diminish and marketing of specialized items will become more common.

CONSUMER SOVEREIGNTY

The above discussion has indicated the important function of the agricultural and food marketing system in providing nutrition and other services to the world's population on a continuing basis. Basic agricultural products are converted into edible foods with varying amounts of services by various activities in the marketing system. Through the mutual interaction of buyers and sellers, information about consumer preferences is communicated to farmers and agribusiness marketing firms. In order to survive, firms must respond to the signals about consumer preferences they receive through the marketing system. Essentially this process can be viewed as consumers effectively determining how resources are used in those sectors of economies that are not centrally planned or controlled by government. As a sociopolitical institution, the marketing system transmits information about consumer preferences to producers. This is the basic concept known as **consumer sovereignty.** This implies that production of basic agricultural commodities and the incorporation of value-added services at various stages of marketing respond to changes in consumer behavior. It is a mutually interdependent process, however, as producers and agribusiness firms make independent decisions about how available resources are used to respond to the specific kind and strength of signals they receive from consumers. Although consumers' demand for food is a constant recurring phenomenon, the retail food market is affected by many factors and consequently is very dynamic.

IMPORTANT FACTORS AFFECTING U.S. FOOD MARKETS

Many factors have caused and are likely to continue to cause changes in the U.S. demand for food at the retail level. These are (1) changing demographic characteristics, (2) new production technologies, (3) increased female labor force participation rates, (4) higher incomes, (5) increased interest in food safety and nutrition, and (6) changes in domestic food and international trade policies. A few comments about each of these topics will help set the stage for the presentation and the discussion in the next chapter about some changes in retail food expenditures in the United States that have occurred in recent years.

Demographics

Change in the size and composition of the U.S. population has affected the retail food market in several ways. First, the rate of increase in total U.S. population has not been as high in recent years as in some earlier periods. This means that the annual rate of change, in total demand for food because of a changing number of domestic consumers, has not been as large as it was in earlier periods. In 1997, the annual rate of increase in the U.S. population was 0.92% (*Statistical Abstract of the United States* 1998). For the last 30 years, the annual growth rate in total population has been close to 1%. This is a smaller rate of increase than what occurred during the 1950s and most of the 1960s.

There are many ways in which changing characteristics of the population other than just the total number of people can affect the retail food market. For example, differential changes in the type of products purchased can occur depending on whether changes in total population are primarily the result of changes in birth, death, or net immigration rates. In recent years, birthrates have fluctuated, death rates have dropped sharply, and the number of immigrants has increased. From 1989 to 1993 the total number of births in the United States was greater than 4 million peaking at 4.148 million in 1990, the highest since 1964 (*Statistical Abstract of the United States* 1998). Since 1990, the number of births have trended downward.

Changes in the age distribution of the population has affected the demand for various products because of differences in consumption patterns associated with age. An increasing number of elderly people have affected the demand for institutionalized food service, since a larger share of the population requires special care. Also, average household size has decreased from 3.33 individuals in 1960 to 2.64 individuals in 1997 (Manchester 1960; *Statistical Abstract of the United States* 1998). Over the last century, average household size in the United States has decreased by more than two individuals. A decrease in the average size of households has created increased interest in smaller food packages to accommodate typical household size.

An increase in ethnic diversity of the population resulting from additional immigrants has also affected the demand for specialized foods. For example, the Hispanic populations in the U.S. more than doubled between 1980 and 1997 (*Statistical Abstract of the United States* 1998). In fact, by the year 2005, the Hispanic component of the U.S. population is expected to be the largest minority. The geographical distribution of the U.S. population has also been changing with increasing proportions in the West and South. Changes in the geographical distribution means that distributional networks for food must adapt to changing locations of consumers.

Production Technologies

Changes in technology at all stages of the marketing process have affected retail food markets. Good evidence of this is to consider the changes that have occurred in the kind of items on grocery shelves and in frozen food cases over the last 5 to 10 years. For example, more than 13,000 new food products were introduced at the retail level each year between 1994 and 1996 (Gallo 1998). Many new food products are similar to existing products. A few new products introduced each year catch on and are successful, but most do not survive very long. New food processing technologies are responsible for a large number of new retail food products introduced into the market each year. New techniques to extend or improve the shelf life of products or to alter other characteristics, and to make meal preparation easier are some of the major kinds of new product development.

Another place where technology has had a substantial influence on the demand for food is in typical household kitchens. The number of new household appliances including microwaves, food processors, bread and pasta making equipment, and so forth, has altered household food preparation activities. Some effects of this equipment are similar to how the introduction of refrigeration and freezing technologies altered the marketing of food in earlier periods.

The increased use of computers has been responsible for many changes in the way food-marketing firms do business. The use of computers for scanning product information has affected checkout procedures at grocery stores and other retail food outlets. Computers have helped in many ways to improve efficiency and thereby influence certain costs of production and marketing activities that ultimately determine the price consumers pay for food products.

Female Labor Force Participation

Perhaps one of the most dramatic factors that has affected the demand for marketing services and different kinds of retail food products is the increasing number of females that work outside the home. In 1997, the female rate of labor force participation was 59.5% compared with 75.9% for males (*Statistical Abstract of the United States* 1998). The female rate of labor force participation was up from 51.2% in 1980. An increasing proportion of females working outside the household has decreased the amount of time available for typical household activities including meal preparation and increased interest in products that are essentially ready to eat or that require minimal preparation time. For example, the results of a survey by Yankelovich Partners, Inc. indicated that half of U.S. households spend less than 45 minutes cooking the evening meal compared to an average of 2 hours, 30 years ago (*Food Processing* 1997). Increases in the female labor force participation have also affected decisions about eating food at home or away from home.

Higher Incomes

Increases in U.S. per capita income over time have definitely influenced the type and quality of individual food products and marketing services purchased and consumed. Between 1990 and 1997, per capita purchasing power increased by 8.3% (Table 1–1 and Figure 1–1). Per capita income actually increased by 31.4% between 1990 and 1997, but a good bit of the change was offset by higher prices meaning a lower rate of growth in purchasing power. Some increase in per capita purchasing power has resulted from the increase in the number of workers associated with the increased female labor force participation discussed earlier. Income has also increased because of increases in human capital and improved labor productivity.

TABLE 1-1

U.S. PER CAPITA DISPOSABLE INCOME IN CURRENT DOLLARS AND 1992 PURCHASING POWER, 1970–1997

Year	Current Dollars	1992 Purchasing Power
1970	$ 3,545	12,022
1975	5,367	13,404
1980	7,730	14,813
1985	12,629	16,654
1990	16,721	17,996
1991	17,242	17,809
1992	18,113	18,113
1993	18,706	18,221
1994	19,381	18,431
1995	20,349	18,861
1996	21,117	19,116
1997	21,969	19,493

Source: Table 722, *Statistical Abstract of the United States*, 1998.

Health, Safety, and Nutritional Interests

Considerable evidence exists that an increasing proportion of consumers make food purchasing and consumption decisions in terms of perceived implications about how healthful or safe certain foods might be (Caswell 1998; Senauer 1989). There appears to be increased attention on the amount of fat, cholesterol, and other nutritional components of foods. Consumers have also become more concerned about food-related illnesses and food safety issues. Safety concerns are also related to pesticide residues on food products. In addition, concerns about the safety of food produced in other countries and what happens to food during various steps in the marketing process are of increased importance to consumers.

Domestic and International Agricultural Policies

A number of domestic and international policies have influenced agricultural and food marketing over time. There has been a mixture of policies designed to restrict, expand, or modify the quantities and prices of agricultural and food products in some way. Government policies that have affected the prices for selected

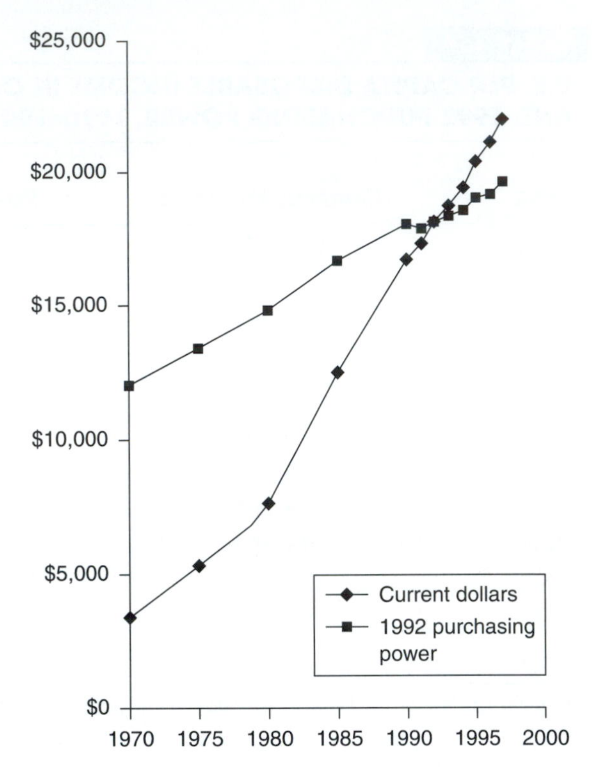

FIGURE 1–1 U.S. per capita disposable income in current dollars and 1992 purchasing power, 1970–1997.

agricultural commodities have had some effect on the relative prices of different food items at the retail level. Differences in relative prices result in differences in food consumption to the extent that consumers respond to price. Also, many governmental programs have been implemented to reduce the cost of food to certain subgroups of the population. This has occurred through the National School Lunch Program, Food Stamps, or direct distribution of food to low-income households.

In addition to domestic agricultural policy decisions that have affected the availability and prices of food products in the United States, there are a number of ways that countries have altered the international movement of agricultural and food products through international trade decisions. More discussion regarding the role of international trade policies on agricultural and food markets will be included in Chapter 12.

SUMMARY

The material in this chapter provides an introduction to the subject matter presented and developed in more detail in subsequent chapters of the book. The chapter also attempts to indicate the breadth of the topics and why they are relevant to every person in the world because of their daily dependence on agriculturally based products. Incomes of agricultural producers and other agribusiness firms also depend on alternative market opportunities.

The important role of invisible market forces in coordinating production and consumption activities is often taken for granted and not fully appreciated until analyzed in detail. A study of agricultural and food markets involves all of the multiple exchanges of ownership of products that occur between initial production and ultimate consumption. From this perspective, the subject matter of this book is about how, when, and where exchanges of ownership occur between farmers and consumers.

The fact that the words marketing and market can be used as nouns as well as verbs can be confusing. Sometimes a very specific geographical location is used in describing buying and selling activities and other times a much wider geographical connotation is implied when referring to interactions between buyers and sellers distributed throughout a region, state, country, or the world. In general, the abstract concept of a market can be viewed as a sociopolitical institution in terms of describing mechanisms by which people interact and thereby make decisions about how resources are used to produce goods and services.

Value-adding physical activities consisting of transforming, transporting, and making items available at times that purchasers desire are part of coordinating production with consumption. This categorization of economic activities leads to viewing market activities in terms of how form, spatial, and temporal characteristics of products are modified. Facilitating functions that help the exchange and physical value-adding processes operate are another set of important activities.

Exchange, physical, and facilitating functions of markets occur for all types of nonagricultural as well as agricultural products throughout the world, but certain attributes of the latter products lead to some special characteristics. For example, the biological nature of agricultural and food production in conjunction with the global distribution of production and consumption activities lead to some interesting marketing phenomenon. Also the bulkiness and relatively low value per unit of weight and homogeneity of many agricultural products cause the markets for these products to be somewhat different from markets for many nonagricultural products.

Finally, the fact that agricultural and food markets are likely to continue to change means that participants need to be aware of recent and potential changes to adapt to a changing environment. One example is the U.S. retail food market that has been affected by changing demographic characteristics of the population and other factors. Among the latter are changes in production technologies, household incomes, female labor force participation rates, food safety and nutritional interests, and governmental domestic food and international trade policies.

QUESTIONS

1. What is meant by value-adding marketing activities?
2. Identify and briefly describe the three major kinds of physical functions performed by agricultural marketing systems.
3. Identify and briefly describe the facilitating functions of agricultural marketing systems.
4. Briefly explain what is meant by **consumer sovereignty** and how it influences what gets produced in the agricultural sector of the U.S. economy.
5. Briefly explain how the biological nature of agricultural products adds some complexity to marketing activities that many nonagricultural or nonfood products do not necessarily face.

REFERENCES

Bromley, D. W. 1993. Revitalizing the Russian food system: Markets in theory and practice. *Choices* 8(4), 4th quarter: 4–8.

Caswell, J. A. 1998. How labeling of safety and food attributes affects markets for foods. *Agricultural and Resource Economics Review* 27(2):151–158.

Clayton, K. C., and B. Claffey. 1993. No news is bad news . . . for markets. *Choices* 8(4), 4th quarter: 18–20.

Davidson, J., and A. Weersink. 1998. What does it take for a market to function? *Review of Agricultural Economics* 20(2): 558–572.

Food Processing. 1997. Who rattles the pots and pan? Chicago, IL: Putnam Publishing Company. 58(5):19.

Gallo, A. E. 1998. *The food marketing system in 1996.* Agric. Info. Bull. 743. Washington, DC: U.S. Department of Agriculture.

International Food Policy Research Institute. 1997. *IFPFI report* 19(2). Washington, DC.

Manchester, A. C. 1960. *Data for food demand analysis: Availability, characteristics, options.* Agric. Econ. Rep. 613. USDA, Economic Research Service, Commodity Economics Division, Washington, D.C.

Rhodes, V. J., and J. L. Dauve. 1998. *The agricultural marketing system.* 5th ed. Scottsdale, AZ: Holcomb, Hathaway Publishers.

Senauer, B. 1989. *Food safety: A growing concern.* Staff Paper P. 89–38. Department of Agricultural and Applied Economics, University of Minnesota. Institute of Agriculture, Forestry and Home Economics. St. Paul, Minnesota.

U.S. Bureau of the Census. 1998. *Statistical Abstract of the United States: 1998.* (118th Edition) Washington, D.C., U.S. Government Printing Office. http://www/census.gov/prod/3/98pubs/98statab/cc98stab.htm

NOTES

1. The extent to which marketing activities require fewer resources to provide the same quantity of value-added attributes to agricultural and food products means that additional resources can be used to produce other goods and services in an economy.

2. An article by Davidson and Weersink (1998) further discusses the concept of a market and the necessary conditions for markets to function.

3. Bromley (1993) provides additional discussion about the role of various social and political institutions that contribute to an appropriate business climate.

4. Clayton and Claffey (1993) discuss the importance of public market information.

5. See pp. 5–6 in Rhodes and Dauve (1998).

CHAPTER

2

MEASURING THE ECONOMIC IMPORTANCE OF AGRICULTURAL AND FOOD MARKETING ACTIVITIES

This chapter introduces and describes some statistical measures frequently used in discussions about the economic importance of marketing agricultural and food products in the United States. An understanding of how these measures are calculated is useful for knowing how to interpret the numbers correctly.

The major points of the chapter are:

1. Conceptual and measurement issues associated with quantifying the economic importance of agricultural and food marketing activities.

2. Major sources and types of annual retail food and alcoholic expenditure and per capita food consumption data for the United States.

3. Recent changes in U.S. food expenditure and consumption patterns.

4. Alternative measures of marketing services including **retail-farm price spreads,** the **marketing bill,** and contributions to the nation's **gross domestic product** and total employment.

INTRODUCTION

The first part of the chapter discusses fundamental issues associated with conceptualizing the economic importance of agricultural and food markets and issues that are confronted in using available information to quantify particular economic concepts. The second part describes major sources of annual U.S. retail food and alcohol expenditures and consumption data associated with one end of the marketing spectrum for agricultural and food products. Changes that have occurred in the composition of U.S. food expenditures and consumption are also described. The final segment of the chapter indicates how different types of information about retail and farm values of agricultural and food products are used to calculate alternative measures of market activities.

Conceptualization

A simple way of thinking about the economic importance or value of agricultural and food markets consistent with the discussion in Chapter 1 would be to consider the difference between the retail value of food consumption and what farmers receive for agricultural products. This difference could be interpreted as the amount of value added to products as they move through the marketing system. Differences between retail and farm values could also be interpreted as the costs of resources required to perform various marketing functions. Although at first glance a difference between retail and farm values of products may appear to be a rather straightforward concept, many difficulties arise in using appropriate data to calculate such a measure.

One of the difficulties encountered in calculating meaningful differences between the retail and farm value of products is variation in the composition or the lack of symmetry between agricultural products at the farm and retail levels. Another difficulty is associated with the global nature of markets.

Comparability of Agricultural and Food Products. Often the words agriculture and food are used almost interchangeably when discussing the marketing system that links agricultural production with food consumption. There are important differences, however, in the form as well as composition of products at each end of the spectrum. For example, most aggregate data about the value of agricultural production usually include information for a number of products that are definitely not food items. For example, the value of cotton and tobacco production are usually part of agricultural production statistics, but the values of these products as purchased by consumers are not usually part of retail food expenditure information. Similarly, forestry and aquacultural products are often included in agricultural production and farm income statistics while the value of offshore fishing is usually excluded. Marketing firms that handle fish and forestry products often perform similar functions as firms specializing in transforming, transporting, or storing other agricultural and food products. Also many of the marketing activities associated with nonfood agricultural products often are similar to the activities for other agricultural products and legitimately could be considered part of agricultural marketing, but generally are not included as part of the agricultural-food marketing system.[1]

Other product compatibility difficulties in the agricultural-food spectrum are associated with the composition of retail food expenditure information. In particular, the way in which expenditure and quantity information for pet food and beverages is treated can lead to complications in making comparisons between retail and farm value data.[2] If retail food expenditure data are based on the amounts of food used only for human consumption, then the value of agricultural production data would need to be adjusted to exclude the quantity of products used for the preparation of pet foods for comparisons between aggregate value measures at retail and farm level to be meaningful.

Additional adjustments of farm level data may be necessary to account for other nonfood uses of particular agricultural products. Also, expenditure and consumption information for alcoholic and some other beverages (including bottled water) are often excluded from retail food expenditure data, or at least treated separately, even if some of the excluded beverages depend heavily on basic agricultural products as ingredients.

Comparability of Individual Products. An alternative to using aggregate data to compute a measure of the importance of marketing activities is to compare the retail and farm values or prices of individual products. If the difference between retail and farm prices is to be a meaningful measure of the value (or costs) of marketing services for individual products, the quantity units need to be comparable. For example, comparing the retail price of bread to the price of wheat is not very meaningful. However, if it is known how many loaves of bread can be produced from 1 bushel of wheat or alternatively how many ounces or grams of wheat are used to make each loaf of bread a price comparison for comparable units is possible.

Similarly, a comparison between a weighted average price per pound of all beef products in a retail meat counter with the price per pound of live cattle is not a meaningful measure of the value of marketing services. This is because more than 1 pound of a live steer is required to produce a pound of steak or any other retail beef product. Additional discussion about converting weights between basic agricultural commodities and retail food products will be included later.

Another problem often encountered in comparing retail and farm values of individual products is difficulty in identifying all of the retail food products that originate from a given agricultural product. This is especially true these days in view of all the transformations and combinations of basic agricultural products used to manufacture various retail food products. Combining the value of several different retail products may be required before a comparison with the value of a basic agricultural commodity provides a meaningful measure of the value of marketing services. Similarly, if one starts with the value of a specific retail food item, it is not always easy to determine the appropriate implicit farm value of the basic commodity. For example, T-bone steaks or hams are not the only products that are derived from live cattle or hogs, respectively. Consequently, differences between the retail value of T-bone steaks and the farm price of cattle or the retail price of ham and the price of live hogs do not represent the value or costs of marketing services for several reasons. One reason is that the quantity units may not be directly comparable as noted earlier. Another reason this kind of comparison would not be very meaningful is because the farm value of live animals represents a composite evaluation of the implicit values of several different components of a commodity used to produce a variety of final retail products. Consequently, even after making an appropriate adjustment for differences in live and dressed weight, comparisons between the retail prices of selected cuts of meat and the farm value of animals are not valid measures of the value of marketing activities for selected retail products.

The market values of live cattle and hogs represent weighted averages of what buyers are willing to pay to obtain components required for higher valued final products as well as what they think some of the lower or less desirable final products are worth. Once animals are slaughtered and the different-valued components are separated, differential prices reflecting relative values in specialized market channels may be observed.

The same type of problem exists for many agricultural products for which different components are used to produce various retail products. In fact one way of describing changes that have occurred in retail food markets over the last 50 years is that at one time consumers primarily purchased foods with a few added services whereas now they purchase services with limited food components.

Relating retail to farm values for purchases of food for away-from-home consumption is an even greater problem than for purchases of food products at retail grocery stores. For example, if a consumer pays $15 for a restaurant meal consisting of several food items, it is not easy to determine the quantities of basic agricultural commodities used to determine how much of the retail price should be attributed to marketing services and how much represents the farm value of basic commodities. The large volume of agricultural commodities used to produce food for away-from-home consumption creates a major problem in estimating marketing costs for individual agricultural commodities.

Complications Because of International Trade. Another difficulty in estimating the value of marketing services is the international or global nature of markets. Consequently, all agricultural products and the added value of marketing services produced in a given country are not necessarily consumed or used in the same country. Agricultural production data include products used by domestic consumers as well as products eventually consumed in other countries. Thus the difference between gross retail and farm level data would not include the value of services the marketing system contributed to food products that are exported to consumers in other countries. Similarly, aggregate retail food expenditures statistics for a given economy usually include some value of agricultural products and marketing services from other parts of the world. Consequently, adjustments of retail and farm level aggregate data are often necessary before a difference in these two measures can be interpreted as any kind of representative measure of the aggregate value of the marketing services associated with agricultural products for a particular economy.

RETAIL FOOD EXPENDITURE INFORMATION

There are two major sources of information about annual U.S. retail food and alcoholic beverage expenditures: the U.S. Department of Agriculture and U.S. Department of Commerce (Manchester 1987, 1990). Although at first glance, it might seem like the U.S retail food expenditures data for a given year ought to be identical regardless of the source, it is important to know why information from different sources varies. The following discussion provides a brief overview of some of the differences in these two alternative sources about U.S. retail food expenditures.

USDA aggregate retail food (excluding alcoholic beverages) expenditure data represent the value of all food purchased (with money or food stamps), donated, or home produced (including sport fish and game) for human consumption. The total value of food includes items purchased by individuals and families using their own resources as well as purchases paid for by businesses and governments for use by their employees. The latter includes purchases of food as part of employee expense accounts and food provided on military bases and similar facilities for employees. The value of food provided residents in all types of institutions is also included in the total value of food expenditures reported by the USDA. Separate estimates of the sales of alcoholic beverages and the value of nonfood items sold by grocery stores are also available.

USDA's retail food expenditures information for selected years is indicated in Table 2–1. The data in the second column of Table 2–1 indicate that the total retail food expenditures in 1997 were approximately $715 billion. This is equivalent to more than $1.9 billion every day or a little less than $100 million per hour. Purchases of alcoholic beverages at all types of retail outlets accounted for another $94.5 billion in 1997. Retail grocery stores also sold more than $106 billion of nonfood items in 1996 (Gallo 1998).

Another type of USDA information about food expenditures represents personal expenditures on food. It differs from the previous series of total expenditures by excluding the value of food purchased by business and government employees on expense accounts because the latter items are not considered to be part of total personal income for national income accounting purposes. Personal food expenditures also exclude the value of donated food, sport fish and game, and nonfarm households' home-produced food. The USDA

TABLE 2-1

AGGREGATE AND PER CAPITA FOOD EXPENDITURES IN THE UNITED STATES

Year	Food Expenditures: Total	Food Expenditures: Value of Food At Home[a]	Food Expenditures: Value of Food Away from Home[a]	Consumer Price Index for Food[a]	Real Values: Food at Home[b]	Real Values: Food Away from Home[c]	Total Population July 1[a]	Real Per Capita Food Expenditures: At Home[d]	Real Per Capita Food Expenditures: Away from Home[e]
	(billion dollars)			(1982–84=100)	(1982–84) dollars		million	(1982–84) dollars	
1970	117.1	77.5	39.6	39.2	197.7	101.0	205.1	963.9	492.4
1975	188.0	119.9	68.1	59.8	200.5	113.9	216.0	928.2	527.3
1980	306.1	185.8	120.3	86.8	214.1	138.6	227.7	940.3	608.7
1985	404.5	235.7	168.8	105.6	223.2	159.8	238.5	935.8	670.0
1990	559.8	311.6	248.2	132.4	235.3	187.5	249.9	941.6	750.3
1994	638.8	349.1	289.7	144.3	241.9	200.8	260.6	928.2	770.5
1995	663.0	364.7	298.3	148.4	245.8	201.0	263.0	934.6	764.3
1996	688.3	380.1	308.2	153.3	247.8	201.0	265.5	933.3	757.1
1997	714.9	394.6	320.3	157.3	250.9	203.6	266.8	940.4	763.1

Source: [a]Data from Putnam, J. Jones and J.E. Allshouse. *Food Consumption, Prices, and Expenditures, 1970–97*. Statistical Bulletin 965, USDA, Economic Research Service, 1999.

[b](Column 3 ÷ Column 5) × 100

[c](Column 4 ÷ Column 5) × 100

[d](Column 6 ÷ Column 8)

[e](Column 7 ÷ Column 8)

TABLE 2-2

PERSONAL FOOD EXPENDITURES AND SHARE OF DISPOSABLE INCOME IN THE UNITED STATES

Year	Personal Food Expenditures (USDA)	Food Expenditures (Commerce)	Disposable Personal Income	Share of Income on Food: USDA Data	Share of Income on Food: Commerce Data
	(billion dollars)			(percent)	
1970	100.6[a]	122.5[b]	727.1[a]	13.8	16.8
1975	161.4[a]	189.1[b]	1,159.2[a]	13.9	16.3
1980	264.4[a]	303.5[b]	1,973.3[a]	13.4	15.3
1985	359.3[a]	415.6[c]	3,002.0[a]	12.0	13.8
1990	483.6[a]	532.0[d]	4,166.8[a]	11.6	12.8
1994	559.3[a]	633.6[e]	5,052.7[a]	11.1	12.5
1995	583.1[a]	649.1[f]	5,355.7[a]	10.9	12.1
1996	606.2[a]	669.0[f]	5,608.3[a]	10.8	11.9
1997	629.4[a]	692.4[f]	5,895.2[a]	10.7	11.7

Sources:

[a]Putnam, J. J. and J.E. Allshouse. *Food Consumption, Prices, and Expenditures,* 1970–97. Statistical Bulletin 965, USDA, Economic Research Service, April 1999.

[b]U.S. Department of Commerce. *The National Income and Product Accounts of the United States,* 1929–82. A supplement of the Survey of Current Business, Bureau of Economic Analysis, September 1983.

[c]U.S. Department of Commerce. *Survey of Current Business.* Bureau of Economic Analysis, July 1989.

[d]U.S. Department of Commerce. *Survey of Current Business.* Bureau of Economic Analysis, August 1993.

[e]U.S. Department of Commerce. *Survey of Current Business.* Bureau of Economic Analysis, August 1997.

[f]U.S. Department of Commerce. *Survey of Current Business.* Bureau of Economic Analysis, October 1999.

personal food expenditures data in Table 2–2 account for 86% to 88% of the total food expenditures values in Table 2–1.

USDA's series of data on personal food expenditures is similar, but not identical to estimates of total personal food expenditures compiled by the Department of Commerce. The major difference is that the Department of Commerce information includes purchases of pet food, ice, and prepared feeds (Putnam and Allshouse 1999). Also, the procedure used by the USDA to estimate at-home food expenditures uses a different source of information from the Department of Commerce resulting in a little bigger adjustment of retail sales for such things as drugs and household supplies purchased at grocery stores.

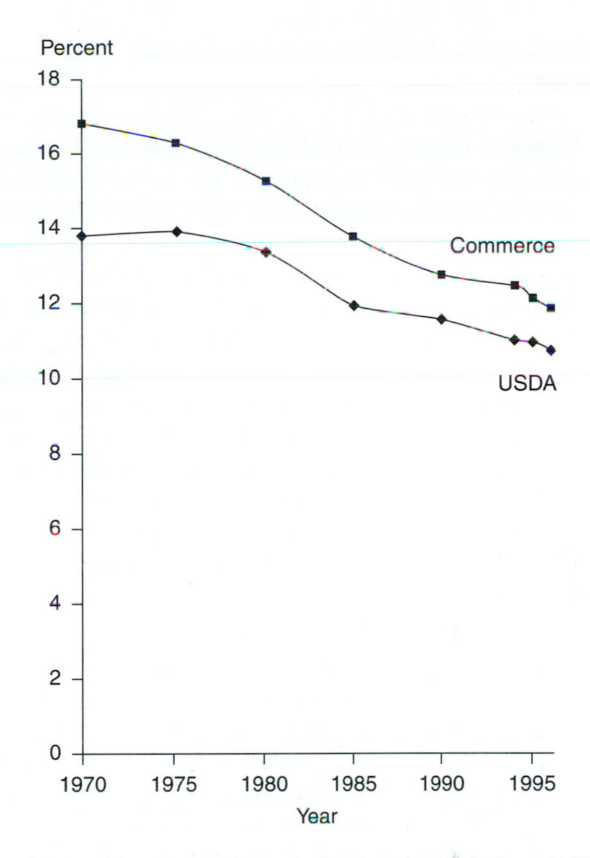

FIGURE 2–1 Share of income spent on food estimated from USDA and Commerce data

Estimates of personal food expenditures compiled by the Commerce Department are indicated in Table 2–2. These numbers are larger than the USDA personal food expenditure estimates because of differences in how the numbers are constructed. The relative size of the difference in the two series has decreased over time. The Department of Commerce's estimate of personal consumption expenditures for alcoholic beverages for 1997 was $88.6 billion, $5.9 billion less than USDA's figure.

Aggregate food expenditure data are often used to track the proportion of disposable household income spent on this category of products. The last two columns of Table 2–2 indicate the percent of total disposable income that U.S. households spent for food. For 1997, the result is either 10.7% or 11.7% of total disposable income depending on which measure of personal food expenditures is used. Similar calculations for the early and mid-1970s resulted in higher proportions (Figure 2–1).

Changes in Composition of U.S. Food Expenditures.

As noted earlier, the USDA's aggregate retail value of food expenditure information is intended to represent the value of food used at home as well as the value of food consumed away from home. The value of away-from-home food consumption in the United States increased much more rapidly than the value of food consumed at home between 1970 and 1997 according to the data in the third and fourth columns of Table 2–1.[3] Over this period of time, the value of the away-from-home food market increased by eight times. This compares to an expansion of five times in the value of food for at-home use. In 1997, the value of food used at home exceeded the total value of food consumed away from home by almost $74.3 billion. If recent trends in food expenditure patterns continue, the size of the away-from-home food market is likely to surpass the value of food consumed at home sometime in the near future.

One of the reasons the total values of food for at-home and away-from-home uses were higher in 1997 than earlier years is because of increases in food prices. Values of the food component of the **Consumer Price Index (CPI),** presented in the fifth column of Table 2–1, indicate how the average retail price of selected food items has changed. Interpretation of price indexes and methods used to calculate them will be discussed in more detail in Chapter 3. For the time being, it is sufficient to note that the value of a price index for a given year indicates how prices for a set of commodities in that year compare on average to prices of the same commodities for another period of time. For example, the value of 39.2 for the food component of the CPI for 1970 in Table 2–1 implies that prices of food items in 1970 were on average 39.2% of what they were in 1982–1984 (for which the corresponding value of the price index is 100). Similarly in 1997, the food price index was 157.3 indicating prices were 57.3% higher than 1982 to 1984 values.

A price index can be used to adjust aggregate expenditure data to determine what expenditures "really" would have been if prices had remained unchanged. Procedures used to adjust aggregate food expenditures for price changes are indicated by the b and c footnotes to Table 2–1.[4]

After adjusting for price changes, food expenditure data provide a better measure of the actual changes in quantities of food. Consequently, the adjusted values are often referred to as "real" expenditures. The data in columns 6 and 7 of Table 2–1 indicate that food sales for off-premise consumption increased from $197.7 billion in 1970 to $250.9 billion in 1997 or an expansion of $53.2 billion after adjusting for changes in prices. Expenditures for the away-from-home market, however, more than doubled over the same period of time from $101.0 billion to $203.6 billion. These numbers confirm that expansion in the away-from-home market has been much greater than food expenditures for at-home use during the last couple of decades.

Another factor responsible for increases in total retail food expenditures in the United States is an expanding population as noted in Chapter 1. The values

FIGURE 2–2 Real per capita food expenditures

in column 8 of Table 2–1 indicate that total U.S. population increased from 205.1 million in 1970 to 266.8 million in 1997 or approximately 30%. The effects of changes in population can also be removed from aggregate expenditure data in order to examine per capita changes in food expenditure and consumption patterns. This can be done by dividing total expenditures (after adjusting for changes in prices) by total population for that year. The resulting values represent a measure of real expenditures on a per person (or per capita) basis.

The next to last column in Table 2–1 indicates that per capita expenditures for at-home food use in terms of 1982–1984 dollars decreased some during the 1970s, but has remained relatively stable during the 1980s, and 1990s. On the other hand, the values in the last column of Table 2–1 indicate that the amount of money per person spent for food consumed away from home has been generally increasing during the same period of time even after removing the effects of higher prices. The relative changes in per capita expenditure for food at home and away from home are illustrated in Figure 2–2.

Changes in Consumption of Individual Foods

Aggregate retail food expenditure data indicate how total expenditures for food consumed at home and away from home have been changing, but do not indicate what has been happening to purchases and consumption of individual food products. Fortunately, several other sources of information can be used to analyze these types of changes.

A major source of information about consumption of individual agricultural commodities is national disappearance data. The USDA calculates this information based on production, storage stocks, and import/export data. Essentially, the process involved in developing these estimates assumes that whatever amount of output of a given product that is not accounted for by changes in storage stocks or international trade flows disappears or is used domestically. For example, suppose that annual domestic production for a particular commodity was 500 units. Also assume that 50 units of this product were in storage at the beginning of the year and 40 at the end of the year. Further assume that 30 units of this product were exported and 10 units imported during the year. The above values can be substituted into the following equation to determine how much of the commodity disappeared or was consumed:

$$\text{Production} + \text{beginning stocks} + \text{imports} = \text{disappearance} + \text{ending stocks} + \text{exports}$$

The terms on the left-hand side of the equation indicate potential sources of availability of an item. The terms on the right-hand side indicate potential uses of an item meaning that disappearance or consumption is essentially a residual measure. Thus, $500 + 50 + 10 = 560$, the total quantity that was available. If there were 40 units in stock at the end of the year and 30 units exported, it would mean that 490 of the 560 total units disappeared or were consumed.

Adjustments in disappearance data are made to take into account industrial and other nonhuman uses of certain commodities in order to get better estimates of the actual annual amounts of particular commodities available for human consumption. Disappearance (or consumption) data for various commodities are usually expressed in retail weight equivalent units. Aggregate disappearance data for various products also is usually divided by total population to facilitate analysis of trends in per capita consumption.[5] Per capita disappearance data is more useful for identifying trends in consumption patterns than for measuring absolute levels because of slippage that can occur between disappearance and actual consumption. In the case of fats and oils, however, even trends in disappearance data may not be totally valid indicators of trends in consumption because of changes in recovery and recycling methods (Putnam and Allshouse 1999).

Percentage changes in U.S. per capita food supply between 1970 and 1996 for several products are indicated by Figure 2–3. The data suggest sizable decreases in per capita consumption for coffee, eggs, milk, and red meat. On the other hand, nine food categories had marked increases in consumption over the same period of time. Per capita cheese consumption increased by 143%, followed

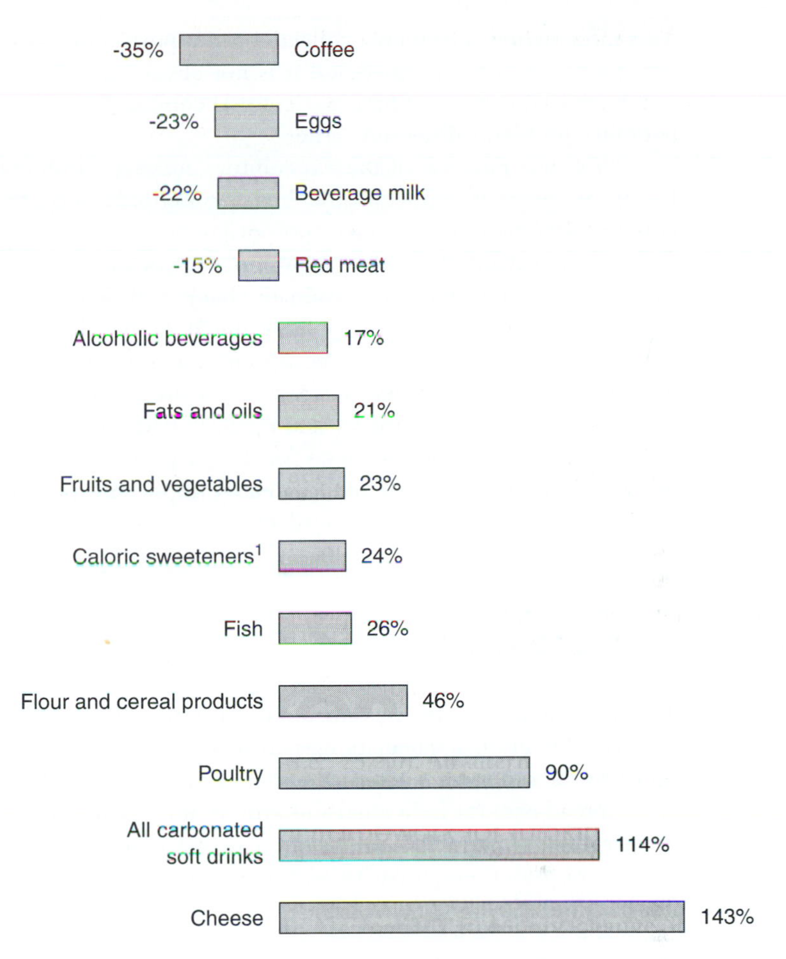

FIGURE 2–3 Percentage changes in U.S. per capita food supply for selected products between 1970 and 1996. (Data from J. Putman and S. Gerrior, Figure 1 of *Food Review,* Vol. 20, No. 3 (Sept.–Dec.), USDA, Washington, D.C. Economic Research Service, 1997.)

[1]Includes caloric sweeteners used in soft drinks.

very closely by soft drinks and poultry. The contrasts in beverage consumption as indicated by decreases in coffee and milk, but increases in alcoholic beverages and soft drinks are especially striking. Similarly among animal products, the decrease in per capita consumption of eggs and red meat are in sharp contrast to increases for fish and poultry products.

Other sources of retail expenditure and consumption data for individual products are based on periodic national surveys of household spending patterns conducted by the Bureau of Labor Statistics and the U.S. Department of Agriculture.[6] These surveys are designed to collect specific information about food consumption and expenditures among different kinds of households and individuals.

This information is generally collected and reported in terms of retail food products purchased or consumed and it is not always easy to translate this information back to quantities of basic agricultural commodities because of product compatibility problems discussed earlier.

A major purpose of the expenditure surveys conducted by the Bureau of Labor Statistics is to determine how households allocate their expenditures among different items to obtain appropriate weights for calculating price indices. On the other hand, the major purpose of the consumption surveys conducted by the USDA is to monitor and evaluate changes in dietary intakes. Consequently, in the one case, emphasis is on how much money is spent on different items whereas in the other survey there is more emphasis on actual quantities of food items consumed regardless of whether the items were purchased, donated, or home produced. Market survey and research firms also collect a variety of other information about food purchases for the purposes of monitoring the movement of individual food products and to identify important market trends. Much of the data however are not publicly available.

ALTERNATIVE MEASURES OF MARKETING SERVICES

The discussion will now be directed toward how price and food expenditure information is used to calculate certain measures of the economic importance of agricultural and food marketing activities in the United States. First, retail-farm price spread data for individual and groups of agricultural and food products will be introduced. Later the contribution of marketing services to gross domestic product and total employment will be presented.

Retail-Farm Price Spreads

The Economic Research Service within USDA uses retail and farm level price information to calculate the price spread for several individual agricultural and food commodity groups. The retail-to-farm price spreads are based on monthly retail prices of a subset of food items collected by the Bureau of Labor Statistics and imputed farm level values. The farm values for individual commodities are intended to represent the value of an appropriate quantity of the basic commodity required to produce a unit of the final product at the retail level. The farm level values for different commodities are based on prices at the first point of sale.[7]

One of the difficulties in calculating retail-to-farm spreads is determining appropriate quantity-conversion factors. For some products like eggs this is not a problem, but for other commodities appropriate conversions are necessary. The comparison of price data can be made meaningful by using conversion factors that account for changes in weight or quantities that occur as basic agricultural products are converted into retail food products. For example, if two units of an agricultural product are required to make each unit of retail product, the farm price needs to be multiplied by 2 before comparing it to the retail price to determine the value of marketing services. For instance, if the farm price of a commodity was $1.25 and the retail price was $5.00, marketing costs would represent $2.50 or 50% of the retail value if two units of the raw commodity were required for each unit of retail product. Conversion factors used by USDA are presented in Table 3 of Putnam and Allshouse (1999).

The average retail price and farm value for a number of agricultural commodities for 1993, 1994, and 1995 are indicated in Table 2–3. Information in Table 2–3 indicates the proportion of each dollar spent at the retail level that producers receive for basic agricultural commodities. The proportion of retail expenditures reflecting the value of marketing services is also implied. These proportions are frequently used to compare the relative costs of marketing different food products as well as how the proportional split between marketing costs and what farmers receive has changed over time. For example, according to data in Table 2–3, the farm value of eggs represented 60% of the retail price in 1995. This means that marketing activities for eggs accounted for 40% of retail value. On the other hand, the farm value for a 16-oz bottle of corn syrup amounted to only 4% of its retail price in 1995. This means that 96% of the latter product's retail value in 1995 was associated with marketing activities.

Values in Table 2–3 indicate variation in the farm and marketing shares for various food products. Products that involve more processing, other kinds of transformations, or generally more marketing services have larger marketing (or smaller farmer) shares of retail values. Over time as more and more processing and additional services have been added to basic agricultural products, the farmer's share of the retail food expenditures has declined and the marketing share has increased.

One of the major shortcomings of price spread information is that the retail price or value information reflects the values of food items purchased only at retail outlets for at-home consumption. As noted earlier, there has been an increasing proportion of total food expenditures allocated to the away-from-home market over the last couple of decades. Consequently, if the costs of marketing agricultural products for away-from-home consumption outlets differ from that of traditional grocery store outlets, price spread information may not be totally representative of all the marketing costs for individual commodities. For most food items, the farmer's share of retail expenditures in the away-from-home market is probably lower than what is indicated in Table 2–3 because of additional marketing services. This suggests that retail-farm price spread data like those in Table 2–3 likely overestimate the share of retail value of food products actually received by farmers.

In addition to calculating the above information for individual and groups of food commodities, the USDA computes another statistic that is frequently used as kind of a summary measure of retail-farm price spread information. This statistic is based on a **"market basket"** concept to track changes in the value or costs of marketing a group of food products relative to farm value. The market

TABLE 2-3

RETAIL FARM PRICE SPREAD INFORMATION FOR SELECTED PRODUCTS, 1993-1995

Product	Retail Price			Farm Value			Farm Value Percentage of Retail Price		
	1993	1994	1995	1993	1994	1995	1993	1994	1995
	(dollars)			(dollars)			(percent)		
Animal products:									
Eggs, Grade A large, 1 doz.	0.91	0.86	0.93	0.53	0.50	0.56	58	58	60
Beef, choice, 1 lb.	2.93	2.83	2.84	1.64	1.46	1.38	56	52	49
Chicken, broiler, 1 lb.	.89	.90	.92	.48	.49	.49	54	54	53
Milk, 2 gal.	1.39	1.44	1.43	.58	.61	.58	42	42	41
Pork, 1 lb.	1.98	1.98	1.95	.73	.63	.67	37	32	34
Cheese, natural cheddar, 1 lb.	3.34	3.35	3.39	1.15	1.17	1.16	34	35	34
Fruit and vegetables:									
Fresh—									
Lemons, 1 lb.	1.08	1.11	1.14	.29	.27	.30	27	24	26
Apples, red delicious, 1 lb.	.83	.80	.84	.19	.17	.21	23	21	25
Potatoes, 10 lb.	3.48	3.74	3.79	.78	.77	.80	22	21	21
Oranges, California, 1 lb.	.59	.56	.62	.13	.11	.12	22	20	19
Grapefruit, 1 lb.	.53	.51	.55	.10	.10	.10	19	20	18
Lettuce, 1 lb.	.66	.61	.80	.12	.12	.18	18	20	23
Frozen—									
Orange juice conc., 12 fl. oz.	1.22	1.21	1.21	.40	.46	.48	33	38	40
Broccoli, cut, 1 lb.	1.15	1.16	1.19	.26	.25	.22	23	22	18
Corn, 1 lb.	1.06	1.12	1.09	.13	.13	.14	12	12	13
Peas, 1 lb.	1.02	1.01	.96	.13	.14	.14	13	14	15
Green beans, cut, 1 lb.	.97	1.02	1.00	.11	.11	.11	11	11	11

Canned and bottled—									
Peas, 17 oz. can	.48	.51	.45	.10	.11	.11	21	22	24
Corn, 17 oz. can	.44	.48	.40	.10	.10	.11	23	21	28
Applesauce, 25 oz. jar	1.02	1.01	1.05	.16	.15	.16	16	15	15
Pears, 29 oz. can	1.23	1.21	1.22	.21	.20	.23	17	17	19
Peaches, cling, 29 oz. can	1.15	1.13	1.13	.18	.18	.16	16	16	14
Apple juice, 64 oz. bottle	1.47	1.37	1.45	.30	.24	.26	20	18	18
Green beans, cut, 17 oz. can	.42	.44	.39	.06	.06	.06	14	14	15
Tomatoes, whole, 17 oz. can	.48	.50	.54	.05	.05	.04	10	10	7
Dried—									
Beans, 1 lb.	.71	.71	.63	.20	.25	.25	28	35	40
Raisins, 15 oz. box	1.53	1.50	1.64	.49	.47	.40	32	29	24
Crop products:									
Sugar, 1 lb.	.39	.38	.38	.13	.13	.13	33	34	34
Flour, wheat, 5 lb.	1.17	1.16	1.23	.33	.36	.43	28	31	35
Shortening, 3 lb.	2.40	2.55	2.66	.70	.84	.80	29	33	30
Margarine, 1 lb.	.80	.82	.83	.19	.24	.23	24	29	28
Rice, long grain, 1 lb.	.51	.55	.53	.08	.12	.11	16	22	21
Prepared food:									
Peanut butter, 1 lb.	1.84	1.85	1.80	.48	.48	.48	26	26	27
Pork and beans, 16 oz. can	.38	.39	.40	.06	.07	.08	16	18	20
Potato chips, regular, 1 lb. bag	1.96	1.93	1.95	.29	.30	.35	15	16	18
Chicken dinner, fried, froz., 11 oz.	1.14	1.15	1.17	.16	.17	.17	14	15	15
Potatoes, french fried, frozen, 1 lb.	.86	.86	.86	.10	.10	.10	12	12	14
Bread, 1 lb.	.75	.76	.79	.05	.05	.05	7	7	8
Corn flakes, 18 oz. box	1.54	1.76	1.75	.09	.09	.09	6	5	6
Oatmeal, regular, 42 oz. box	2.58	2.56	2.56	.17	.16	.16	7	6	7
Corn syrup, 16 oz. bottle	1.56	1.59	1.63	.05	.06	.06	3	4	4

Source: Elitzak, H. "Food Marketing Costs Rose Less Than the Farm Values in 1995," *Food Review,* USDA, September–December, 1996, p. 6.

basket of commodities is based on the average annual quantity of food purchased per household for at-home use for a selected period of time (currently 1982–1984). The change in the retail value of the fixed set of commodities (excluding imports, seafood, and nonalcoholic beverages) is recalculated each year using retail price information collected by the Bureau of Labor Statistics. Changes in the farm level value for an equivalent quantity of basic agricultural commodities is calculated using prices received by agricultural producers. The difference between the retail and farm value for this set of products provides a measure of the value of marketing services associated with converting basic agricultural commodities into edible products. This decomposition of total retail value is often used to calculate a ratio representing the farmer's share of the retail value of the market basket of commodities. Changes in the values of this ratio between 1950 and 1994 indicate that the share of each dollar spent for food purchased at the retail level for at-home consumption received by farmers decreased from 47% to 24% (Table 2–4). This means that the share accounted for by the marketing system increased from 53% to 76%. In the early 1970s when world grain prices increased sharply, the farmer's share increased and the marketing share decreased as a result, but within a few years the downward trend in the farmer's share began again.

TABLE 2-4

ALTERNATIVE MEASURES OF FARMERS' SHARE OF RETAIL FOOD EXPENDITURES IN THE UNITED STATES FOR SELECTED YEARS

YEAR	MARKET BASKET[a]	MARKETING BILL[b]
	(percent)	
1950	47	41
1955	41	35
1960	39	33
1965	38	33
1970	37	32
1975	40	33
1980	37	31
1985	32	25
1990	30	24
1991	27	22
1992	26	22
1993	26	22
1994	24	21

Sources:

[a]Table 5 of *Food Cost Review, 1995*, Agricultural Economic Report, No. 729.

[b]Table 18 of *Food Cost Review, 1995*, Agricultural Economic Report, No. 729.

Aggregate Measures of Marketing Services

A closely related, but somewhat different measure of the value or costs of marketing food products also calculated by the USDA is the marketing bill. This measure produces alternative estimates of the shares of retail expenditures for food received by farmers and marketing firms (Table 2–4). Calculating the marketing bill starts with an estimate of the total amount of retail food (including nonalcoholic beverages) expenditures. The share of these expenditures attributable to imported food and seafood products is eliminated to estimate the retail value of all domestically produced food. The retail value of domestically produced food purchased for away-from-home consumption as well as products purchased for at-home use is included and hence the information is based on a different set of food commodities than those included in the market basket. These calculations produced an estimate of $546.5 billion for retail value of domestically produced food for 1996 (Table 2–5). This value is $144.7 billion less than the total value of retail food expenditures for 1996 reported in Table 2–1. The imputed gross value of domestically produced food products at the farm level was estimated to be $122.8 in 1996 (Table 2–5). This means that the difference

TABLE 2–5

COMPONENTS OF MARKETING BILL FOR FOOD FOR SELECTED YEARS

COMPONENT	1980	1985	1990	1995	1996
			(billion dollars)		
Labor[a]	81.5	115.6	154.0	196.6	206.3
Packaging materials	21.0	26.9	36.5	47.8	46.9
Rail and truck transportation[b]	13.0	16.5	19.8	22.3	22.9
Fuels and electricity	9.0	13.1	15.2	18.6	19.3
Pretax corporate profits	9.9	10.4	13.2	22.8	24.0
Advertising	7.3	12.5	17.1	20.0	20.8
Depreciation	7.8	15.4	16.3	18.7	19.4
Net interest	3.4	6.1	13.5	11.7	12.1
Net rent	6.8	9.3	13.9	19.6	20.2
Repairs	3.6	4.8	6.2	8.0	8.3
Business taxes	8.3	11.7	15.7	19.4	20.1
Total marketing bill	182.7	259.0	343.6	415.7	423.7
Farm value	81.7	86.4	106.2	113.8	122.8
Consumer expenditures	264.4	345.4	449.8	529.5	546.5

Source: Table 1 of *Food Review*, September–December, 1997, USDA, p. 28.

[a]Includes employee wages/salaries and health and welfare benefits.

[b]Excludes local hauling charges.

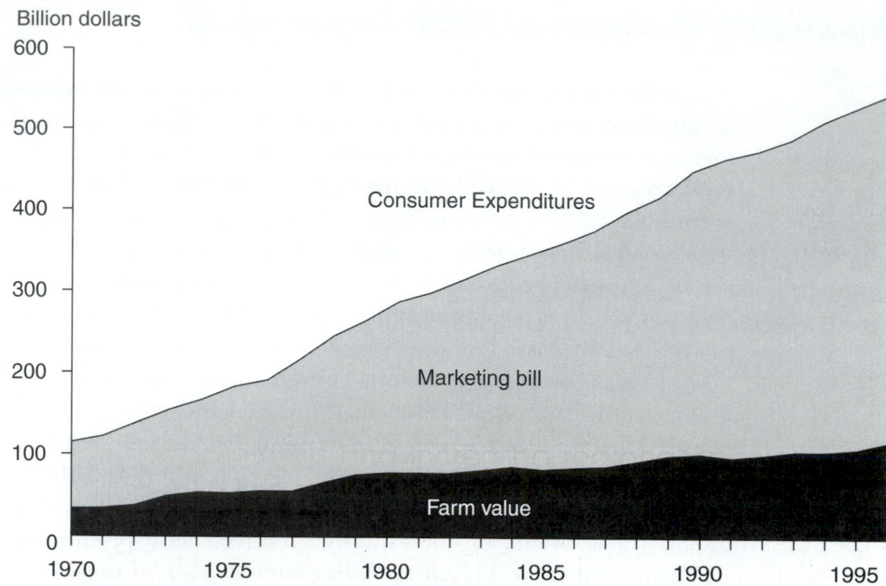

FIGURE 2–4 Marketing bill and farm value components of U.S. retail expenditures for food, 1970–1996. (Data from H. Elitzak, *Food Cost Review, 1995,* Agric. Econ. Rep. 729, USDA, 1996 and Table 9–29 in *Agricultural Statistics,* USDA, 1998.)

of $382 billion between the retail and farm values of domestically produced food represents the value of the marketing bill. A graphical depiction of changes in the marketing bill and farm value of food commodities between 1970 and 1996 is presented in Figure 2–4.

The separation of total retail food expenditures into the marketing bill and aggregate farm value of domestically produced food commodities provides information that can also be used to compute the proportion of consumer food expenditures received by farmers. For example, the marketing bill for 1996 indicates that farmers received 22.5 cents of every dollar consumers spent for domestically produced food products. Conversely, this means that 77.5 cents of every retail food dollar represented the amount of money for various marketing services. Between 1970 and 1994, the farmers' share decreased from 41% to 21% according to the calculations of the marketing bill (see Table 2–4). This compares to a decrease from 47% to 24% indicated by the market basket measure. Both measures indicate generally the same trend in proportional shares and their values are now more similar than they used to be (Figure 2–5).

Changes in the proportion of retail food expenditures received by producers (as implied by the market basket or marketing bill calculations) are often used in making comparisons of changes in the amount and/or costs of providing marketing services relative to the costs of producing basic agricultural products. When interpreting such data, however, it is important to remember how the

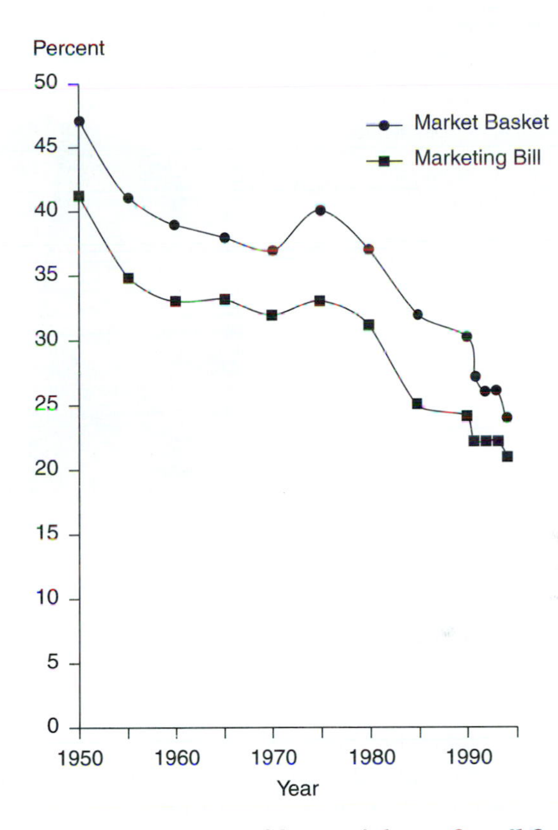

FIGURE 2–5 Alternative measures of farmers' share of retail food expenditures.

values are calculated. For example, changes in composition of the products purchased by consumers can affect the share of retail food expenditures attributable to marketing versus the proportion received by farmers. To illustrate this point, consider the following scenario. Suppose consumers decide to purchase products that have a greater quantity of marketing services instead of products with a smaller quantity of marketing services. This would result in an increase in the marketing bill even though there might not be any change in the aggregate value of underlying basic agricultural products. Thus the magnitude of the marketing bill can be affected by changes in the composition of retail food purchases over time. This cannot occur with calculations based on the market basket concept, however, because the underlying mix (or basket) of products is assumed to be the same over time. This measure, however, does not include purchases of food for away-from-home consumption and consequently is not representative of all retail expenditures for food products.

In addition to indicating the properties of total retail food expenditures attributable to marketing services, information in Table 2–5 and Figure 2–6 indicates the composition of the marketing bill. For example nearly half of the

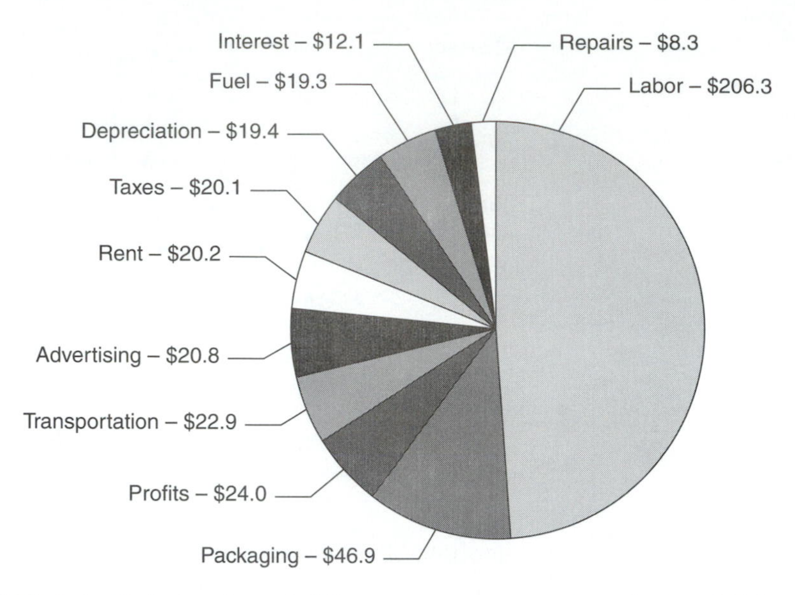

FIGURE 2–6 Relative size of different components of marketing bill for food in 1996 (in billion dollars).

total marketing bill for 1996 consists of labor costs. The next largest component of the marketing bill is the cost of packaging materials, followed by pretax corporate profits, transportation, advertising, rent, and business taxes.[8]

Contribution to Gross Domestic Product (GDP)

The two previously discussed measures of the U.S. agricultural-food marketing system are essentially residual measures. They are residuals in the sense that each measure is derived by starting with different values of domestic retail food expenditures and working backward after calculating gross values of the basic commodities at the farm level. An entirely different approach to estimating the value of food and fiber marketing and production in terms of contributions to **gross domestic product** and total employment is also used by the USDA. The estimated net values of agricultural production and marketing services based on this procedure include economic activities that occur in the United States associated with exports and imports of agricultural and food products. For example, if U.S. grown animal or crop products are processed in the United States before being exported, the net value associated with the processing activities is included in the value the agricultural marketing sector contributes to the U.S. economy. Similarly, whatever value is added to imported food products after they come into the United States is also considered part of the agricultural marketing activity.

According to these calculations, manufacturing and the distribution of food and fiber products accounted for $631 billion in 1996 (Table 2–6). The total value associated with manufacturing and distribution of food and fiber products was nine times greater than the value added at the farm level. One reason why the value for farming based on this procedure is smaller than the farm value of domestically produced food products used in calculating the marketing bill is that the values for the farm sector in Table 2–6 represent only the amount of value added to products over and above the cost of purchased inputs. The value of domestically produced food used in calculating the marketing bill represents the final value of agricultural commodities at the farm level including the cost of all purchased inputs.

The three activities that account for most of the value added by the manufacturing and distribution of agricultural commodities are wholesaling and retailing, food service, and food processing. These three components account for nearly 84% of the total value added by manufacturing and distribution of agricultural products in 1996.

The data in Table 2–6 indicate that the total value of inputs purchased by farmers and marketing firms amounted to $295 billion in 1996. The biggest component of the total is for services including financial services, insurance, communication, and other supporting services.

One way of expressing the relative importance of the food and fiber marketing and production is to consider each sector's proportional contribution to the nation's gross national product. The data in Table 2–6 indicate that food and fiber production and marketing accounted for just over 9% of gross domestic product in 1996. If the contributions of purchased inputs are added the food and fiber sector accounted for 13.1% of gross domestic product in 1996.

Employment

An alternative way of expressing the economic importance of food and fiber marketing and production in an economy is to consider how many individuals are employed in various activities. In 1996, 18.3 million people were employed in farming, manufacturing and distribution activities related to food and fiber (Table 2–6). The number employed in manufacturing and distribution activities was approximately ten times greater than the number of people employed at the farm level. Another 4.3 million people were employed in producing inputs purchased by farmers and marketing firms.

The production and marketing of food and fiber in the United States accounted for about 13.7% of total employment in the United States. This proportion is a little larger than the share of gross domestic product discussed earlier. One of the reasons for this difference is because the average wage (and presumably the value) of some of the jobs in the food and agricultural sectors tends to be less than in some other areas of the economy. This should not be interpreted to mean that all of the food and fiber jobs have low wages. Opportunities to increase labor

TABLE 2-6

CONTRIBUTION OF THE FOOD AND FIBER SYSTEM TO THE U.S. ECONOMY, 1996[a]

Industry	Value Added to GDP	Share of GDP	Number of Workers	Share of Total U.S. Employment
	($ Billion)	(Percent)	(Thousands)	(Percent)
Farming	71.3	0.9	1,637	1.2
Total manufacturing and distribution:	631.4	8.3	16,716	12.5
Manufacturing—				
Food processing	108.0	1.4	1,316	1.0
Textiles	48.2	0.6	1,352	1.0
Leather	0.3	—	7	—
Tobacco	18.4	0.2	45	—
Distribution—				
Transportation	33.8	0.4	602	0.4
Wholesaling and retailing	283.1	3.7	6,519	4.9
Foodservice	139.2	1.8	6,874	5.1
Total inputs:	295.4	3.9	4,343	3.2
Mining	13.4	0.2	60	—
Forestry, fishing, and agricultural services	8.7	0.1	315	0.2
Manufacturing	94.4	1.2	1,186	0.9
Services	178.9	2.3	2,782	2.1
Total food and fiber system	997.7	13.1	22,694	16.9

Dash is less than 0.1 percent.
[a]Numbers may not add to totals due to rounding.

Source: Table 1 from Lipton et al. (1998).

productivity in certain aspects of food service have been somewhat limited relative to other segments of the economy. This is especially true for many labor-intensive jobs in the expanding away-from-home market.

Eating and drinking establishments account for the largest number of jobs among different food marketing activities. In 1996, more than 40% of total employment in manufacturing and distribution of food and fiber was associated with food service. Retailing and wholesaling was the next largest source of employment. Food processing and the textile industry had almost the same number of employees as the number at the farm level.

SUMMARY

Several measures of the economic importance of agricultural and food markets in the United States were introduced in this chapter. Conceptual difficulties associated with comparability of agricultural and food products at the farm and retail levels and complications involving availability of retail and farm level information about individual and groups of commodities were noted. Differences in how agricultural and food products are defined and globalization of markets mean that subtracting the total farm value of agricultural products from retail expenditures for food does not provide a meaningful measure of the value of marketing services. Difficulties also arise in making meaningful comparisons between retail and farm level prices of individual commodities for a couple of reasons. The variety of retail food items produced from the same animal or crop means it is not easy to select a single representative retail price to compare to the value of the basic commodity at the farm level. Another problem is that many prepared retail food products consist of a mixture of different agricultural commodities.

Another factor affecting the calculation of meaningful empirical measures of marketing activities is the availability of appropriate data. Differences in definitions and the procedures used to estimate annual U.S. retail food expenditures by the U.S. Department of Agriculture and the U.S. Department of Commerce are summarized. The difference between the U.S. Department of Agriculture's estimates of total retail food expenditures and total personal food expenditures is also noted. Procedures used for estimating national per capita consumption (or disappearance) of individual agricultural and food commodities from production, inventory, and international trade data were also discussed.

Recent trends in aggregate and per capita expenditures for food indicate a greater expansion in away-from-home food consumption relative to expenditures for food eaten at home. USDA data also indicate that changes in per capita consumption of different food products have not been uniform. For example, the consumption of coffee, eggs, milk, and red meat has decreased between 1970 and 1996 whereas nine other food categories including cheese, soft drinks, fish, and poultry increased during the same period of time.

The manner in which retail-farm price spread information is calculated by USDA and certain limitations of the resulting measures for selected products are also described. Similarly, the way aggregate value measures of marketing services associated with a representative "market basket" of commodities as well as an aggregate "marketing bill" are computed is indicated. Depending on which measure is used, the estimated proportions of total annual retail food expenditures received by producers and marketing firms differ. Both measures indicate an increasing share of retail food expenditures allocated to marketing activities and a decreasing share to producers over time.

Recently, food production and marketing activities have accounted for about 9% of U.S. gross domestic product. The total value of food marketing activities is more than ten times the value of food products at the farm level. Food production and marketing account for about 13.7% of total U.S. employment. Employment in the marketing sector is about ten times greater than the number of people involved in producing food. Nearly 40% of total employment in the manufacturing and distribution of food and fiber are associated with food service.

QUESTIONS

1. Briefly explain the conceptual differences between USDA's measures of total food expenditures and personal food expenditures.
2. What does the following information indicate about the disappearance of a given commodity during a particular time period?
   Beginning inventory—125 units
   Ending inventory—95 units
   Production—650 units
   Exports—70 units
   Imports—35 units
3. Briefly explain how a retail-farm price spread for a particular commodity is calculated and what it represents.
4. Briefly explain why USDA retail-farm price spread information for a particular commodity likely overestimates the proportion of retail food expenditures on that product actually received by farmers (i.e., producers).
5. Suppose that the retail price of a food product is $4.00/lb and the price of the major agricultural ingredient required for that retail food product is $1.00/lb at the farm level. If 2 pounds of the farm product is required to produce each pound of retail product, what does the above information indicate about the farmer's share of retail expenditures on this product?
6. Assume that marketing costs account for 60% of the price of a retail food product and each unit of this product requires three units of a basic agricultural commodity produced by farmers. If the retail food product is currently selling for $3.00/unit, what does the preceding information imply about the price received by producers per unit of the basic agricultural product? Explain the logic supporting your answer.
7. Briefly describe one of the ways the calculation of marketing costs based on the market basket differs from that of the marketing bill.

REFERENCES

Elitzak, H. 1996. *Food cost review, 1995*. Agric. Econ. Rep. 729. USDA, Economic Research Service. Washington, D.C.

Elitzak, H. 1997. Farm value grew more than marketing costs in 1996. *Food Review* 20(3), (Sept.–Dec.): 28–32. USDA, Economic Research Service. Washington, D.C.

Gallo, A. 1998. *The food marketing system in 1996*. Agric. Info. Bull. 743. USDA, Food and Consumer Economics Division, Economic Research Service. Washington, D.C.

Lipton, K.L., W. Edmondson, and A. Manchester. 1998. *The food and fiber system: Contributing to the U.S. and world economies*. Agric. Info. Bull. 742. USDA, Economic Research Service. Washington, D.C.

Manchester, A.C. 1990. *Data for food demand analysis: Availability, Characteristics, Options,* Agric. Econ. Rep. 613. USDA, Economic Research Service. Washington, D.C.

Manchester, A.C. 1987. *Developing an integrated information system for the food sector.* Agric. Econ. Rep. 575. USDA, Economic Research Service. Washington, D.C.

Putnam, J.J., and J.E. Allshouse. 1999. *Food consumption, prices, and expenditures, 1970–97.* Statistical Bull. 965. USDA, Economic Research Service. Washington, D.C.

Putnam, J., and S. Gerrior. 1997. Americans consuming more grains and vegetables, less saturated fat. *Food Review* 20(3) (Sept.–Dec.): 2–13. USDA, Economic Research Service. Washington, D.C.

Schrimper, R.A. 1989. Cross-sectional data for demand analysis. In R.C. Buse (ed.), *The economics of meat demand* (pp. 74–92). Proc. of the Conference on The Economics of Meat Demand, Charleston, South Carolina.

Smallwood, D.M., J.R. Blaylock, S. Lutz, and N. Blisard. 1995. Americans spending a smaller share of income on food. *Food Review* 18(2): 16–19. USDA, Economic Research Service. Washington, D.C.

U.S. Department of Agriculture. 1998. *Agricultural statistics.* Washington, D.C.

U.S. Department of Commerce. *Survey of current business.* Bureau of Economic Analysis, various issues.

U.S. Department of Commerce. 1983. *The national income and product accounts of the United States, 1970–1995.* A supplement of the Survey of Current Business, Bureau of Economic Analysis.

Weimer, J. 1996. USDA's role in nutrition education and evaluation. *Food Review* 19(1): 41–45. USDA, Economic Research Service. Washington, D.C.

NOTES

1. Differences in how food and agricultural products are defined not only complicate interpreting differences between retail food expenditure and agricultural production data, but are also why variation occurs among different

measures of farm income, agribusiness employment, and other indicators of agricultural activities for a given geographical area. More comprehensive definitions of agricultural activities result in bigger numbers.

2. Pet foods can be even a bigger problem when aggregate food production or retail information is used to analyze changes in human nutrition intakes. Another issue is how expenditures and consumption of vitamins and mineral supplements are handled in different retail data series. Vitamins and mineral supplements are usually not included as part of the agricultural-food marketing system even though they affect nutrient intake.

3. The Commerce Department also provides total values of food purchased for off-premise consumption, purchased meals and beverages, food furnished to employees including military personnel, and food produced and consumed on farms. The first two of these components include the value of alcoholic beverages and consequently are not directly comparable to USDA's estimates of at-home and away-from-home food consumption. The total personal food expenditures data from the Commerce Department in Table 2–2, however, exclude alcoholic beverages.

4. Adjusting expenditure information for price changes results in adjusted levels of expenditures that are higher than the unadjusted values for years for which prices are less than the base year. Similarly, the values of total expenditures are adjusted downward for those years for which prices are greater than the base year. The adjustment procedure will be more fully explained in the next chapter.

5. Additional details about the calculations of per capita disappearance data can be found in Putnam and Allshouse (1999).

6. Additional details about continuing national surveys of household expenditure and food intake surveys conducted by the Bureau of Labor Statistics and the USDA are available in Schrimper (1989), Smallwood et al. (1995), and Weimer (1996).

7. The farm level values are based on prices received by farmers as tabulated by the National Agricultural Statistics Service less allowance for any byproducts and may include some grading and packing services. In some cases, a value at the farmer's level is imputed based on wholesale price information because of the integrated structure of producing and marketing certain commodities.

8. The USDA also calculates a price index that summarizes changes in the price of the various inputs used for marketing food products. This index is published in *Agricultural Outlook* and can be used as an indicator of what might happen to the size of the marketing bill because of changes in the prices of certain inputs used to provide a given quantity of marketing services. Further details about the items included in this index are available in Elitzak, *Food Cost Review,* 1995.

CHAPTER

3

INDEX NUMBERS

This chapter describes how index numbers are constructed and used for summarizing and analyzing changes in market conditions.

The major points of the chapter are:

1. Key decisions that are involved in constructing any type of price index.
2. The meaning and interpretation of **price relatives.**
3. Difference between **Laspeyres** and **Paasche price indices.**
4. How the **base year** of a price index can be modified to facilitate comparisons.
5. What the Consumer Price Index measures and some of its uses.

INTRODUCTION

Frequently it is useful to characterize how market conditions change over time by comparing prices. If the comparison involves only a single product, it is easy to compare prices for different time periods provided relevant information for the product is available. If the comparison involves several commodities, however, some way of combining changes in prices of several products into a summary measure is required. Various kinds of statistical measures called price indices are often constructed for purposes of making such comparisons.

The first part of this chapter discusses three basic issues associated with any type of index representing a composite or summary measure of changes in prices of a multiple set of goods and/or services over time. Alternative types of price indices that differ in the way price and quantity information are combined are introduced in the middle part of the chapter. The next part of the chapter indicates how price indices can be modified and used to adjust economic data to remove the effect of changes in prices. The final section of the chapter includes information about the Consumer Price Index that is frequently used as a measure of inflation.

BASIC ISSUES

Whenever any kind of price index is calculated, there are essentially three issues that have to be faced. The first issue is what items to include in the index. For example, if an index of prices received by agricultural producers is to be constructed, one set of commodities might be selected. On the other hand, if an index of prices paid by consumers is being constructed, an alternative set of products would be relevant. Determining the items to include in a price index is dictated primarily by what set of price changes are going to be measured and summarized or how the index is going to be used. Another factor influencing selection of items can be availability of price information about particular commodities.

A second major issue in constructing a price index is what sort of weights are going to be attached to the individual prices or price changes for various commodities included in an index. In other words how is price information for a number of commodities going to be combined into a single number? Is the price information for each item going to be weighted equally or are different weights going to be applied to prices of different products?

A third and closely related issue to weighting concerns is what base year to use for reporting values of the index. Later there will be a discussion of alternative ways in which the term **base year** is used in conjunction with price indices.

To illustrate the preceding three issues, consider constructing a price index for three major U.S. field crops to determine how prices changed between 1988 and 1997. For example, suppose the following three commodities are selected for inclusion in a price index: cotton, corn, and soybeans. The 1988 and 1997

TABLE 3–1

AVERAGE PRICE RECEIVED BY U.S. FARMERS

	Marketing Year	
Item	*1988*	*1997*
	$	
Cotton (per lb)	.566	.674
Corn (per bu)	2.54	2.60
Soybeans (per bu)	7.42	6.50

Source: Data from *Agricultural Statistics, 1998*, USDA.

prices and quantities of these three commodities are indicated in Tables 3–1 and 3–2. The values in Table 3–1 indicate that substantially greater (absolute as well as proportional) changes in the prices of cotton and soybeans than in the price of corn occurred between 1988 and 1997. Cotton and corn prices increased between 1988 and 1997 whereas the price of soybeans decreased by almost $1 per bushel Figure 3–1. The purpose of a price index is to be able to calculate a single summary measure of price changes over some interval of time for the commodities selected.

Average of Price Relatives

One type of price index that could be calculated would be the average of price relatives between 1988 and 1995 for the three commodities. A price relative for

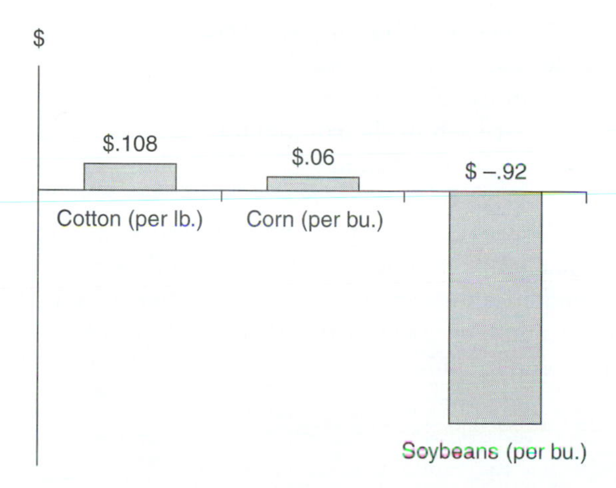

FIGURE 3–1 Changes in prices between 1988 and 1997. (Data from Table 3–1).

TABLE 3–2

QUANTITY OF U.S. PRODUCTION

	Marketing Year	
Item	*1988*	*1997*
	(billion)	
Cotton (lb)	7.403[a]	9.113[a]
Corn (bu)	4.927	9.366
Soybeans (bu)	1.549	2.727

Source: Data from *Agricultural Statistics, 1998,* USDA.
[a]Based on values of production divided by average price per pound.

an individual commodity is equivalent to $(P_1/P_0) \times 100$ where P_1 is the price for one particular year and P_0 is the price in an initial (or base) year. Based on the data in Table 3–1, the price relatives for the three commodities in 1997 relative to 1988 are 119.1 for cotton, 102.4 for corn and 87.6 for soybeans. These price relatives result from dividing the 1997 price of each commodity by its price in 1988 and multiplying the result by 100. Comparing each of the price relatives to 100 (the price relative for the base year) indicates that the price of cotton increased 19.1%, the price of corn increased 2.4%, and the price of soybeans decreased 12.4% between 1988 and 1997.[1] A price relative indicates the percentage change for the respective commodities because percentage changes are always measured from a base of 100. Since the numerator and denominator of the price relative for any commodity in a base year are identical, a price relative for the base year will always be equal to 100. If a price relative is 110 for some year, it would indicate a 10% increase from the base year. A price relative that is less than 100 indicates a decrease in price from the base year.

In order to calculate a price index using the average of price relatives, the price relatives for each of the commodities included in an index are added and the total divided by the number of commodities. This would be equivalent to weighting the price relative for each commodity equally. For the above example, the simple average of the price relatives (or weighting each of the three price relatives equally by one-third) is 103.0. This indicates that the 1997 prices of the three commodities on average were 3.0% higher than in 1988. The 3.0% change represents a simple average of the 19.1%, 2.4%, and −12.4% changes in prices of the three commodities.

Real Prices

When analyzing changes in price, it is important to distinguish between nominal and real changes. A nominal change refers to the actual change in prices or a price index at two different points in time. Real changes refer to changes after the effects of general inflation (or deflation) have been removed from nominal price changes. One of the most commonly used measures of inflation in the United States is the Consumer Price Index for all urban consumers (CPI-U). More information about how this index is constructed will be provided later, but for the time being it is sufficient to note that the CPI-U index indicates an increase in the general price level of 35.6% between 1988 and 1997. In other words, the CPI-U index for 1997 was 135.6 relative to a base value of 100 in 1988. Dividing the average of the price relatives for the three field crops in the previous example by the 1997 value of the CPI-U based on a value of 100 for 1988 is one way that the change in prices for the three commodities could be expressed in real terms relative to 1988. In other words, dividing 103.0 by 135.6 and multiplying by 100 yields 76.0. This means that the ***real*** prices for the three commodities in 1997 were on average 76.0% of what they were in 1988. Another way of interpreting this number is that the purchasing power of the three commodities decreased by 24.0% between 1988 and 1997 in terms of the items included in the CPI.

An alternative deflator for measuring the purchasing power of the agricultural commodities is an index of prices paid by U.S. farmers for various kinds of **inputs.** This index is calculated by the USDA. The index of prices paid by U.S. farmers increased by 28.6% between 1988 and 1997 or approximately 80% of the rate of change in the consumer price index during the same period. Using the index of prices paid by farmers for deflating the average of the price relatives for the three commodities would involve dividing 103.0 (the average of the price relatives) by 128.6 (the index of prices paid by U.S. farmers on a 1988 = 100 base) and multiplying by 100. This results in a value of 80.1 which represents a decrease in the purchasing power of the three commodities of almost 20% between 1988 and 1997. The difference between a 20% or 24% decrease in the real prices or purchasing power of the three commodities between 1988 and 1997 is the result of using different groups of items for measuring purchasing power. In the first case, the comparison is made to items included in the Consumer Price Index and in the second case, the comparison is relative to the prices of various kinds of inputs purchased by U.S. farmers.

WEIGHTING ISSUES

One objection to using an average of price relatives as a representative measure of **nominal price** changes for a set of commodities is that the way it is calculated assumes that the price change for each commodity is equally important. In the previous example, the price relatives of cotton, corn, and soybeans each had an equal weight of one-third in terms of influencing the overall average change in prices even though the total values of corn and soybean production were far greater than the value of cotton production in 1988 as well as in 1997 based on the data in Tables 3–2 and 3–3. This gets to the heart of the second

TABLE 3-3

1988 AND 1997 VALUES OF PRODUCTION

	Marketing Year			
Item	*1988*		*1997*	
	Billion $	Proportion	Billion $	Proportion
Cotton	4.190[a]	.148	6.142[a]	.127
Corn	12.515[b]	.444	24.352[b]	.505
Soybeans	11.494[b]	.408	17.726[b]	.368
	28.199	1.000	48.220	1.000

Sources: [a]Data from *Agricultural Statistics, 1998.* USDA.
[b]Based on values in Tables 3-1 and 3-2.

issue mentioned initially in discussing the construction of any price index (i.e., what sort of weights should be applied to the price changes of different commodities). Equally important issues are what commodities to include in an index and what base period to use, but for the moment assume that a choice has been made in terms of what products to include in the index and that 1988 has been selected as the base period for comparison purposes. This leads to consideration of alternative price indices that use different weighting procedures.

Laspeyres Index

One of the most widely used procedures for calculating a price index is the Laspeyres formula. What this amounts to is applying different weights to the price relative of each item included in an index. The weights are determined by the relative importance (in terms of value) of each item in a base period. This requires calculating the proportion of the total value of each item like those in Table 3–3 or Figure 3–2 for a base period. The price relative for each commodity is multiplied by the proportion that each item contributes to the total value of all items in the index. The value of the price index is obtained by summing the values that result after multiplying each price relative by the appropriate weight.

Multiplying the 1988 proportions in Table 3–3 by the price relatives calculated earlier for each commodity and then summing produces a value of 98.8 or $(.148 \times 119.1) + (.444 \times 102.4) + (.408 \times 87.6)$. This indicates that the Laspeyres index suggests that the prices of the three commodities in 1997 were 1.2% lower than they were in 1988. This is in contrast to the 3.0% increase indicated by the average of the price relatives. The difference in the two results is because the price increase for cotton receives less weight and the decrease in soybean prices receives greater weight in calculating the Laspeyres index than when the price change for each commodity was weighted equally.

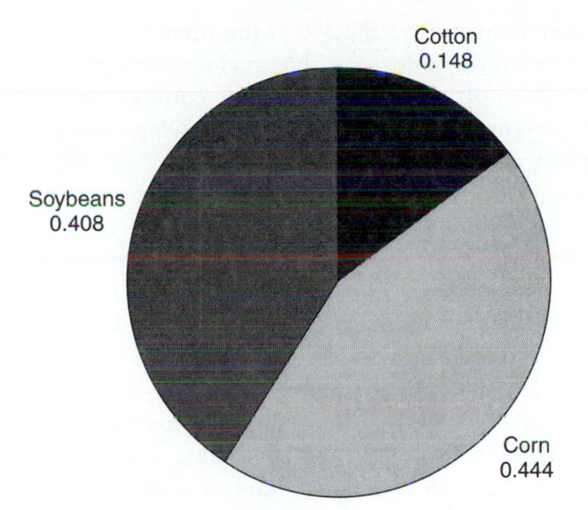

FIGURE 3–2 Proportion of total value of three commodities in 1988. (Data from Table 3–3.)

A mathematical representation of the above calculations is one way of expressing a formula for a Laspeyres index. It can be written as the summation (Σ) of the weights for each item multiplied by the respective price relative for all items included in the index. Thus the Laspeyres index (I_L) can be represented as follows:

$$I_L = \sum_{i=1}^{n} w_i \text{ (price relative of } i^{\text{th}} \text{ commodity)} = \sum_{i=1}^{n} w_i \left(\frac{P_{1i}}{P_{0i}} \times 100 \right) \quad \textbf{(Eq. 3–1)}$$

where $w_i = \dfrac{P_{0i}Q_{0i}}{\sum_{i=1}^{n} P_{0i}Q_{0i}}$ with i referring to each of the particular items included in the index and the first subscript on P and Q referring to a particular time period. The zero subscript refers to the base period and 1 refers to the year for which the index is being calculated. Thus P_{0i} is the price of the ith commodity in the base period and P_{1i} represents the price of the ith commodity for a subsequent year for which the price index is being calculated.

An alternative and more frequent way of expressing and calculating a Laspeyres price index results by substituting the expression for w_i into the above formula and doing some algebraic simplification to obtain:

$$I_L = \frac{\sum_{i=1}^{n} P_{1i}Q_{0i}}{\sum_{i=1}^{n} P_{0i}Q_{0i}} \times 100 \quad \textbf{(Eq. 3–2)}$$

The denominator of the Eq. 3–2 is the total base year's value of the commodities included in the index. For the previous example, the denominator would be $28.199 billion or the value of the three commodities in 1988 (see Table 3–3) in that it represents the summation of multiplying each price times the quantity of the respective item in the base year (in this case 1988). The numerator of the second expression for the Laspeyres formula represents a measure of what the total value of the quantities in the base period would have been using the prices that existed at a later time. To compute the numerator of the second formula, the 1997 price for each commodity in Table 3–1 needs to be multiplied by the 1988 quantity of that commodity indicated in Table 3–2 and the results summed as illustrated in the second column of Table 3–4. The value of $27.869 billion indicates what the total value would have been if producers had received 1997 prices for the same amount of each of the three commodities produced in 1988. Dividing $27.869 billion by $28.199 billion and multiplying by 100 results in a value of 98.8, exactly the same value as multiplying the price relatives by the 1988 proportions in Table 3–3 and summing the results. This illustrates that a Laspeyres price index can also be calculated as the ratio of two aggregate value (revenue or expenditure) measures.

The calculation of a Laspeyres index in either of two ways leads to identical results. If the price relatives for individual commodities are readily available it may be easier to calculate proportional weights for a base period (like those in Table 3–3), multiply by the respective price relatives, and then sum the results to get the value of the index. If price relatives for individual commodities have not been calculated, it may be easier to calculate the aggregate value of the numerator of Eq. 3–2, divide by the total value of the commodities in the base year, and multiply by 100. Either way results in the same numerical value of a Laspeyres price index. In fact, doing the calculations both ways is a good way of checking to make sure no errors have been made.

TABLE 3–4

ALTERNATIVE COMPONENTS FOR CALCULATING LASPEYRES AND PAASCHE PRICE INDICES

Item	1988 Quantities Valued at 1997 Prices	1997 Quantities Valued at 1988 Prices	Proportional Shares of 1997 Quantities Valued at 1988 Prices
	Billion $	Billion $	
Cotton	4.990	5.158	.105
Corn	12.810	23.790	.484
Soybeans	10.069	20.234	.411
Total	27.869	49.182	1.000

Based on data in Tables 3–1 and 3–2.

Paasche Index

An alternative weighted price index is the Paasche index (I_p). Mathematically this index can be expressed as follows:

$$I_p = \frac{\sum_{i=1}^{n} P_{1i}Q_{1i}}{\sum_{i=1}^{n} P_{0i}Q_{1i}} \times 100 \qquad \textbf{(Eq. 3–3)}$$

At first glance, the formula for the Paasche index looks very similar to the formula for the Laspeyres index. The only difference between the two formulae is different subscripts on the quantity variables used in calculations of the numerator and denominator. The Paasche index uses quantities for the most recent period (1) whereas the Laspeyres index uses quantities for a base period (0). The Paasche index requires quantity information for the latest period for which index numbers are being calculated, but uses the same price information as the Laspeyres formula. For an example, using the same three commodities as before, the numerator of the Paasche index amounts to the actual total value of production in 1997 because the 1997 price of each commodity is multiplied by the 1997 quantity and added as in the fourth column of Table 3–3. The denominator of the Paasche index is what the total value of the three commodities would have been in 1988 had production been the same as the 1997 quantities (see Table 3–4). Dividing \$48.220 billion by \$49.182 billion and multiplying by 100 yields 96.8. This value suggests a slightly greater decrease in prices between 1988 and 1997 than what is reflected by the Laspeyres index. The reason for the difference in values of the two price indices is that cotton is not weighted as heavily in the Paasche calculation as the Laspeyres index because it did not have as large an increase in the total value between 1988 and 1997 compared to the other two commodities (see the proportions in Tables 3–3 and 3–4). Situations with substantial change in the weighting factors over time will lead to a greater difference in values for the Laspeyres and Paasche indices than in the previous example.

An alternative way of considering the difference in the two indices is that the Paasche index weights price relatives using the proportions of total value based on 1997 quantities and 1988 prices, whereas the Laspeyres weights price relatives by the relative importance of each commodity in the base year. One way to remember the difference in the formulae for the Laspeyres and Paasche indices is to recall that the Laspeyres index uses Q_{0i} and the Paasche index uses Q_{1i} in the calculations. A mnemonic scheme that may be useful for remembering this is that in the alphabet the letter *L* (for Laspeyres) comes before *P* (for Paasche), just like 0 precedes 1. Both indices use the same price information.

Recently, governmental price indices have begun to calculate year-to-year changes in price based on weights for the current and preceding years.[2] These kind of indices are referred to as "chain-weighted" indices which eliminate the use of fixed weights. Using chain-weighted indices means that the weights change gradually over time with changes in the quantities of items included in an index.

BASE PERIOD

Choosing between a Laspeyres or Paasche index is essentially deciding between using different quantities (Q_{0i} or Q_{1i}) for weighting purposes. In the case of a Laspeyres index, selecting the particular set of Q_{0i} values determines what period of time will be the base or reference point for the index. This is because the index will automatically be equal to 100 because the numerator and denominator components of the index are identical in the base year. For all other time periods, the index will deviate from 100 unless each and every price remains unchanged, or there are a sufficient number of price increases offset by price decreases to maintain exactly the same value of the numerator over time. Therefore selecting the appropriate set of Q_{0i} to use is essentially determining the base period for a Laspeyres index in that it is equivalent to setting the index equal to 100. The values of the index for other time periods indicate percentage changes from the base period.

The fact that a percentage difference between any two numbers (including index numbers) can be calculated, leads to an alternate way in which values of a price index can be converted or rescaled to a different base period. This is done by selecting one of the original values of the index and expressing all other values of the index relative to the selected value similar to what is involved in computing a price relative for an individual commodity. This procedure rescales all values in the index by selecting one of the years to have a value of 100 and expressing all other values relative to that point.

For example, each of the CPI values in the second column of Table 3–5 indicates the percentage change in prices from 1982 to 1984 for which the base value of the index is 100. For example, the values indicate that in 1998, consumer prices were on average 63.0% higher than they were in 1982 to 1984. The values of this index can also be used to calculate percentage changes between other points in time. In fact, when converting the average of the price relatives for the three commodities into real terms earlier it was noted that the CPI increased by 35.6% between 1988 and 1997. The value can be observed directly if each of the CPI values in the second column of Table 3–5 is divided by 118.3 (the 1988 value) and multiplied by 100. This process would lead to a new series of CPI values (the third column of Table 3–5) with a value of 100 for 1988 and 135.6 for 1997 reflecting a 35.6% increase from 1988. The converted series of CPI values is especially convenient for making comparisons for any year relative to 1988 instead of from 1982 to 1984. Without additional information, a value of 100 for a Laspeyres index does not necessarily mean that the weights used in its construction refer to that period. The original index could have been rescaled to a different base year for ease of making comparisons.

TABLE 3-5

U.S. CONSUMER PRICE INDEX AND DISPOSABLE PERSONAL INCOME PER CAPITA FOR SELECTED YEARS[a]

Years	CPI-U[a] 1982-84 = 100	CPI-U 1988 = 100	Disposable Personal Income per Capita[b]
1988	118.3	100.0	13,896[b]
1990	130.7	110.5	16,721[b]
1991	136.2	115.1	17,242[b]
1992	140.3	118.6	18,113[b]
1993	144.5	122.1	18,706[b]
1994	148.2	125.3	19,381[b]
1995	152.4	128.8	20,349[b]
1996	156.9	132.6	21,117[b]
1997	160.5	135.6	21,633[c]
1998	163.0	137.8	22,299[c]

Sources: [a]Table 772, *Statistical Abstract of the United States, 1998* and *Agricultural Outlook*, April 1999.
[b]Table 722, *Statistical Abstract of the United States, 1998.*
[c]Table 2.1, *Survey of Current Business*, February 1999.

A similar situation exists with Paasche indices. In general, the Paasche index would equal 100 only in the case where the prices in the numerator are identical to the prices in the denominator except for extremely unlikely circumstances of offsetting price differences among commodities. It is also possible to rescale Paasche indices however, by dividing all index values by the value for a selected year thereby expressing everything in terms of percentage changes from a year for which the index is equal to 100.

Thus the term *base period* can refer to the time period selected for weighting purposes, or a period from which all percentage changes are expressed. It is possible that these could refer to the same year or be different.

CONSUMER PRICE INDEX

As noted previously the Consumer Price Index (CPI) is a widely used measure of inflation or changes in the general price level in the United States. This index is frequently used for making "cost-of-living" adjustments in wages, retirement benefits, and so forth. The CPI is a Laspeyres type of index that is recomputed every month based on new price information, and is further adjusted about every decade with a new set of weights representing more current household expenditure patterns. The weights are based on information from a large national sample survey of households about how much money they spend on various items.[3]

Price quotations for approximately 71,000 goods and services are collected from approximately 22,000 retail establishments each month or bimonthly in urban areas of the United States (Boskin et al. 1998). Additional information on rent and owner's equivalent rent is obtained from about 35,000 households. All of this information is incorporated into a modified Laspeyres type of formula to measure the relative change in consumer prices from a base period.[4]

The CPI is frequently used in analyzing changes in household or per capita real incomes or real prices of individual commodities. If incomes have increased more than the change in the CPI, it would indicate an increase in the purchasing power or real income of consumers. For example, nominal values of disposable per capita personal income in Table 3–5 indicate an increase from $13,896 in 1988 to $22,299 in 1998 or an increase of 60.5%. Given that the number is more than the 37.8% increase in the consumer price index over the same period, an increase in real income or purchasing power is indicated. If income had increased by a smaller percentage change than the CPI, it would have indicated a decrease in the purchasing power of consumer incomes.

In order to calculate how much of a change has occurred in real incomes between 1988 and 1998 or any other period, nominal income data must be converted to real income by removing the effects of inflation. This can be done by dividing nominal income for each year by a CPI value for that year (assuming some particular base year has been selected) and multiplying the result by 100. This process will lead to measures that represent the real purchasing power in terms of prices of the base year of the price index.

If the disposable income numbers in Table 3–5 are deflated by the first column of CPI data in Table 3–5, the resulting values of real income would be in terms of 1982 to 1984 purchasing power. If changes in real income in terms of 1988 purchasing power were of interest, it would be necessary to use the converted CPI series in the third column of Table 3–5 to deflate the income data. This would result in $\left(\frac{22{,}299}{137.8} \times 100\right)$ or a value of $16,182 for 1998 in terms of 1988 dollars. The latter value could then be compared to the $13,896 of disposable personal income per capita in 1988 indicating an increase of 16.4% between 1988 and 1998. Consequently, measures of real income in terms of 1982 to 1984 or 1988 purchasing power can be obtained using essentially the same information depending on which base year is selected for representing the price index.

Hourly wage rates, aggregate expenditures on food or other commodities, or any other nominal variable can be put on a real basis by deflating or removing the effects of general changes in price levels using price indices. In the case of deflating food expenditures, a component of the overall consumer price index that measures only food prices might be more appropriate than the overall CPI index. This would remove the effects of price changes for just food products when examining what has been happening to trends in real expenditures for food (i.e., the amount of money that would have been spent on food quantities in the absence of price changes). The process would be identical to the way information about disposable income and the CPI in Table 3–5 can be used to compute an adjusted series of real income. Dividing total food expenditures for each year by the value of a food price index for the same year and multiplying by 100 (as in Table 2–1 of Chapter 2), produces a revised series of expenditure information in terms of the prices that existed during the base period of the price index used for deflation. The adjusted expenditure information is often referred to as deflated data because of generally having to adjust for increasing prices over time. The same process would be applicable for adjusting expenditure data if prices had been decreasing.[5]

SUMMARY

Price indices are frequently used to summarize changes in prices for various kinds of goods and services. The three basic issues encountered in constructing such measures are (1) what particular set of items should be included, (2) how price information for different items should be weighted, and (3) what base period is to be used for comparison purposes.

The simplest kind of price comparison for an individual item is a price relative. A price relative is the ratio of prices of an item at two different points in time multiplied by 100. If prices for the two observations used for calculating a price relative are identical, the value of the price relative is equal to 100. The extent to which a price relative differs from 100 indicates the percentage difference (or change) in price from the value in the denominator (or base period).

Once decisions are made about what base period to use for calculating price relatives and what items to include in an index, different composite or summary measures can be constructed depending on how information for each price relative is weighted. The simplest price index is an unweighted average of price relatives for the items of interest. This kind of index permits a given percentage change in price for each item to have the same impact on the final value of the index regardless of the "relative economic" importance of the item.

Alternative ways of calculating a price index by differentially weighting price relative information yield Laspeyres or Paasche price indices. Laspeyres price indices use the proportional share or relative value of each item in a base period for weighting purposes. In this way, the influence of each item's price relative is weighted by its relative value in a base year. In contrast, Paasche price indices use the most recent quantity information of each item for weighting purposes. If quantities change very little over time, the values of Laspeyres and Paasche indices will be very similar. Changes in the relative quantities of items over time, however, can result in different values of Laspeyres and Paasche price indices for the same set of items. Quantities for successive years are used for weighting year-to-year changes in prices when calculating chain-weighted price indices.

The term base year has a couple of different interpretations when dealing with index numbers. One use of the term is to indicate the time period selected for weighting purposes. Once a price index has been calculated for several time periods, however, it is easy to convert or rescale the values of the index by expressing each value as a percentage change from any of the original observations. This is accomplished by dividing each of the original values of an index by the value of the index selected as a new base. Thus, regardless of how a price index is originally constructed, it can be rescaled or adjusted to different base periods indicating percentage differences from any predetermined observation to facilitate comparisons.

One of the most frequently used price indices is the Consumer Price Index which is recomputed each month in the United States. It measures changes in retail prices of goods and services purchased by U.S. households and is widely used as a measure of inflation. The CPI is often used to adjust household income and expenditure data to analyze changes in real purchasing power or expenditures. The adjustment process consists of dividing income or expenditure information for various periods of time by a corresponding value of the CPI for each observation and multiplying by 100. This process essentially adjusts income or expenditure data to values corresponding to prices of the **base year** of the price index. The same process can be used to compare any variable expressed in monetary terms relative to the prices of items included in the CPI.

QUESTIONS

1. Suppose that the 1998 retail price of a product is $6.00/unit and the price of the same identical product in 1975 was $4.00/unit. Show how you would determine the numerical value of the 1998 price relative based on the above information.

2. Assume that the following data represent the average prices and quantities of tomatoes and lettuce purchased in 1985 and 1995 by a typical U.S. household.

	1985		1995	
Commodity	**Price**	**Quantity**	**Price**	**Quantity**
Tomatoes	$.30/lb	20 lb	$.45/lb	25 lb
Lettuce	$.40/head	10 heads	$.80/head	12 heads

a. What is the 1995 average of the price relatives for these two commodities?

b. Indicate how you would interpret your answer to part a.

c. Based on the above data, calculate a Laspeyres index indicating the 1985 to 1995 changes in prices of the two commodities.

3. Would the inclusion of more commodities in a Laspeyres price index tend to increase or decrease its value if prices of the additional commodities increased less rapidly than the prices of the other commodities included in the index? Briefly explain the reasoning for your answer.

4. Suppose that the Consumer Price Index on a 1982 to 1984 = 100 base changes from 120 to 204 between 1989 and 2009.

   a. What would the values of the Consumer Price Index be for 1989 and 2009, if you wanted to use 1989 as a base period to facilitate making comparisons of price changes?

   b. Using the above information, what hourly wage rate would be required in 2009 to maintain the same purchasing power (or real value) that an $8.00/hr rate of pay had in 1989? Show your work.

5. Assume the Consumer Price Index had a value of 150 in 1996 (based on 1982–1984 = 100) and 154.5 in 1997. If you paid your workers $15.00/hr in 1996, how much higher should your wage rate be in 1997 if you agreed to adjust your workers' wage rate each year to keep up with the rate of change in the CPI?

REFERENCES

Boskin, M.J., E.R. Dulberger, R.J. Gordon, Z. Griliches, and D. Jorgenson. 1998. Consumer prices, the consumer price index and the cost of living. *Journal of Economic Perspectives* 12(1): 3–26.

Dalton, K.V., J.S. Greenlees, and K.J. Stewart. 1998. Incorporating a geometric mean formula into the CPI. *Monthly Labor Review* 121(10): 3–7.

Economic Research Service. 1999. *Agricultural outlook.* Washington, DC: USDA. http://www.econ.ag.gov/epubs/pdf/agout/ao.htm

National Agricultural Statistics Service. *Agricultural statistics, 1998.* Washington, DC: USDA. http://www.usda.gov/nass/pubs/agr98/acro98.htm

U.S. Bureau of the Census. 1998. *Statistical abstract of the United States* (118th Edition) 1998. www.census.gov/prod/3/98pubs/98statab//cc:98stab.htm

NOTES

1. Price relatives are always positive values because the numerators and denominators are always positive numbers. A value less than 100, however, indicates a decrease in price from the base year.

2. For the most recent year, the weights reflect the composition of goods and services in the preceding year. For all other years, the weights reflect the composition of the goods and services in the preceding and current year (*Statistical Abstract of the United States, 1998,* pp. 486–487).

3. In January 1998, weights based on 1993 to 1995 expenditure patterns replaced those based on 1982 to 1984 expenditures.

4. Since January 1999, the CPI is no longer based totally on a set of fixed quantity weights, but instead uses a procedure recommended by the Boskin Commission to more accurately take into account some of the substitutions that consumers make in response to changes in relative prices. The process involves calculating geometric averages using expenditure proportions as weights to calculate changes in prices within subcategories of the overall index (Dalton et al. 1998).

5. In this case, the revised series would consist of higher (or more inflated) values than the original data.

Part

II

Analytical Tools and Framework

Part II of the book contains five chapters that are designed to help students develop a useful framework for thinking about the way markets link production and consumption activities. The framework uses fundamental economic concepts to illustrate the outcomes of buyer and seller decisions. The objective of this section of the book is to demonstrate how demand and supply concepts can be used to analyze implications of changes in market conditions on producers, consumers, and marketing firms.

Chapter 4 describes key concepts that are useful for describing potential buyers' interests in acquiring food and agricultural products. Individual and aggregate consumer demand schedules provide a mechanism for illustrating how buyers respond to prices and other factors influencing their willingness to purchase items. Elasticities are introduced to characterize the way in which consumers respond to factors that are important for understanding the dynamic nature of markets.

Chapter 5 introduces concepts useful for understanding the willingness of producers to sell varying quantities of food and agricultural products at alternative prices. Many of the ideas underlying supply relationships are similar to those introduced in Chapter 4. A key difference, however, is that considerable time lags often occur between when decisions are made to produce items and the results are visible in the marketplace. Upon closer examination, however, the ways in which buyers and sellers behave over varying time intervals have some common dimensions.

Chapter 6 develops a framework of market activity by combining the concepts of the previous two chapters to show how changes in prices and quantities of products at the farm and retail levels are linked together. This framework differs from what is typically introduced in elementary economics courses in that it includes multiple demand and supply schedules. Vertical distances between primary and derived demand and supply schedules at the retail and farm levels represent the value of marketing services. The framework is developed initially assuming a competitive market structure that is especially relevant for analyzing global and national markets.

Chapter 7 discusses how noncompetitive market structures alter buying and selling behavior relative to the competitive market structure assumed in the previous chapter. The abstract nature of market structure is initially discussed before considering how the existence of just one seller (monopoly) or a single buyer (monopsonist) could affect market behavior. These, cases are especially relevant for analyzing what happens when governments become the sole provider or buyer of products. Brief descriptions of several other types of market structures are also included in this chapter.

Chapter 8, the final chapter in Part II, describes the variety of ways buyers and sellers discover a mutually satisfactory price at which they are willing to exchange ownership of commodities. This chapter provides a summary of the variety of ways individual and group activities affect the manner in which buying and selling occurs. A key idea presented in the chapter is that each type of marketing institution involves transaction costs. As new technology changes the relative costs of communication and transportation, buyers and sellers are likely to continue to change the way in which they interact and discover mutually satisfactory prices for exchanging ownership of commodities.

CHAPTER

CHAPTER 4

DEMAND AND ELASTICITY CONCEPTS

This chapter reviews and extends basic economic concepts about demand for goods and services that students usually get exposed to in introductory economics courses. Emphasis will be placed on helping students become comfortable in using the concepts for analyzing markets. Consequently, the theoretical underpinnings of the concepts will not be developed in detail.

The major points of the chapter are:

1. Important characteristics of individual and **market demand** schedules.

2. The difference between a change in demand and a change in quantity demanded.

3. How **price elasticities of demand** can be calculated and the implications of different numerical values.

4. Calculation and implications of **cross-price** and **income elasticities of demand.**

5. Effects of changes in population on demand.

INTRODUCTION

The first part of the chapter discusses the basic concept of an inverse relationship between the quantity demanded of any product and its price and different ways of representing these relationships. The second part of the chapter indicates how individual demand schedules can be aggregated to obtain market demand schedules. Then attention is turned to how price elasticities of demand can be calculated and used to describe how consumers respond to alternative prices. The final parts of the chapter indicate how elasticity concepts are used to quantify the sensitivity of demand relationships to changes in prices of other products, income, and population.

BASIC DEMAND CONCEPTS

An inverse relationship between the price and quantity demanded per unit of time is an important abstract concept to summarize potential buyer interest in any commodity. Demand schedules are assumed to represent the outcome of individual consumers (or households) making choices among products to maximize their satisfaction subject to resource constraints. An inverse relationship between price and quantity demanded is known as a demand schedule. It can be expressed in a tabular form indicating specific quantities of a product desired at alternative prices, in a graphical format, or as a mathematical relationship.

Information such as that in Table 4–1 indicates the quantities of an item a representative individual (or household) would be willing to buy per unit of time at alternative prices. As the price increases the quantity demanded gets smaller. The four combinations of price and quantity in Table 4–1 can also be graphically represented in a two-dimensional diagram like Figure 4–1 that illustrates the inverse relationship between price and quantity. The unit of time is crucial in specifying demand relationships in that the quantities associated with alternative prices would be expected to vary depending on what length of time is being considered (e.g., week, month, or year).

Considering more than the four prices in Table 4–1 is like thinking about a continuous line indicating the quantities desired at any particular price along

TABLE 4–1

HYPOTHETICAL DEMAND FOR A COMMODITY

PRICE PER UNIT	QUANTITY PURCHASED
$ .25	75
$ .50	50
$ .75	25
$1.00	0

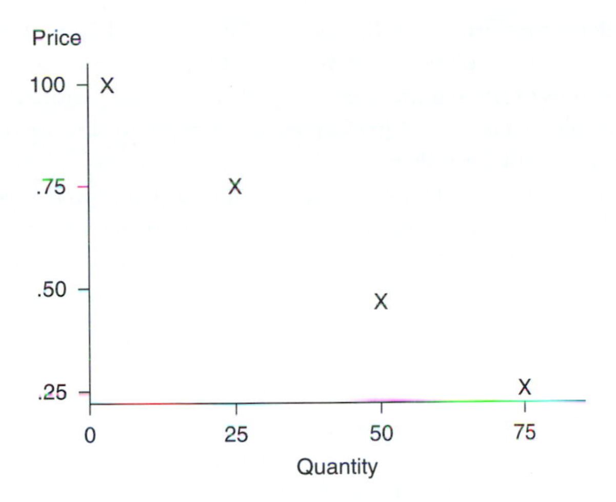

FIGURE 4–1 Graphical representation of demand.

the vertical axis of Figure 4–1. Graphical representations of demand relationships usually involve continuous lines even though it might be more realistic to consider respective quantities associated with only a finite set of prices. Furthermore, linear relationships often are used for ease of illustration even though there is no reason that demand relationships necessarily are restricted to that kind of mathematical functional form. The basic idea of inverse relationships between price and quantity, however, can be illustrated with a linear line just as well as more complicated configurations.[1]

Tabular and graphical representations of demand relationships are alternative ways of representing information from an explicit mathematical functional representation that indicates how quantity is related to alternative prices. For example, $Q = 100 - 4P$ provides information about quantities associated with alternative values of P between 0 and 25.

In considering the fundamental concept of a demand schedule associated with an individual person or household, a number of things are assumed to be constant. Among the things held constant for a particular demand schedule are purchasing power (or income) of the individual or household, the prices of other goods, and tastes and preferences.

Purchasing power and prices of other goods can be explicitly introduced as additional variables in a linear demand equation like Equation 4–1.

$$Q = 50 - 4P_1 + 2P_2 + .01I \qquad \textbf{(Eq. 4–1)}$$

where P_1 = price of the commodity of interest
P_2 = price of another commodity[2]
I = income

Substituting specific numerical values for P_2 and I into the above equation produces a simple linear inverse relationship between quantity and price of an item.

Considering different combinations of P_2 and/or I in Equation 4–1 would result in a set of linear relationships between Q and P_1 with different intercepts.

Values of the parameters in Equation 4–1 are assumed to be consistent with a given set of tastes and preferences. Changes in any of the coefficients in the equation would be indicative of a change in tastes and preferences causing the demand relationship to be different, holding prices and income constant.

The list of things assumed constant when specifying a demand relationship is very useful when thinking about what causes changes in demand. For example, a change in a demand function for an individual or household will occur if and *only* if a change occurs in income, prices of other goods, or tastes and preferences. This implies that the demand for a commodity doesn't increase or decrease just because of a change in its own price. For example, the change from point A to point B along demand schedule (D_1) in Figure 4–2 does not represent a change in demand. On the other hand a change from point A to point C is the result of a change in demand.

A demand schedule encompasses the complete set of quantities desired at each and every alternative price that might exist under a given set of circumstances. It is not just the quantity that would be purchased at a given price, but the entire set or schedule of quantities that would be purchased at alternative prices. When economists refer to a change in demand, it is interpreted as a change in the whole demand schedule, not just a movement from one point to another along the same curve. A change in demand refers to a shift in the entire price-quantity relationship. In graphical terms, it represents a move of the orig-

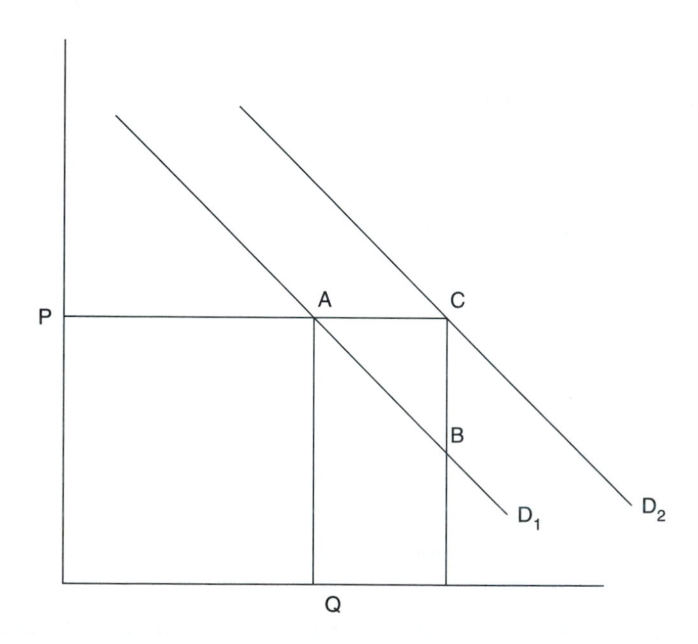

FIGURE 4–2 Two demand schedules.

inal demand schedule to the right or left such as D_1 to D_2 in Figure 4–2 as a result of a change in one of the three variables discussed previously.

A change in the price of a commodity will result in a *change in quantity demanded* of that commodity, not a *change in demand.* It is only when there has been a change in income, the price of some other good, or tastes and preferences, that a change in demand would occur. It should not be surprising that a consumer would alter the quantity demanded of a commodity as its price changes. For example, assume an individual would be willing to buy Q_1 of a commodity if its price is P_1, but a smaller quantity Q_2 if the price of the commodity increases to P_2. There are two underlying reasons why such a change in quantity demanded is reasonable.

First, it is clear that if the price of any good increases, a consumer cannot continue to buy everything that previously was being purchased without a compensating increase in income (or reduction in savings). Thus, an increase in any price without any change in income effectively results in a reduction in the purchasing power of existing income. Some adjustment in consumption patterns would be required because of the change in available purchasing power associated with a higher price of an item. Individuals simply are not able to continue to purchase everything they did under the previous set of prices for all commodities. Some reduction in quantity is required as price increases. Similarly, if the price of a commodity is reduced, a consumer could continue to purchase all the original quantities and have some income left over. Thus, a decrease in price is likely to lead to an increase in quantity demanded because of the resulting increase in purchasing power. The above kind of adjustment in quantities associated with a change in price is the so-called ***income effect*** that accounts for part of the change in quantity demanded associated with different prices.

The other part of the adjustment in quantities associated with a change in price is the ***substitution effect.*** To see how the substitution effect results in quantity adjustments, think about what happens to the purchase and use of different products when there is a change in relative prices. If the price of one commodity increases relative to the prices of other items, there is an incentive to reduce purchases of the higher priced commodity and replace it with goods whose prices have not changed. Similarly, if the price of a good decreases there would be an incentive to use more of the cheaper good in place of more expensive items, other things equal. This is the substitution effect associated with a change in relative prices that is separate and apart from the income effect.

As a result of these two responses associated with a price change, quantity adjustments along a demand curve consist of two parts: a substitution effect and an income effect. As long as the income effect is positive, that is, people would buy more of the good if they had more income, the substitution and income effects tend to reinforce each other in terms of effects on quantities demanded. For example, if the price of a good decreases, more of the good would be used in place of other goods because of the substitution effect. Furthermore, part of the effective additional purchasing power would also be allocated to that good if more of these goods would be purchased with higher incomes. This means there would be a positive effect on quantity of the good whose price decreased because of both the substitution and income effects.

A practical illustration of these effects is to consider how consumers react to a decrease in the price of poultry. Some of the increase in quantity of poultry demanded results from substituting poultry for other meat products. Some of the increase in quantity of poultry demanded also occurs because consumers would have a little extra money left over in response to a reduced price permitting them to buy more poultry as well as other products. This effect is what happens if incomes increase, but prices remain the same. This is an example of where the substitution and income effects work in the same direction. There can be situations where the income and substitution effects work in opposite directions but they are rather rare.

MARKET DEMAND SCHEDULES

In the previous section, the demand schedule for an individual consumer or household was considered for the purpose of emphasizing the inverse relationship between price and quantity as well as to note the difference between a change in demand and a change in quantity demanded. For purposes of market analysis however, it is necessary to aggregate individual demand curves to consider the total market demand for a commodity.

The conceptual development of an aggregate market demand consists of a horizontal summation of all the individual demand curves relevant for a particular market. That is, market demand represents the total quantity of a commodity demanded at each and every price by every individual in a market. Aggregating (or adding) quantities demanded by all individuals (or households) in a specific market at each price indicates the appropriate quantities for the market demand curve. For example, if a market consisted of three individuals A, B, and C and they each were willing to purchase 10, 20, and 25 units of a commodity respectively at a given price, the total quantity demanded at that price in the market would be 55 units. Aggregating the quantities that each of these three individuals would be willing to purchase at each respective price would lead to a representation of the demand relationship for a market comprised of A, B, and C. Repeating the process for each and every alternative price, leads to a market demand schedule. If the demands are represented in mathematical form, the aggregate market demand is the algebraic sum of the equations. For example, the algebraic sum of two demand equations $Q = 10 - 4P$ and $Q = 5 - 2P$ would be $Q = 15 - 6P$. The latter equation indicates the total quantity demanded at any price between 0 and 2.5.

Obviously, anything that causes individual demand schedules to shift will be a potential shifter of market demand. This means that income, the prices of other goods, and tastes and preferences are shifters of market demand functions because they shift individual demand functions. An additional factor that can cause changes in aggregate market demand, however, is how many individuals are in a particular market at a given time. If there is an increase in the number of people demanding a particular good, it will shift the aggregate demand for the commodity. This additional shifter of market demand relationships is usually referred to as changes in population.

Sometimes a broader concept than just the total number of individuals is used to specifically note that the composition of the population as well can be a potential shifter of market demand curves. The composition of the population has particular relevance for certain agricultural commodities to the extent that different age or ethnic groups consume different kinds of products. This is not necessarily the result of changes in individual tastes and preferences, but reflects a different composition of aggregate tastes and preferences in the population.

Adding size and/or composition of the population to income, prices of other products, and tastes and preferences provides a useful checklist of things that can be responsible for changes in the demand for products. Anytime consideration is given to whether the demand for a product has or will change; the first thing that must be determined is if there has been or will be a change in any of the four items in the above list. If there has been no change in any of the items in the list, demand has not shifted. If something occurs to cause a change in one of the demand shifters, then it may alter demand for a product. Once a potential shifter of demand has been identified, the next thing is to decide if it is going to shift the demand to the right or to the left. Some changes may lead to increases in demand while others may cause decreases in demand. Analytical concepts to quantify these shifts will be developed using some of the same principles that are widely used to measure the responsiveness of quantities demanded to changes in price.

PRICE ELASTICITY OF DEMAND

With tabular representations of market demand relationships it is fairly easy to determine how large of an adjustment in quantity demanded is associated with particular changes in price. Similarly with a mathematical representation of demand behavior, the difference in quantity implied by two alternative prices can be compared to calculate the relative change in quantity per unit change in price. A graphical representation or measure of the responsiveness of quantity to changes in price is the "slope" of the demand function ($\Delta Q/\Delta P$). A demand function that indicates very little change in quantity as price changes, would be more vertical than a demand function representing considerable responsiveness to price changes. Thinking about the responsiveness in terms of the above "slope" concept initially can be a little confusing because it is not the usual mathematical concept of a slope in terms of a change in vertical distance relative to changes in horizontal distance in Figures 4–1 or 4–2. The reason why $\Delta Q/\Delta P$ is considered to be a slope measurement by economists is because quantity is a dependent variable and price is an independent variable from the perspective of individual consumers even though it is conventional to plot quantity on the horizontal axis and price on the vertical axis. However, it is reasonable to assume that price responds to alternative quantities offered for sale in a market.

Although the $\Delta Q/\Delta P$ (slope) measurement of a demand relationship is a valid empirical representation of the relationship between quantity and price, it is not entirely satisfactory. One problem with such a measurement is that its magnitude depends on the units used for measuring quantities and prices, which creates problems in making comparisons among commodities. For example, the slope of the demand for watermelons might be 50 and a similar ratio for the demand for beef might be 100. This would not necessarily imply that the demand for one commodity is more responsive to price changes than the other unless comparable measurements of quantities and prices were used for both commodities.

Similarly, it would be difficult to compare the slopes of demands for the same commodity in two countries that used different monetary measurements of prices. In the case of comparing the demands for the same commodity in Mexico and the United States, the numerator of the two slopes might have the same units, but the denominator for the U.S. slope would be dollars while for Mexico it would be pesos. Differences in measurement units of quantities and prices among commodities and across economies create difficulties in using slopes for comparisons of responsiveness of quantity demanded to price changes.

Fortunately, a way around this problem is to multiply the slope by a ratio of price to quantity. The units associated with the price to quantity ratio would be exactly the reciprocal of the units for the slope. Thus, when the two components are multiplied together, the product is invariant to the units used to represent quantity and price because the units of the two components of the elasticity formula offset each other. This procedure produces values that can be readily compared regardless of whether the demands for different commodities or the demands for the same commodity in different countries are being considered. The mathematical representation of this formula is

$$e = \frac{\Delta Q}{\Delta P} \frac{P}{Q} \qquad \textbf{(Eq. 4–2)}$$

where e represents price elasticity and P and Q represent price and quantity, respectively. Once the price elasticities of demand for different products are calculated, they can be compared to tell which demand is more responsive to price.

If a particular price-quantity combination on the demand function is used for selecting values for the second component of the price elasticity formula, the result would be referred to as the point elasticity. The data in Table 4–1 can be used to illustrate how the elasticity formula works. For example, the change in quantity for each \$.25 decrease in price is 25. Thus, the ratio of $\Delta Q/\Delta P$ would be $\frac{25}{-.25} = -100$. Different points on that demand function, however, have different ratios of P/Q. If one selects the point corresponding to a price of \$.50 the ratio of P/Q is .01 and the price elasticity would equal -1. On the other hand, a price of \$.75 yields a ratio of P/Q of .03. Multiplying the slope of −100 by .03 results in a price elasticity value of −3. Consequently, different values of price elasticity of demand can occur depending on which point on the demand curve is selected even if the slope of the demand curve is constant.

If the mathematical functional form for a demand relationship is known, the change in quantity associated with a given change in price can be obtained in a couple of ways. One way is to determine the difference in quantities associated with two different prices.[3] Even though $\Delta Q = -4$ for each unit change in P_1 in Equation 4–1, particular values for all prices and income would be required to determine a particular value of Q to use for the second part of the elasticity formula.

The price elasticity formula can also be expressed as $\frac{\Delta Q}{Q} \div \frac{\Delta P}{P}$ by algebraically manipulating the components. If the numerator and denominator of the preceding expression are each multiplied by 100, the formula has a very natural interpretation. After multiplying by 100, the numerator of the formula represents a percentage change in quantity. Similarly, multiplying the denominator by 100 indicates percentage change in price. This means that the **coefficient of price elasticity** can be interpreted as the percentage change in quantity associated with each 1% change in price along a demand schedule. It is important to emphasize the 1% change in price because dividing a given percentage change in quantity associated with a particular percentage change in price represents the average percentage change in quantity for each 1 percentage change in price. For example, a price elasticity of −.5, means that on average a .5 percentage *decrease* in quantity demanded is associated with each 1% *increase* in price. Similarly, it implies a .5 percentage *increase* in quantity demanded for each 1% *decrease* in price. The negative sign attached to the price elasticity is consistent with the inverse relationship between price and quantity that is inherent in demand relationships.

One way of overcoming some of the variation in measures of price elasticity associated with different P and Q combinations along a demand schedule with a constant slope, is to calculate a measure of price elasticity over a given range of the demand schedule. In this case, the first part of the formula stays the same but two points on the demand function are selected to compute an average quantity and average price to use in place of a single price and quantity for the second component. For example, to obtain a representative elasticity over the range of prices from \$.50 to \$.75 in Table 4–1 average prices and quantities for those two points could be used to calculate the elasticity. That is, the slope of −100 could be multiplied by a ratio of .625/37.5 resulting in an elasticity value of −1.67. The result could be interpreted as a representative elasticity for a range of prices on the demand schedule. This type of calculation is known as the *arc elasticity* as opposed to *point elasticity.*

One way to determine the slope of any linear demand function is to use the differences in prices and quantities associated with two points on the demand function. This would produce values for ΔQ and ΔP that when multiplied by the average price and quantity for the two points produces a value for the price elasticity of demand. Thus an alternative way of writing the formula for arc price elasticity would be as follows:

$$\frac{Q_1 - Q_2}{P_1 - P_2} \times \frac{P_1 + P_2}{Q_1 + Q_2}$$

where Q_1 and P_1 refer to one point on the demand schedule and Q_2 and P_2 refer to an alternative point. The second component of the above elasticity formula is a ratio of sums rather than averages of prices and quantities for two points on the demand schedule. It produces the same ratio of P/Q as using averages because calculating average prices and quantities requires dividing the sum of prices and the sum of quantities by two. Dividing the numerator and denominator of the second component by two produces the same value as the ratio of the sums.

The point and arc formulas are both appropriate for evaluating price elasticities of demand. Both formulas represent the concept of responsiveness of a demand schedule and differ only in terms of whether a point or an arc is used for computing percentage changes. The only issue in using the arc price elasticity formula is which points to select. A general rule of thumb is that points should be selected that represent the relevant range of the demand curve over which price responsiveness is of interest.[4]

One word of caution when dealing with the price elasticity of demand is that the slope or the ratio of ΔQ to ΔP will always have a negative sign as long as prices and quantities are inversely related. It is important to include the minus sign for price elasticities of demand to differentiate them from **price elasticities of supply** which will be discussed in Chapter 5. The sign of the price elasticity will always be automatically correct using the arc price elasticity formula if the same subscript is used for the price and quantity of a given point on the demand schedule. For example, if point A in Figure 4–2 is designated as P_1 and Q_1 and point B as P_2 and Q_2, the correct sign of price elasticity will always occur regardless of whether the change in quantities and prices are computed using the values at point A minus the values at point B or the values at point B minus the values at point A. Sometimes the absolute values of price elasticities of demand are used (i.e., $|e|$) to simplify communication and avoid having to always write a minus sign. In such cases the negative sign of price elasticities of demand is suppressed and only the numerical value is used.

If the absolute value of price elasticity of demand is equal to 1, the demand schedule is said to be ***unitary.*** If the absolute value of price elasticity of demand is less than 1, demand is called *inelastic.* If the absolute value of price elasticity of demand is greater than 1 demand is ***elastic.*** This classification of demand relationships based on price elasticity is very important. Each of the three categories implies different changes in total revenue (or total expenditures) for movements along demand curves.[5]

The reason for the relationship between price elasticity of demand and total revenue is because relative changes in Q and P along a given demand function affect the product of P times Q. The inverse relationship between P and Q underlying a demand function means that changes in each of the two components tend to cause total revenue to change in different directions as different P and Q combinations are considered. The net effect on directional changes in total revenue depends on the relative magnitude of changes in the two components, which is what the coefficient of price elasticity represents. For example, if one considers a given percentage change in price (either an increase or decrease), the only thing that is going to determine whether total revenue (TR) increases or decreases is whether the relative change in quantity is greater or smaller than the corresponding relative change in price.

In the case of a demand schedule with unitary elasticity, the percentage changes in price and quantity tend to offset each other producing no change in revenue. This can be illustrated by representing the directional changes in P and Q with arrows of the same magnitude.

$$\begin{aligned} TR &= PQ \\ &= \uparrow\downarrow \text{ if price increases or} \\ &= \downarrow\uparrow \text{ or in case of a decrease in price} \end{aligned}$$

In the case of a demand relationship that is elastic (i.e., $|e| > 1$), however, percentage changes in quantities are greater than percentage changes in price. Using directional arrows of different magnitudes to represent the magnitude of changes, it is clear that total revenue will change in the same direction as quantity. That is,

$$\begin{array}{lll} TR & = & PQ \\ \downarrow & = & \uparrow \Big\downarrow \text{ if price increases or} \\ \uparrow & = & \downarrow \Big\uparrow \text{ in case of a decrease in price} \end{array}$$

This means that higher prices will produce lower total revenue and lower prices will be associated with increases in total revenue.

If a demand function is inelastic (i.e., $|e| < 1$), then the results will be just the opposite of the elastic case because of the responsiveness in quantity to price adjustments. In this case, the relationship can be illustrated as:

$$\begin{array}{lll} TR & = & PQ \\ \uparrow & = & \Big\uparrow \downarrow \text{ if price increases or} \\ \downarrow & = & \Big\downarrow \uparrow \text{ in case of a decrease in price} \end{array}$$

which indicates that revenue changes in the same direction as prices change or inversely with quantities. The results of the preceding three cases are summarized in Table 4–2.

TABLE 4-2

SUMMARY OF CHANGES IN TOTAL REVENUE FOR DIFFERENT PRICE ELASTICITIES OF DEMAND

ABSOLUTE VALUE OF PRICE ELASTICITY OF DEMAND	DESCRIPTION	Δ IN TR WITH Δ IN P	Δ IN TR WITH Δ IN Q
1	Unitary	No change	No change
>1	Elastic	Opposite direction	Same direction
<1	Inelastic	Same direction	Opposite direction

Instead of thinking about movements along a demand function because of a changing price, the effects on total revenue or expenditures associated with a change in quantity demanded could also be of interest. It is important to know the directional changes in total revenue for upward or downward movements in price or quantity along demand functions with different price elasticities.

Examples of price elasticities for some selected food commodities are presented in Table 4–3. All of the price elasticities in Table 4–3 are **inelastic.** Meat products, tomatoes, carrots, and bananas appear to be a little more responsive to price than most of the other items.

Coefficient of Price Flexibility

It was noted previously that changes in total revenue associated with changes in quantity along demand curves were just the opposite of the changes associated with price adjustments because of the inverse relationship between price and quantity. In some situations, it is more convenient to analyze quantity adjustments associated with particular changes in price and in other situations an adjustment in price resulting from specified changes in quantity may be of more interest. The latter situation is frequently encountered in the case of agricultural commodities because of interest in how much the market-clearing price would adjust in response to particular changes in quantity supplied. That is, what kind of adjustment in price would be required to clear the market for each 1% change in quantity? Conceptually the answer to this question is just the reciprocal of the definition of price elasticity of demand. The percentage change in price associated with each 1% change in quantity along a given demand function is defined to be the coefficient of price flexibility. A quick and easy way to obtain an approximate value for the coefficient of price flexibility is to take the reciprocal of the coefficient of price elasticity.[6]

Multiplying the coefficient of price flexibility by a particular percentage change in quantity provides an estimate of the percentage change in price that

TABLE 4.3

PRICE AND INCOME ELASTICITIES FOR SOME FOODS

	Price Elasticity	Income Elasticity
Beef	−.62	.39
Pork	−.73	.66
Chicken	−.37	.08
Turkey	−.53	−.13
Eggs	−.11	.29
Cheese	−.25	.42
Milk	−.04	.12
Butter	−.24	.54
Margarine	−.01	−.34
Apples	−.19	−.36
Bananas	−.50	.09
Lettuce	−.09	.37
Tomatoes	−.62	.92
Carrots	−.53	.68
Sugar	−.04	.01
Coffee	−.17	.82

Source: Adapted from Table 3 of Huang, *American Journal of Agricultural Economics,* 1996b.

might be anticipated under competitive market clearing conditions assuming no change in demand. For example, if the coefficient of price flexibility for a particular product was −2.0, one would expect a 2% increase in price for each 1% decrease in quantity. This means that a 10% increase in price might be expected assuming no change in demand if a 5% decrease in output were anticipated.

Factors Affecting Magnitude of Price Elasticities

The major thing that influences the size of the price elasticity of demand for a particular product is the product's degree of substitutability for other products. This is consistent with earlier discussion about the quantity response along a given demand curve depending primarily on the substitution effect. Thus the response reflected by price elasticity coefficients is largely the movement along a demand curve associated with the degree of substitutability. If the product of interest has several good substitutes or is a good substitute for other products, then there is likely to be a bigger quantity response to any given price change and the price elasticity is likely to have a larger absolute value. If a product has few substitutes, the price elasticity is likely to be small (in absolute value) reflecting the fact that there is not likely to be much response in quantity purchased to alternative prices.

Applying this concept to the case of the demand for all food products, it should not be surprising that the price elasticity of the demand for total food is quite low because of the lack of **substitutes** to satisfy nutritional needs. On the other hand, the price elasticity of demand for an individual food item is likely to be greater than the price elasticity for all food because of the **substitution** possibilities that exist among alternative food products. This leads to a general rule that the price elasticity of demand for a group of commodities tends to be smaller than the price elasticity of demand for many (but not necessarily all) of the individual commodities making up the group. At first glance this generalization sounds counterintuitive like some other basic economic concepts but makes sense when closely examined.

Another point about the price elasticity of demand is that the income effect associated with a change in price is going to be small for products that account for a small share of total expenditures. The smaller the proportion of total expenditures that a product accounts for, the smaller the income effect is going to be for any change in price of that commodity. This suggests that price elasticities should be smaller for commodities on which a small percentage of income is spent because of the income effect associated with changes in price. For example, how much extra purchasing power will the typical consumer realize if the price of salt changes by 100%? The change in purchasing power resulting from this change in price is not likely to cause much of a change in the amount of salt purchased. On the other hand, if the price of housing changes, the income effect would be much greater than for many other commodities and is likely to lead to a larger price elasticity of demand.

Another important dimension of price elasticities is the need to differentiate the responsiveness in terms of how long a time period is being considered for consumers to adjust to a given price change. A particular change in price is very likely to lead to different quantity responses per unit of time depending on how long an adjustment period is considered. Another way of stating this is that the slopes of demand relationships for a given product are likely to vary for different lengths of time consumers have to adjust their purchasing patterns. The initial adjustment in quantity of a product purchased per unit of time to a change in price from P_1 to P_2 might be different than if the price change persists for a longer period of time. This suggests that the demand curve may pivot around an initial point consistent with the quantity adjustment to a given price change being larger the more time consumers have to adjust purchasing patterns, Figure 4–3. The usual argument therefore is that price elasticities will be bigger the longer the period of time allowed for consumers to adjust to a price change. This is an important principle, but a minor qualification needs to be incorporated for short periods of time. This occurs to the extent that the timing of purchases can be changed and the degree to which commodities can be stored. A price change for a storable item may result in a larger rate of adjustment in purchasing be-

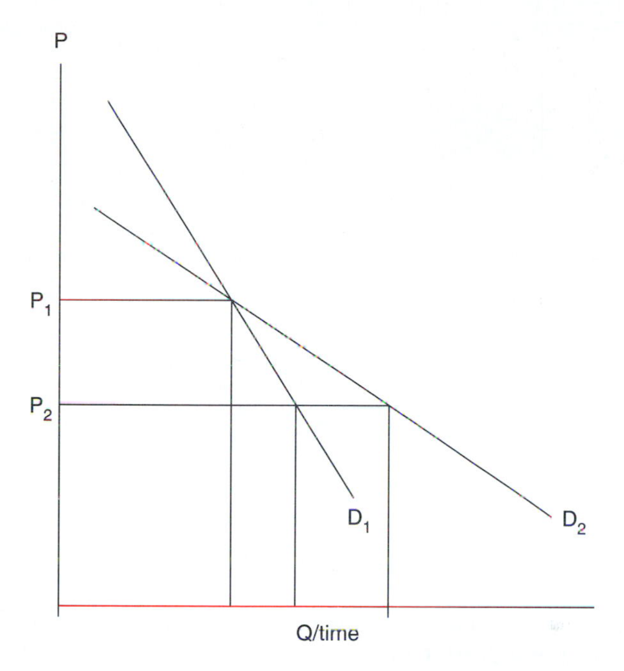

FIGURE 4–3 Demand relationships for different lengths of run.

havior than what will be observed for longer time periods if consumers modify their inventories of goods in response to short run changes in price.

Combining the short and the longer run effects on purchasing behavior means that a U-shaped curve might depict how the absolute value of the price elasticity of demand varies with length of adjustment time (Figure 4–4). The first kind of adjustment described suggests that the elasticity increases the longer the length of the adjustment period. The second change indicates the response in the short run may be larger than for intermediate lengths of time. In each case the same relative change in price is being considered with the only difference being the amount of time consumers have to adjust to the given change in price.

A good example of these effects is to see how people respond to changes in gasoline prices. In the very short run, people tend to buy more than usual when the price decreases especially if they anticipate the price change is a temporary phenomenon. Once it is realized that the change in price is more of a permanent adjustment, purchases and inventories will adjust to new equilibrium levels. Thus, the percentage change in purchases for an intermediate time period is not likely to be as great as the short run. In the short run, people can change their purchases by postponing or adding to their stocks depending on whether the price has gone up or gone down. If purchases can be postponed or stocks altered, the response observed in terms of quantities purchased reflect these adjustments and consequently a U-shaped curve can characterize price elasticities for different lengths of time.

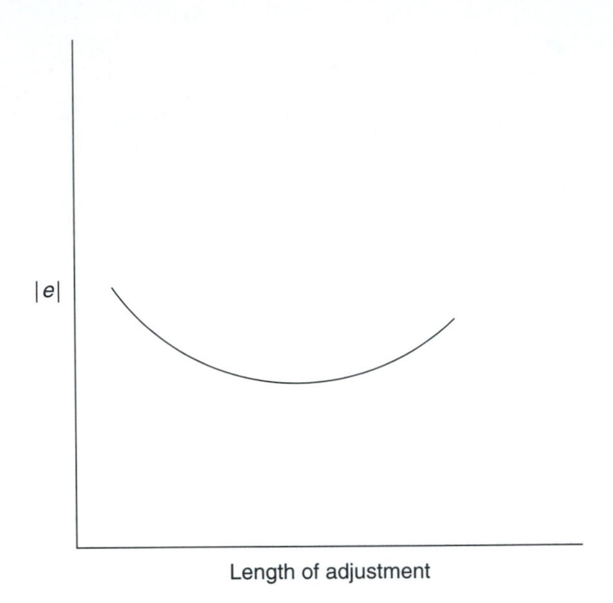

FIGURE 4–4 Differences in price elasticity of demand for different lengths of adjustment.

CROSS-PRICE ELASTICITY

Analogous concepts to price elasticity can be used to quantify shifts in demand relationships. Instead of considering movement along a given demand curve from one price-quantity combination to another one, interest might exist in how much the entire demand curve shifts to the left or right in response to one of the demand shifters. If a change in the demand for a commodity is the result of a change in the price of some other commodity, the effect can be measured by the concept of cross-price elasticity. This elasticity measure indicates how much the demand for one commodity shifts when the price of some other commodity changes by one percent. It measures the responsiveness in the quantity of one commodity to a change in the price of another commodity. The cross commodity relationship is why the concept is referred to as *cross-price elasticity* or sometimes just as cross elasticity of demand.

Essentially the same kinds of formulae used to compute direct price elasticities are used for calculating cross-price elasticities. The only difference is that the quantity of one commodity is considered along with a change in the price of another commodity. All other prices and income are assumed to remain constant. If subscripts are used to differentiate commodities, the appropriate formulas are very similar to those used for calculating price elasticities of demand. Thus, the formula for cross-price elasticity is

$$\frac{\Delta Q_i}{\Delta P_j} \cdot \frac{P_j}{Q_i} \qquad \textbf{(Eq. 4–4)}$$

where Q_i and P_j represent the quantity of the ith good and the price of the jth good, respectively.

Cross-price elasticities can be calculated using either a point or an arc formula similar to the options for calculating direct price elasticities. That is, a specific price and quantity combination or average prices and quantities associated with the old and new price can be used. For example, if the price of the jth commodity changing from $.75 to $1.00 causes the quantity of the ith item to increase from 10 to 15, the cross-price elasticity could be calculated as $\frac{5}{.25} \cdot \frac{.75}{10} = 1.5$. The cross-price elasticity indicates that the quantity of the first commodity increases by 1.5% for each 1% increase in the price of the second commodity. If information about the demand relationship is specified in mathematical form, it would be necessary to evaluate the change in Q that is associated with two values of the price of another commodity in order to determine $\Delta Q_i/\Delta P_j$. After obtaining a value for this part of the formula, it could be multiplied by one of the P_j/Q_i values or a ratio of the average of the P_j's divided by the average of the Q_i's to compute the cross-price elasticity.[7]

The sign of the cross-price elasticity indicates whether the direction of the change in demand for the ith good is to the right or left. If the sign of the cross-price elasticity is positive it indicates that there is an increase in the demand for the ith good as the price of the jth good increases. For example, if the price of beef goes up, people are likely to demand more chicken or pork to replace some of the beef and one would expect positive cross-price elasticities among these products. If two products are *substitutes,* an increase in the price of one good will cause the demand for the other product to increase (consistent with a positive cross-price elasticity). As the demand for the second good increases, the quantity demanded of the first good decreases (assuming a negative price elasticity of demand). Assuming income effects are small for commodities on which a small part of the budget is spent means that the cross-price elasticity is largely influenced by whether a substitution or complementary relationship exists between commodities. If the cross-price elasticity is negative it suggests the two goods are **complements** or tend to be purchased and/or used together. Some examples of *complements* would be cranberries and turkeys, hot dogs and rolls. Examples of substitutes are butter and margarine or beef and pork. If the demand for one commodity is unaffected by changes in the price of another commodity, the cross price elasticity between the two goods would be zero and the goods would be said to be *independent.*[8]

Cross-price elasticities are important for considering the effects of changes in prices of goods that influence the demand for other commodities. For example, if you were managing a grocery store and decided to reduce the price of hot dogs, you might want to think about the effect this is likely to have on the demand for hot dog rolls, mustard, ketchup, and other condiments so that you would have enough inventory to satisfy anticipated demand.

INCOME ELASTICITY

Another kind of shift in the demand curve that can be measured by an elasticity measure is the effect of changes in income. Again the same kind of formulas that have been introduced earlier are appropriate except now a change in income is considered rather than a change in price, that is,

$$\frac{\Delta Q}{\Delta I} \cdot \frac{I}{Q} \qquad \textbf{(Eq. 4–5)}$$

where I represents income. Thus, if a change in income from \$25,000 to \$30,000 results in increased purchases of an item from 100 to 110, the **income elasticity coefficient** could be calculated as $\frac{10}{5{,}000} \cdot \frac{25{,}000}{100} = .5$.[9] Income elasticity measures indicate the percentage change in demand of a product associated with each 1% change in income.

Income elasticities represent how responsive demand curves are to alternative levels of income. If the income elasticity is positive, the good is said to be *normal* which implies the demand increases or shifts to the right with higher incomes. If the income elasticity is negative, the good is said to be *inferior.* If income elasticity is zero, it indicates a *neutral* commodity or one for which income does not have any effect on demand.

Some of the effects of income elasticities on various goods can be seen by examining changes in per capita consumption of various goods over time. For example some of the decrease in per capita consumption of sweet potato consumption in the United States can probably be attributed to changes in incomes because most estimates of the income elasticity of demand for sweet potatoes are negative. Thus increases in real income have probably had a negative impact on the demand for sweet potatoes as well as potatoes in general, and probably a few other products.

The coefficient of income elasticity can be interpreted as dividing a percentage change in quantity by a percentage change in income. Conceptually these percentages are evaluated holding the price of the product constant. Thus an alternative way of computing income elasticities (as well as cross-price elasticities) for commodities is to use the percentage change in expenditures because it will be the same as the percentage change in quantity holding prices constant. This alternative way of calculating elasticities is especially useful for commodities for which quantities may be difficult to measure. For example, when considering the demand for clothing, housing, or other product aggregates, defining appropriate quantity units may be difficult, if not impossible. What can be done in these cases is to use changes in expenditures to calculate income elasticities.

This is valid because multiplying the numerator and denominator of the income elasticity formula by a given price essentially converts quantity measurements into expenditures.

Calculating the income elasticity using either quantity or expenditure information should lead to the same estimate of income elasticity provided the price of the good is constant. One problem is that many times it is difficult to make sure that some of the change in expenditures does not reflect purchasing more expensive items within a product group. As income changes there are likely to be changes in the quality as well as quantity of the products purchased. For example, as incomes increase consumers may substitute steak for ground beef instead of just buying more ground beef. Even if the total quantity of beef purchases remained exactly the same following an increase in income, total expenditures might increase because of quality adjustments. If consumers continue to buy exactly the same quality of products when income increased, expenditures would change by the same percentage as quantity. If a change in quality occurs, however, the percentage change in expenditures will produce a bigger measure of income elasticity than if percentage changes in quantity were used. Conceptually using quantity or expenditure data ought to produce the same empirical elasticity estimate provided the price is held constant, but the latter assumption may be very difficult to maintain considering the possibilities of changing quality of items purchased.

The magnitude of income elasticities has important implications on how the proportions of total income spent on different items (budget shares) change with income. For example, if expenditures on each item purchased changed at the same rate as income (i.e., income elasticities for all products were 1) there would be no change in budget shares. An income elasticity greater than 1, however, indicates that expenditures on that item increase more rapidly than income implying an increasing budget share. Items with income elasticities of less than 1 have decreasing budget shares as income increases.

In general, income elasticities for agriculture products are fairly small. (For example, see the values in Table 4–3.) In fact, a reasonable estimate of the income elasticity for all agricultural products is often considered to be somewhere in the neighborhood of .2 or .3. This means that for every 1% increase in income, only a .2 or .3 percentage change in the demand occurs. One implication of income elasticities for agricultural products being less than 1.0 is that the relative size or importance of the agricultural sector is going to decline as increases in real incomes occur. This is because the demand for agricultural products will be increasing at a slower rate than some other parts of the economy. The implication of a low income elasticity of demand for agricultural products is that the agriculture sector will not expand as fast as incomes rise in the overall economy. In fact, one indirect measure of economic development of various economies that is sometimes used is the size of the agricultural sector. If the agricultural sector is a large part of an economy, it probably indicates that the economy is not very highly developed in that a large part of the country's total economic activities involve fulfilling basic nutritional needs of the population. On the other hand, if the

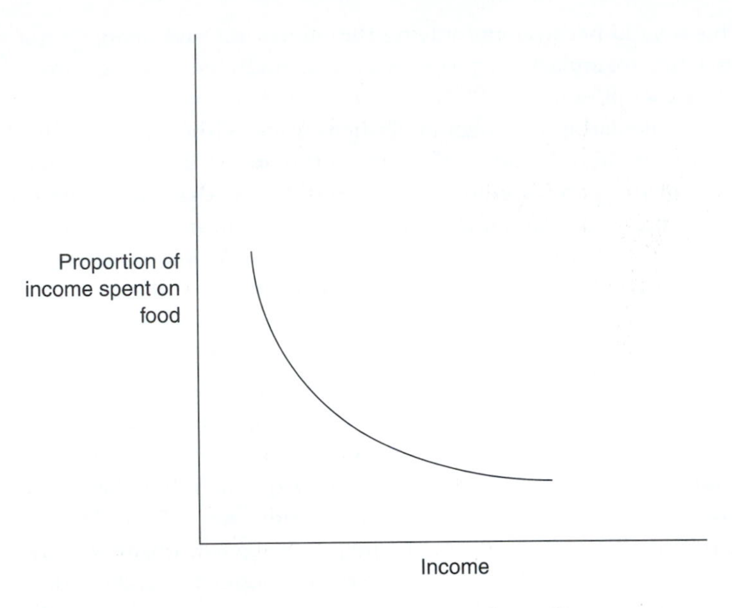

FIGURE 4–5 Relationship between proportion of income spent on food for different levels of income.

agricultural sector is a small part of an economy, it may indicate a substantial level of economic development. This indicator of course ignores the extent to which export and import activities may affect the size of various sectors.

A low income elasticity of demand for agriculture products also means that people with higher incomes will be spending a smaller percent of their total income for food products than people with lower incomes. This relationship can be illustrated by a graph that shows how the percent of total income spent on food varies inversely with income (Figure 4–5). The relationship between these two variables is an implication of **Engel's law** that is associated with a low income elasticity for food.

POPULATION

In projecting the rate of change in market demand for a commodity, one of the important variables that must be considered is what is expected to happen to population. In order to quantify the effects of this shifter of demand relationships, one more elasticity is often used. Usually it is assumed that each 1% change in population results in a 1% change in demand. That is, if there are 10% more people to feed, a 10% increase in demand for products will occur, holding prices and real purchasing power per capita constant. Assuming a population elasticity of 1 is a good first approximation when considering demand for individual products, but additional considerations of possible changes in the age distribution of the population must also be taken into account. For example, changes in the birthrate influencing the age composition of populations can be quite significant on the demand for baby food or milk products. Also, to the extent that elderly individuals consume a different set of food products than people in other age groups in the population, a changing proportion of older individuals can affect the demand for agricultural products. Changes in the age distribution of a population or an increase in population because of migration from other countries can have different implications on demand for particular products.

SUMMARY

The key concept used to represent potential interest in purchasing commodities is inverse relationships between price and quantity demanded per unit of time. Demand relationships can be expressed in tabular, graphical, or mathematical terms. An important distinction about demand relationships is the difference between a change in demand (a shift in the entire relationship) and a change in quantity demanded (the change in quantity associated with a change in the price of the commodity). A demand relationship for a given item is defined as holding income, the prices of other products, and tastes and preferences constant. Changes in any of these variables can cause changes in demand for products.

The extent to which the quantity demanded of a given product varies with price along a demand function is primarily because of substitution effects, although there may also be an income effect. Individual or household demand relationships can be horizontally aggregated to obtain market demand relationships representing the interests of all potential consumers or buyers in a market. Consequently, an additional factor that can shift market demand relationships is the size and composition of a market's population.

Price elasticity of demand provides a measure of the responsiveness of the quantity demanded to changes in price along a demand relationship. A particular point on a demand function or an average of points over a given region of the demand function can be used to calculate elasticities. The most important aspect about a price elasticity of demand is whether its absolute value is greater than (**elastic**), less than (inelastic), or equal (unity) to 1. Each of the three price elasticity categories has different implications about directional changes in total revenue (or expenditure) for price and quantity adjustments along a demand curve.

Another concept used to represent adjustments along a demand relationship is the **coefficient of price flexibility** that indicates the relative responsiveness of price to changes in quantity. Price flexibilities are especially used in forecasting how market prices will respond to anticipated changes in quantities available for sale.

Cross-price and income elasticities indicate the direction and amount of change in demand relationships in response to changes in the prices of other products and incomes, respectively. A positive cross-price elasticity indicates that the demand for one product increases as the price of another product increases indicating that the two goods are substitutes. A negative cross-price elasticity implies a complementary relationship between goods. A zero cross-price relationship implies that a change in the price of one good has no effect on the demand of the other good and the goods are considered to be independent.

A positive income elasticity indicates that the demand for an item changes in the same direction as income. Items with positive income elasticities are considered to be **normal goods.** Negative income elasticities are associated with **inferior goods.** A zero income elasticity indicates a *neutral* good. If the income elasticity for an item is less than 1, expenditures on this item will change at a slower rate than income implying a decreasing share of total expenditures as incomes increase. This is one reason why the proportion of total income spent on food has decreased over time with increasing average levels of income and is generally lower among higher income households in any given market.

Another important shifter of demand relationships is the size and composition of a market's population. Assuming that the demand for most products will change at approximately the same rate as population changes implies an elasticity of 1 for population changes. Changing age and ethnic composition of populations, however, can have important differential effects for individual commodities.

QUESTIONS

1. Identify and briefly describe the two components of a price effect in the context of why the quantity of pizzas demanded at low prices is likely to be greater than the quantity of pizzas demanded at higher prices.

2. Assume that the following two equations represent the demands for sweet potatoes for immediate consumption (fresh market) and processing purposes, respectively, for a given market area.

Fresh market $Q = 100 - 4P$
Processing $Q = 25 - P$

where Q = million hundredweight of sweet potatoes
P = dollars/hundredweight

Based on the above information, what would be the appropriate algebraic equation representing aggregate demand for sweet potatoes in the given market?

3. If the price elasticity of demand for sweet corn from your roadside market outlet is -2.5, would you want to raise or lower the price you are charging in order to increase total revenue? Explain.

4. Last month U.S. retail cigarette prices increased by approximately 20%.
   a. If the price elasticity of demand for cigarettes is approximately $-.4$, how much of a change in the quantity of cigarette sales would you anticipate to occur, other things being equal?

b. Would you expect total revenue from the sales of cigarettes to increase, decrease, or stay about the same as a result of the 20% increase in price? Briefly explain the logic supporting your answer.

5. Recently a number of food processing companies have introduced several new kinds of frozen potato products in retail food outlets. How is this likely to affect the price elasticity of demand for frozen french fries in retail food stores? Briefly explain the logic behind your answer.

6. Briefly indicate how each of the following events would be expected to affect the U.S. demand for strawberries and explain your reasoning.

a. A 15% increase in the price of strawberries because of increased costs of production.

b. A 10% decrease in the price of substitute products.

c. A report indicating that the current supply of strawberries is going to be larger than normal because of favorable growing conditions.

7. Assume that the income elasticities for soft drinks and beer are .5 and .8, respectively.

a. How much would total expenditures on soft drinks change as a result of a 10% increase in income assuming no change in the prices of soft drinks or beer?

b. How would a 10% increase in income be expected to affect total expenditures on beer, assuming no change in the prices of soft drinks or beer?

c. Given your answers to part b, would you expect the typical household with $40,000 annual income to have a *larger* or *smaller* share of their total expenditures allocated to beer compared to a typical household with $60,000 annual income? Explain the logic underlying your answer.

8. Suppose that the following equation represents annual household **demand** for broilers.

$$Q = 160 - 25P_1 + 10P_2 + .001I$$

where Q = annual pounds of broilers
P_1 = price/lb of broilers
P_2 = price/lb of beef
I = annual income

Calculate the direct and cross-price elasticities for broilers when the prices of broilers and beef are $.80 and $3.00/lb, respectively, and annual income is $30,000/yr.

9. Assume that the income elasticity for oysters is 2.0.

a. How much of a change in demand for oysters would you expect to occur if there is a 10% increase in average incomes of all households in a given market?

b. How much of a change in demand for oysters would you expect to occur if there is a 5% increase in population of a given market?

c. How much of a change in demand for oysters would you expect to occur if there is a 5% increase in population and a 10% increase in average income at the same time?

REFERENCES

Eales, J. 1996. A further look at flexibilities and elasticities. *American Journal of Agricultural Economics* 78(4): 1125–1129.

Huang, K. S. 1994. A further look at flexibilities and elasticities. *American Journal of Agricultural Economics* 76(2): 313–317.

Huang, K. S. 1996a. A further look at flexibilities and elasticities: Reply. *American Journal of Agricultural Economics* 78(4): 1130–1131.

Huang, K. S. 1996b. Nutrient elasticities in a complete food demand system. *American Journal of Agricultural Economics* 78(1): 21–29.

NOTES

1. Furthermore, for relatively small variations in price, a linear representation may be a reasonable approximation of curvilinear relationships.

2. Actually, P_2 could be considered to include several variables representing prices of multiple commodities.

3. For those familiar with calculus, the slope ($\Delta Q/\Delta P$) can be found by evaluating the derivative of the mathematical representation of demand at a particular point.

4. One reason the selection of points becomes important is that in the case of a linear demand function if the two extreme points on each axis are selected an elasticity of -1 results regardless of the slope of the demand function.

5. Directional changes in total revenue for a seller is the same thing as changes in total expenditures for buyers associated with different points along a demand schedule.

6. For more discussion about the relationship between these two coefficients see Huang (1994), Eales (1996), and Huang (1996a).

7. If the new price and quantity combination is used, the cross-price elasticity will be $\frac{5}{.25} \cdot \frac{1.00}{15} = 1.33$. If the arc formula is used, the value will be $\frac{5}{.25} \cdot \frac{1.75}{25} = 1.4$.

8. A zero cross-price elasticity could exist among commodities that are substitutes because the formula uses the change in the quantity of one good that results from substitution as well as income effects associated with a change in price of another good.

9. Alternative values of the income elasticity would be $\frac{10}{5{,}000} \cdot \frac{30{,}000}{110}$ or $\frac{10}{5{,}000} \cdot \frac{53{,}000}{210}$ depending on whether point or arc calculations are used to evaluate the elasticity.

SUPPLY CONCEPTS

The purpose of this chapter is to review basic economic concepts regarding the supply of goods and services. Similarities and differences with the concepts discussed in Chapter 4 are emphasized.

The major points of the chapter are:

1. Important characteristics of individual and market supply schedules.
2. Major shifters of supply relationships for food and agricultural commodities.
3. Calculation and interpretation of price elasticities of supply.
4. Differences in price elasticities of supply.

INTRODUCTION

The first part of the chapter discusses supply schedules and the difference between changes in supply and quantity supplied. Next, the chapter summarizes some of the economic assumptions underlying supply schedules. In particular, the difficulty farmers have in making production decisions based on an unknown price anticipated to exist when an item will be ready for sale is considered. The third part of the chapter provides a list of supply shifters. Finally, the remaining parts of the chapter indicate how price elasticities of supply are calculated and expected to vary for different periods of time and level of product aggregation.

BASIC CONCEPTS

The basic concept of a supply relationship refers to the entire schedule of quantities of a product that individual firms, or all firms in an industry, are willing to produce and sell at alternative prices. It is important to keep the schedule concept in mind in order to differentiate between a change in supply and a change in quantity supplied. The difference is whether the entire supply schedule shifts or whether there is just movement from one point to another one along a given supply schedule in response to a change in price. For example in Figure 5–1, a

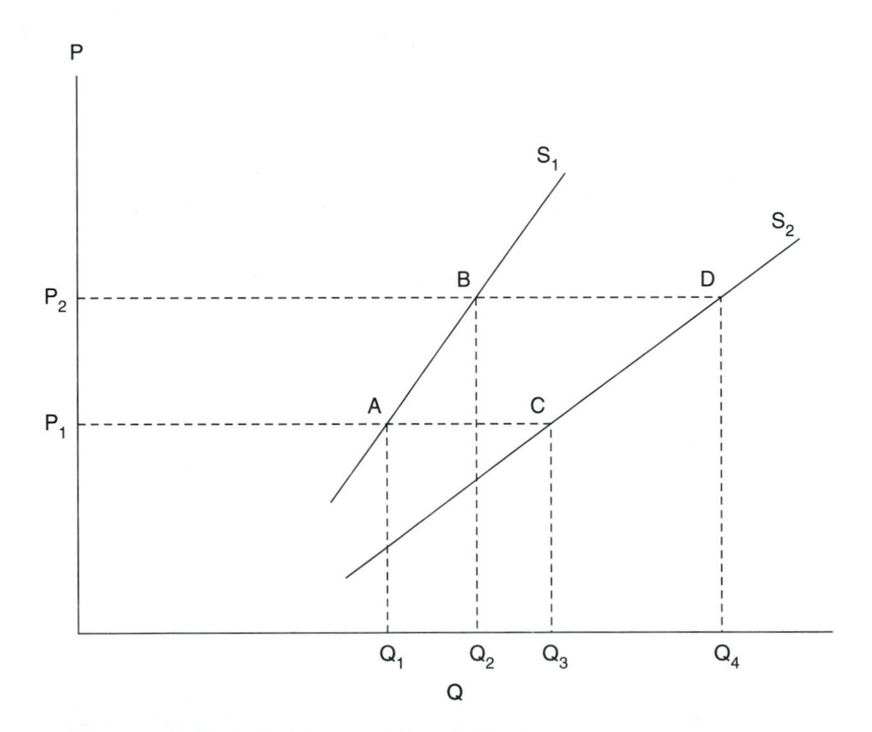

FIGURE 5–1 Alternative supply relationships.

change from Q_1 to Q_2 associated with a change in price from P_1 to P_2 is a change in *quantity supplied* in that it reflects movement along a given supply relationship. A movement from Q_3 to Q_4 associated with the change in price from P_1 to P_2 would also be described as a change in quantity supplied along a different supply function. A shift in the entire schedule from S_1 to S_2 would reflect a *change in supply*. The supply schedule S_2 represents an increase in quantity at each and every price relative to points along S_1. Thus a change from Q_1 to Q_3 (when the price is P_1) or from Q_2 to Q_4 (when the price is P_2) would indicate the effects of a change in supply. In either case a change in quantity occurs without any change in price. A change in price can lead to a change in the quantity supplied but does not alter the supply of the commodity.

ECONOMIC ASSUMPTIONS

Several economic assumptions underlie the way in which alternative quantities are related to different prices as characterized by a supply relationship. First, an underlying production function indicating how various resources or inputs can be combined to produce a product is assumed to represent or describe the transformation process involved in producing goods and services. Second, prices of all inputs are assumed to be known and held constant in defining a supply schedule. These two pieces of information determine the **marginal** or extra **cost** of producing varying quantities by modifying the combination of inputs.

A fundamental concept of basic economics is that producers maximize profits by finding the correct combination of resources to expand production until the **marginal cost** of producing the last unit is equal to the price they expect to receive. If the extra revenue obtained from the last unit produced is equal to the extra or additional cost of producing the last unit, maximum profit is guaranteed. The schedule of optimal quantities of production associated with all possible prices generates a supply function for an individual firm. The horizontal summation of supply curves of all firms producing the same item leads to the industry or aggregate market supply. The horizontal summation process is identical to what is involved in aggregating individual household demand relationships to obtain market demand relationships.

The preceding ideas are extremely useful for thinking about all kinds of production decisions, but some elements require some additional consideration when considering the supply of agricultural products. For example, the above framework assumes that production decisions are made assuming the price of the product is known. When production decisions about agricultural products are made, there is often no certainty about what price will be received when the product is ready to be sold unless a producer has a **forward contract,** uses a futures market to lock in a price, or the government announces a guaranteed or fixed price.[1] Much of the conceptual framework underlying supply curves assumes producers instantaneously respond to price changes. In the case of agricultural products, it is important to realize that producers often have to make decisions based on expected prices. Whether or not an expected price is the price that actually occurs depends on a lot of intervening events between the time a production decision is made and when the product is ready to be sold.

Once a product is produced, decisions of when to put it on the market or offer it for sale can be quite responsive to current market prices, but many production decisions with respect to agricultural products have to be made months or years prior to the realization of actual output. Thus, basic production decisions often are made based on some type of price expectation.

A lot of economic analysis and attention has been given to how producers formulate price expectations. At least for commodities, for which a futures market exists, it is intuitively appealing to argue that futures markets provide some information about expected prices for alternate points in the future. Chapter 15 will indicate how futures markets provide an opportunity for producers to more or less lock in a price when making production decisions.

Futures markets do not exist for all agricultural commodities, however, so it is necessary to also consider how price expectations might be formulated in the absence of a formal futures market. One way to formulate an expectation about price is to think about what demand and supply conditions are likely to be in the future. This essentially means that an expected price is based on the anticipated demand and supply conditions. A more formal way of expressing this kind of reasoning is known as **rational expectations.** Using a rational expectations formulation is a forward-looking approach in that expected price is equal to what economic analysis suggests the price would be for a future time period based on all current information about future demand and supply conditions. The rational expectation theory assumes that producers are rational since they take into account all current information in formulating an expected price for making production decisions. This line of reasoning does not imply that the expected price will necessarily be the realized price. All sorts of new conditions and new information could develop between the time a price expectation is formulated and the time a commodity is ready to be sold. Consequently, the realized price reflecting actual demand and supply conditions could deviate significantly from expected price.

SHIFTERS OF SUPPLY FUNCTIONS

Assuming that the supply curve represents the schedule of quantities that would be produced and sold at alternative (expected) prices, the entire schedule may change in response to a number of different factors. It is useful to have a list or system for categorizing various factors that cause changes in supply schedules similar to the list of shifters of demand schedules developed in Chapter 4. Each of the major shifters of supply functions will be identified and discussed in the following paragraphs.

One reason supply relationships shift is because of changes in marginal costs. Marginal costs can change either because of changes in technology or the price of inputs that alter the cost of production. Either one of these two events can cause a change in marginal costs of production and a firm's willingness to produce and offer a commodity for sale.

A change in technology enables a producer to get more output with the same quantity of inputs or use fewer inputs to get the same quantity of output. Changes in technology will lead to expansions in supply unless the use of a new technology is prohibited by government regulations. Changes in technology result in marginal cost curves shifting downward or to the right, which is probably the most significant factor accounting for shifts in the supply of agricultural commodities over time. Examples of technological improvements include improved crop and vegetable varieties, new production technologies, and so on.

Changes in prices of inputs can also cause a change in marginal costs of production. Depending on whether input prices increase or decrease a supply curve could shift to the left or the right. If input prices decrease, the marginal cost curve would be lower or shift to the right resulting in an increase in supply. If input prices increase, it will cause higher marginal costs and a decrease in supply.

The price of other products can also shift supply curves in terms of altering the opportunity costs of a given enterprise. For example, if the price of corn goes up relative to soybeans, land could be switched out of soybean production, shifting the supply of soybeans to the left. This type of change represents a change in the opportunity cost or value of land in terms of producing one crop relative to another. It may not be easy to shift specialized production facilities from one enterprise to another, but if inputs can be used in the production of multiple commodities, the price of the product can have a significant influence on the supply of other products. In the case of products that are produced jointly an increase in the price of one item may increase the supply of the other commodity. For example, an increase in the price of wool may increase the supply of lambs.

A final point about opportunity cost concerns the effect of returns for nonagricultural uses of factors of production. For example, if the returns to off-farm work changes, the aggregate supply of farm products could shift simply because of the effects on the opportunity costs of labor engaging in agricultural production. The same thing is true about nonagricultural uses of land. If it becomes more profitable to use land for building houses, apartments, or shopping centers, land will shift out of agricultural production. This kind of adjustment simply represents factors of production responding to their best opportunity return. This type of phenomenon has been occurring for many years with labor and land shifting from agricultural to nonagricultural uses in the United States.

The opportunity cost of using entrepreneurial skills to produce agriculture and food products essentially determines the number of firms willing to produce a given product. The number of firms in existence at a given point in time determines the nature of the horizontal aggregation of individual supply relationships required to obtain the market supply for a given geographical region. In this sense, the number of firms in operation plays a similar role in supply analysis that total population does in demand analysis.

Another kind of shifter of supply functions is random effects such as weather conditions, diseases, and so on. Random effects can cause a deviation in what is actually produced relative to the amount of anticipated production thereby shifting the realized supply schedule. This kind of shift has the same effect on market conditions as if conscious decisions to shift supply were made by producers.

Effects of weather and diseases or other random effects on the aggregate supply of individual products are generally not very significant. Random events may have especially large effects on the supply schedule of individual producers or small areas in a country, but they generally do not have large aggregate effects. For example, there may be a terrible drought or flood in one part of the country while at the same time growing conditions may be ideal in other areas of the country, except in special cases such as periodic El Niño weather effects.

Governments can also influence or cause supply schedules of products to be altered. One way is if the government places an explicit limit on the amount of output that can be produced. This would mean that individual as well as aggregate supply curves become vertical at a specified quantity regardless of how high a price is considered. Government programs might also limit the quantity of particular inputs that can be used, for example, so many acres, so many pounds of fertilizer, or so many units of water. Recently, livestock producers have been facing additional restrictions on the management of animal wastes. Government policies also control the kinds of chemicals that can be used in the production and processing of agricultural products. All of these regulations ultimately have implications on the supply schedules of agricultural products.

Another way in which governments indirectly influence the supply of agricultural products is through expenditures for research and extension or other educational activities related to agricultural production. One rationalization for government's role in subsidizing the research and education activities is the competitive nature of the industry. To the extent that research activities are successful in developing new technology and people are made aware of new technologies through education an outward shift in supply will occur. If expenditures on research and extension are increased, an expansion in supply and lower prices will eventually occur, benefiting consumers. The fact that students at state-supported institutions learn about agricultural markets at tuition rates less than full cost is one example of an educational subsidy. Another example is a publicly subsidized extension service.

The above list is a fairly complete list of factors that can cause shifts in supplies. Thus whenever something happens in the world, the first thing to do in predicting its effect on a given market is to consider whether the event will be a shifter of demand or supply relationships. Once the list of different demand and supply shifters is checked, a basic demand and supply framework can be used to analyze the effect of shifting demand or supply functions and translate any event into an appropriate anticipated price effect. This kind of framework will be developed in Chapter 6.

PRICE ELASTICITY OF SUPPLY

When working with supply relationships, it is useful to know how large a movement along a supply curve will occur in response to a particular change in the price of the commodity. For example, if the price should change from P_1 to P_2 in Figure 5–1, how big a change in quantity supplied will occur if S_1 is the appropriate supply relationship? For this kind of measurement, the same kind of formula used for measuring price responsiveness of demand relationships is appropriate (i.e., price elasticity). Thus the percentage change in quantity supplied divided by the percentage change in price can be used to measure the responsiveness of changes in quantity supplied. A price elasticity of supply can be calculated using either a point or an arc formula. The price elasticity of supply using the point formula is:

$$\left(\frac{\Delta Q}{\Delta P}\right)\left(\frac{P}{Q}\right)$$

where the latter component represents a particular price-quantity combination on the supply function.

For example, if P_1 and P_2 in Figure 5–1 are \$2.00 and \$3.00, respectively, a change in price from P_1 to P_2 would be a change of 50%. If the corresponding values of Q_1 and Q_2 are 20 and 25, respectively, the percentage change in quantity associated with the change in price from P_1 to P_2 would be 25%. These values indicate a supply elasticity of .5 (i.e., 25/50). An alternative way of arriving at this result would be to substitute values directly into Equation 5–1, for example,

$$\frac{5}{1.00} \cdot \frac{2.00}{20}.$$

If one desired to use the arc formula, the only change in the formula would be to replace the ratio of P to Q by a ratio of average prices to average quantities. For the previous example, this would mean replacing 2.00/20 with 2.5/22.5. The arc formula for the supply elasticity would produce a value of .56 instead of .5. The reason for the difference is that percentage changes in prices and quantities are calculated with a different base than if the point formula is used.

If an algebraic representation of a supply relationship is available, it is possible to numerically determine the change in quantity supplied associated with a change in the price.[2] This would provide the necessary information for the first part of the price elasticity formula. For the second part of the formula, a particular price-quantity point on the supply function or an average of two points could be used depending on whether the point or arc formula was used. For example, assume that Equation 5–2 represents the supply response for a given commodity.

$$Q = 10 + 2P - 1L$$

where

Q = quantity produced
P = price of commodity
L = price of labor

In this case, Q would change by two units for each unit change in P so $\Delta Q/\Delta P = 2$. It would be necessary to specify a particular value for L, however, to know exactly how much Q would be produced for a given value of P. Once a value for L is selected, the equation would indicate an appropriate value of Q for any particular value of P selected.

Price elasticities of supply will always be positive values as long as supply curves have a positive slope. Also, any increase in price along a positively sloped demand curve will result in greater total revenue. Thus, the direction of changes in revenue associated with different points along a supply function is related only to the direction of price change and not the magnitude of the price elasticity of supply. Any decrease in price will produce less revenue. It does not matter whether the supply curve is fairly flat or fairly steep, or whether the supply elasticity is less than 1 or greater than 1, total revenue changes in the same direction as price along a supply curve. The amount of change in total revenue, however, between different points on a supply curve depends on the price elasticity of supply as well as the magnitude of price adjustment.

In general, there do not appear to be as many empirical estimates of price elasticities of supply of agricultural products readily available as there are price elasticities of demand. Table 5–1 contains a range of price elasticities of supply for several selected field crops and vegetables in the southern part of the United States reported by Shumway (1986). For some of the commodities, the range in the empirical estimates of price elasticity of supply is fairly large.

Differences in Price Elasticity of Supply at the Farm Level

The price elasticity of supply for any commodity is likely to differ depending on how long a change in price remains in effect just like the price elasticity of demand. This concept is especially important because of the inability to instantaneously adjust production to changes in actual or expected price. The time dimension is very important in terms of thinking about what sort of a response in quantity supplied is likely associated with a given change in price.

For purposes of contrasting differences in price responsiveness of supply relationships it is convenient to consider three varying periods of time. The first time period to be considered is a **marketing period.** This is especially appropriate for storable commodities that are produced once a year for which the marketing period essentially exists from one harvest period to the next.[3] Even if it is technically feasible to store a commodity for more than 1 year, there is usually little in-

TABLE 5-1

PRICE ELASTICITIES OF SUPPLY FOR SELECTED FIELD CROPS AND VEGETABLES IN THE SOUTHERN UNITED STATES

COMMODITY[a]	RANGE
Field crops	
Corn	.34 to 1.59
Cotton	.25 to .34
Rice	.06 to .77
Soybeans	.62 to .94
Wheat	.35 to .46
Food grains	.40 to .51
Oil crops	.15 to .34
Vegetables	
Cabbage	.16 to .31
Potatoes	.15 to .22
Tomatoes	.27 to .92

Source: Adapted from Table 5, Shumway, (1986). "Supply Relationships in the South—What Have We Learned?" *Southern Journal of Agricultural Economics* 18(1).
[a] Estimates reported only for commodities with multiple studies and if estimates were significant at 5% level.

centive to store commodities longer than a marketing period because of seasonal price patterns. This topic will be discussed more fully in Chapter 13.

The usual assumption about the supply curve of a commodity for a marketing period is that it is quite inelastic. This implies that a change in price really would not have much effect on changing the total quantity available over the entire marketing period. The basic reason is that there can be no change in quantity resulting from adjustments in production until the next harvest period. Thus the supply elasticity would be quite small until a new crop can be harvested.

Even though a fixed quantity of a crop is available for a marketing period, the supply curve would probably not be perfectly inelastic because if the product's price becomes too low some reduction in the quantity supplied might result simply due to failure to harvest or market some of the commodity. That would give responsiveness to downward price changes. Similarly, if the price increased extra care might be used in the harvesting and marketing process to minimize losses thereby increasing the total quantity available for sale. Another source of responsiveness for some commodities depends on how much is used for at-home consumption. This is not as big a factor in the United States as it previously was, but is still important in some countries. If the price increases enough there would be an incentive to sell more of the commodity rather than using it for household food consumption. If the market price is low the amount used at home could increase and smaller quantities sold on the market, producing a positively sloped supply response. Another reason for some responsiveness to price during a marketing period is that price can influence culling decisions and/or grading practices. If the price got high some inferior quality products that would ordinarily be discarded might be sold.

Consider now a second period of time that is shorter than a marketing period, that is, a period of time that might be a day or a month within a marketing period. In this case, the price elasticity of supply would be greater than for the entire marketing period. At first glance this sounds contradictory because the reason for the initial low price elasticity of supply is that production could not be changed and certainly production cannot be changed during even shorter periods of time. The reason for increased responsiveness for shorter periods of time, however, is because the supply curve is defined as the quantity that people are willing to produce and sell. If a shorter period of time than a marketing period is considered, the timing of the selling decision dominates the production decision as far as supply responsiveness. There may be only a fixed amount of the commodity to sell during an entire marketing period, but there is more flexibility about how much is sold in a given week or month. Thus what is observed is the effect of changes in decisions about when to market or sell the commodity in response to price changes. This means there may be quite a bit of responsiveness to price changes in the short run because of the decision of when to sell. For example, it is reasonable to expect that most of the corn that is produced in a given year will be sold before next harvest, but when will it be sold? If the price goes up in December there is likely to be a sizable percentage response in terms of the amount put on the market. This means that the concept of market supply is not just a schedule of quantities producers are willing to produce, but the schedule of quantities that producers are willing to produce *and sell* at alternative prices. When considering short periods of time, the relevant supply curve is defined by how much of a commodity already produced is likely to appear on the market at alternative prices. In using the economic concept of supply, extra care must be exercised because of the distinction of what is meant by supply for different lengths of time. In some cases, production decisions are the most relevant. In other cases, the marketing decisions about what is going to be offered for sale are more relevant. Even for continuously produced commodities, there is a decision of whether to sell the commodity today or keep it a little longer hoping that the price will increase.

The final period of time for considering different supply responses is one that is longer than the marketing period. It is reasonable to argue that the supply elasticity for this sort of time period would be bigger than what it is for a marketing period. This is because producers would have time to make adjustments in production. If there is time to modify production decisions some responsiveness to changing prices would be observed. It is not possible to say whether the price elasticity for the long run is greater than the short run, but in general a U-shaped curve, like that described in Chapter 4, is again appropriate to characterize how price elasticities of supply for a given commodity might vary with the length of time associated with a given price change.

Another generalization about the price elasticities of supply that is similar to the case of price elasticities of demand is the elasticity for aggregate commodities is likely to be smaller than for individual commodities. This again sounds counterintuitive in terms of the elasticity for the aggregate being different from the sum of the parts. The reason for this generalization is that changing from one enterprise to another enterprise is generally easier than changing specialized inputs from agricultural production to nonagricultural production. Empirical verification of this fact is that if all agricultural prices go up relative to nonagricultural prices, the percentage change in aggregate production will not be as great as the percentage change in output of an individual commodity in response to the same percentage change in price. Also, it is much easier to transfer inputs from the production of one commodity to the production of another than it is to expand production of all commodities. For example, the response in corn production will be much larger if corn prices go up relative to other prices than if all agricultural prices went up.

SUMMARY

The willingness of farmers and agribusinesses to produce and sell varying quantities of agricultural and food products at alternative prices is an important concept underlying market behavior. A basic assumption underlying supply relationships is that producers attempt to maximize profits by selecting an optimum combination of inputs until the marginal cost of the last unit produced is equal to the price or the anticipated increase in total revenue.

Many characteristics of supply relationships are similar to concepts about demand introduced in Chapter 4. In particular, the distinction between a change in quantity supplied (a movement along a supply schedule) and a change in supply (a shift in the entire schedule) is very important. A key difference between demand and supply behavior, however, is that purchasing decisions are often made in response to known prices whereas considerable price uncertainty may exist at the time decisions are made to produce items. Consequently, price expectations play an even more prominent role in production and selling than purchasing decisions. Although individuals presumably use many sources of information in forecasting prices, the use of futures market prices and a rational expectation framework are reasonable approaches to formulate price expectations. The rational expectations framework essentially assumes anticipated demand and supply conditions are used to formulate price expectations.

Things that can shift supply schedules are changes in technology, prices and/or opportunity costs of inputs, governmental policies that affect production decisions, and random events. Government policies may place limits on quantities of production or operate in many indirect ways. Random events resulting in actual production deviating from anticipated output have the same consequences on availability of products as explicit decisions altering quantities of production. The usefulness of the list of supply shifters and the list of demand shifters developed in the previous chapter will be considered more fully in Chapter 6.

The responsiveness in quantity supplied to changes in price along a supply curve is indicated by the price elasticity of supply. The price elasticity of supply is defined as the percentage change in quantity supplied for each 1% change in price of the commodity, holding everything else constant. Price elasticities of supply can be calculated using point or arc concepts similar to the formulas introduced in the previous chapter. Assuming quantities of production always increase with higher prices implies that price elasticities of supply are positive. Also directional changes in total revenue along a supply curve depend only on whether price (and quantity) is increasing or decreasing and not the magnitude of price elasticity.

Price elasticities of supply for a given commodity are assumed to vary for different lengths of time similar to the U-shaped curve described in Chapter 4. The price elasticity of supply for a periodically produced storable commodity would be smallest for the period of time between harvests. Price elasticities of supply for shorter as well as longer time periods would generally be larger than the price elasticity for the entire marketing period. Larger elasticities would be the result of being able to vary the time of sale and/or change production. Also the price elasticity of supply for an aggregate group of commodities would generally be expected to be less than the price elasticity of supply for individual commodities because of the production substitution possibilities.

QUESTIONS

1. Briefly indicate how each of the following events would be expected to affect the U.S. supply of corn and explain your reasoning.
   a. A 15% increase in the expected price of corn because of an increased demand for livestock feed.
   b. A 10% decrease in the expected price of soybeans.
   c. The development of a new variety of corn that increases yields per acre by 8%.
2. Briefly indicate how you would expect each of the following events to affect the U.S. supply of turkeys and explain your reasoning.
   a. A 15% increase in the price of corn and soybeans.
   b. A 10% decrease in the price of turkeys.
   c. Development and adoption of a new synthetic hormone that decreases turkey mortality and improves feed efficiency.
3. Assume that the following equation represents the supply of wheat from all producers in a given region.

$$Q = 34 + 2.5P - 40F$$

where

Q = million bushels of wheat
P = price/bu of wheat
F = price/lb of fertilizer

Calculate the price elasticity of wheat supply if the price of wheat is $4.00/bu and fertilizer is $.10/lb.

4. Briefly explain how the price elasticity of supply of corn for 1 month is likely to compare to the price elasticity of the same product for a 12-month period.

5. How would you expect the price elasticity of supply for peppers to compare to the price elasticity of supply of all vegetables in a given geographical region? Briefly explain your reasoning.

REFERENCES

Heifner, R. G., B. H. Wright, and G. E. Plato. 1993. *Using cash, futures, and options contracts in the farm business.* Agric. Info. Bull. 665. USDA, Economic Research Service. Washington, DC.

Shumway, C. R. 1986. Supply relationships in the South—what have we learned? *Southern Journal of Agricultural Economics* 18(1):11–19.

NOTES

1. Additional discussion about the use of cash, futures, and options contracts is available in Heifner, Wright, and Plato (1993).

2. The change in quantity per unit change in price can be evaluated by substituting alternative values of price into the equation or using calculus.

3. For many continuously produced commodities such as livestock products, the **marketing period** is the time from when initial production decisions are made until the actual product is ready to be sold. This can vary from a few weeks to several months depending on the particular product.

CHAPTER

6

MODELS OF MARKET BEHAVIOR

This chapter indicates how some of the basic concepts of demand and supply introduced in Chapters 4 and 5 can be used to analyze changes in market conditions. Also a framework is developed to show how prices and quantities of food and agricultural products at the farm and retail levels of the **marketing** system are linked together. Emphasis is on showing how the framework can be used for analyzing the effects of changes in demand, supply, and marketing costs of any commodity.

The major points of the chapter are:

1. How basic demand and supply concepts can be used for analyzing changes in market conditions.
2. The development of an expanded framework for considering the linkage between retail and farm prices of commodities.
3. How the expanded framework can be used for analyzing effects of changes in demand, supply, and the cost of marketing products.
4. Additional implications of general framework for analyzing markets.

INTRODUCTION

Chapter 6 illustrates how basic knowledge about demand and supply relationships can be used to analyze markets. The first part of the chapter indicates how changes in observed quantities and prices of an item can be used to determine if changes in unobservable demand and/or supply schedules have occurred in a given market. Even though analysis using single demand and supply relationships is extremely useful for analyzing a particular market it does not provide an adequate mechanism for understanding the linkage between retail and farm prices of commodities. Consequently, the second part of the chapter indicates how basic demand and supply concepts can be extended to show the interrelationship between retail and farm level markets. The third part of the chapter discusses how changes in retail and farm level prices as well as the quantity of a product moving through the marketing system occur in response to changes in retail demand, the supply of products at the farm level, or marketing costs. The final part of the chapter reviews some of the simplifying assumptions and limitations of the general framework.

BASIC FRAMEWORK

An understanding of the basic principles of demand and supply that have been discussed in Chapters 4 and 5 in conjunction with deductive reasoning is the starting point for most market analysis. The process can be used to determine whether observed changes in market conditions have been the result of changes in demand or supply. The process also can be used to forecast how market conditions will be affected by particular shifts in demand and/or supply relationships.

In order to illustrate the simplest kind of analysis, an initial market clearing price and quantity of a commodity can be used to define four quadrants in a price-quantity diagram. In Figure 6–1, point A is assumed to represent an initial market clearing price (P_c) and quantity (Q_c) determined by the intersection of an unobservable negatively sloped demand function and an unobservable positively sloped supply function. The boundaries of the four quadrants in Figure 6–1 are determined by horizontal and vertical lines defined by the price and quantity associated with the initial point (A). All points in quadrant I represent observations with higher prices and greater quantities relative to point A. All points in quadrant II represent higher prices but lower quantities than point A. Similarly, points in quadrants III and IV define alternative combinations of prices and quantities relative to the initial price and quantity associated with point A.

Delineation of the four quadrants in Figure 6–1 provides a mechanism for inferring what is responsible for changes in market conditions as changes in price and/or quantity are observed. For example, a new market clearing price and quantity consistent with a point in quadrant I indicate that there has definitely been an increase in demand for the commodity relative to the previous situation. Without

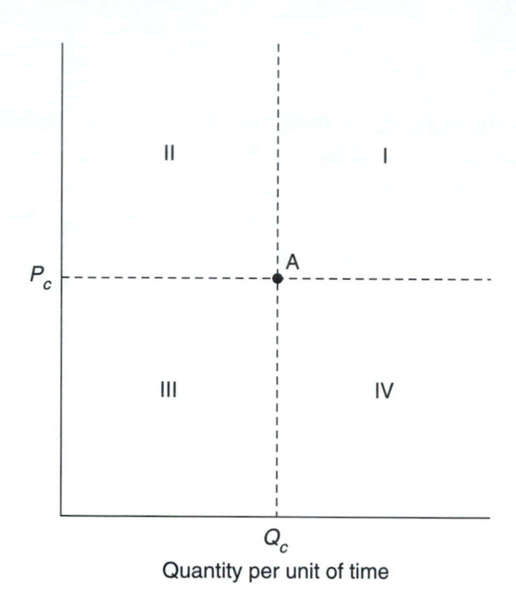

FIGURE 6–1 Alternative quadrants defined by initial equilibrium price and quantity.

additional knowledge about the slope of the supply function, it would not be possible to conclude anything about changes in the supply function. All points in quadrant I can be consistent with the supply curve remaining the same, increasing, or decreasing relative to the relevant supply for point A. Point A and any point in quadrant I could be located on the same supply function or located on an entirely different supply relationship depending on how the quantity supplied responds to changes in price. It is clear, however, that a change from point A to any other point in quadrant I implies that an increase in demand has occurred since it is not possible for both points to be on the same negatively sloped demand function. Similarly, changes in equilibrium prices and quantities into other quadrants could be used to make alternative inferences as indicated in Table 6–1.[1]

TABLE 6-1

ALTERNATIVE QUALITATIVE INFERENCES CONSISTENT WITH CHANGES IN MARKET CLEARING PRICE AND QUANTITY.

Change to	Conclusion
Quadrant I	Increased demand
Quadrant II	Decreased supply
Quadrant III	Decreased demand
Quadrant IV	Increased supply

Although the previous ideas are fairly simple, they are extremely useful to begin explaining changes in market conditions. Such an analysis does not indicate what may have been responsible for a change in demand or supply, but the set of possible influences is at least reduced to either the list of supply or demand shifters discussed in earlier chapters.

The same framework can also be used to make predictions about changes in market conditions. For example, if demand is expected to increase without any change in supply, a subsequent increase in forecasted price and quantity or a move into quadrant I would be implied. How much of a change in price and quantity would occur depends on the amount of increase in demand as well as the price elasticities of demand and supply. This type of information would be required to translate the above kind of qualitative market analysis into more quantitative terms. An example of quantitative analysis is contained in Appendix A.

Manipulation of demand and supply relationships for a given commodity can also be used to illustrate how markets for different commodities are interrelated. For example, a change in the price of one commodity can affect the demand for another commodity as indicated by the cross-price elasticity of demand. Any change in demand resulting in a new market clearing price and quantity will have implications on the amount and directional change in the value of purchases (or sales) in that market.

Limitations of the Basic Framework

It is important that the **demand** and supply relationships be at the same level in the marketing system for the previous kind of analysis to be meaningful. For example, is the price and quantity of the product being considered at the retail or farm level? The kind of market being analyzed would dictate what kind of demand and supply relationships would be relevant.

As noted in Chapter 4, individual consumer demand schedules can be horizontally aggregated to conceptually represent aggregate demand at the retail level for products for some market aggregate (e.g., city, state, region, nation, world). For example, the retail demand for bread in a given market could be obtained by aggregating individual household demand schedules. What is of interest and importance to wheat producers, however, is how the retail demand for bread or other bakery products affects the demand and price of wheat. Even though consumers do not directly buy wheat from producers, changes in the demand for bread and other food products at the retail level have definite implications for the producers of primary agricultural products used to produce retail items. Thus, a mechanism for extending basic concepts about consumer demand for foods is required in order to understand implications about the demand for products that agricultural producers sell. Similarly, factors affecting the supply of primary agricultural products must be linked to the supply of retail food products for an overall conceptual model of the marketing system for agricultural products.

Incorporating Marketing Activities

As noted in Chapters 1 and 2, many activities occur between the sale of products at the farm level and the retail level that are responsible for changes in the form, location, and time products are sold. It is obvious from differences in the nature of products that farmers sell and products purchased by consumers at the retail level that a host of marketing activities add value to products. For example, the value of wheat is enhanced by the transformation into products consumers want to consume. Also, the values of wheat and related products are enhanced as items are transported to locations more convenient to consumers. An example of the value added by changes in time of ownership is that consumers prefer to purchase and consume bread continuously, even though wheat is harvested only during a limited period of time each year. Storage of wheat or flour and related products between harvest periods provides time value in terms of consumers being able to buy bread closer to the time they actually want to eat it. Storage requires the use of economic resources and therefore consumers must be willing to pay for the convenience of obtaining bread throughout the year to offset the cost of the resources used for storage of wheat or flour.

All marketing activities require the use of economic resources and therefore involve certain costs. For simplification, all of the value-enhancing marketing activities can be aggregated conceptually and their total cost considered as the cost of getting a product from producers to consumers.

Assuming a given marketing cost per unit of product, it is possible to derive the demand relationship for basic raw products at the farm level associated with a given level of demand at the retail level. It is convenient to introduce some simplifying assumptions to facilitate illustrating the major relationships. Once a basic framework for linking retail and farm level markets is developed, the simplifying assumptions can be modified to reflect alternative conditions.

Simplifying Assumptions

One of the simplifying assumptions useful for developing an expanded framework is to assume that every retail food product consists of a fixed proportion of inputs of raw agricultural products and other inputs. For example, it is convenient to assume that 1 dozen eggs is required at the farm level for each dozen available for sale at the retail level. Also, a fixed quantity of other economic resources is assumed to be required for all of the marketing activities associated with each dozen of eggs. In essence, this assumes that the **production function** for converting raw farm products into consumer products requires that inputs be used in fixed proportions. The latter assumption can be modified to consider substitution possibilities between marketing inputs and the raw products in producing final retail goods.[2] For example, the use of additional refrigeration services may reduce spoilage and thereby reduce the quantity of raw products required to obtain a given quantity of products available for sale at the retail level. The idea of substitution among inputs implies that a given quantity of products at the retail outlets might be produced from varying quantities of raw products through the use of alternative technologies. The idea of fixed proportions technologies in the marketing system ignores this type of substitution, but is a useful starting point for thinking about **derived demands.**

The assumption of a fixed proportion between basic agricultural products and products sold at the retail level makes it easy to make the conversion between equivalent units of products at different levels of the marketing system. This means quantities at the farm or retail level can be expressed in equivalent units. For example, in the case of bread and wheat, the quantity axis for representing demand and supply relationships can be expressed either in terms of loaves of bread or bushels of wheat provided the conversion factor between the two entities is known. This permits demand functions for both wheat and bread to be represented in a single diagram using the same quantity axis (Schrimper 1995).

Similarly, the assumption of fixed proportions permits prices at the retail and farm levels to be expressed in equivalent terms (e.g., cents per loaf of bread or cents per ounce of wheat). Thus, a retail demand curve for bread can be expressed as an equivalent relationship in terms of bushels of wheat and price per bushel as long as the fixed conversion factor between quantities of wheat and loaves of bread is specified. Alternatively, the farm level demand for wheat could be expressed in terms of the price of bread and associated number of loaves at the farm level.

Another assumption that is useful for developing an expanded framework for linking retail and farm markets is to assume that prices of the resources used in the marketing process do not vary with the quantity being used. Another way of saying this is to assume that the supply of all inputs other than the raw farm products are perfectly elastic as far as the agriculture marketing sector is concerned. For example, this is consistent with assuming that if additional quantities of refrigeration and/or transportation services are used to market agricultural products it would not affect the price of these inputs. Similarly, if more labor is required in the agricultural marketing sector it would not affect hourly wages. Essentially, this assumes that the agricultural sector is sufficiently small in terms of using the type of resources used for marketing activities that variation in demand for these inputs from the agricultural sector will not materially affect the price of these resources.

Assuming a fixed proportion production function along with a perfectly elastic supply of other inputs used in the marketing process at any point in time is equivalent to assuming the marketing cost per unit (measured at the retail or farm level) of product is a constant amount. A constant cost of marketing services per unit of the product means the cost of marketing a product will not vary with quantities of the product moving through the marketing system. Marketing costs per unit are likely to change over time, but at a given point in time the marketing costs per unit are assumed to be the same for alternative quantities of product under the above simplifying assumptions.

DERIVED DEMAND

The retail demand for a product is referred to as the primary demand for the finished product because the demand for products originates at the consumer level. The demand for any basic agricultural product used in producing the retail product is referred to as the derived demand. Watermelons provide a useful illustration to illustrate these concepts because one does not have to deal with converting units of the product at the farm level into retail products. Since there is a one-to-one relationship between the quantity of watermelons at the farm and retail level there is no need to convert quantities. Also with a perishable product, there is little if any storage or inventory aspects to consider.

Assume that it costs $1.25 to get each watermelon from the farm to the retail level. This means that for any given quantity of watermelons demanded at the retail level, the price or value for the same identical quantity at the farm level would be $1.25 per unit less than the point for the corresponding quantity on the demand relationship at the retail level. For example, if D_r represents the retail demand for watermelons, the derived demand at the farm level could be represented as D_f in Figure 6–2. The demand schedule at the farm level would represent the schedule of quantities that retailers (or their purchasing agents) would be willing to buy at alternative farm prices. By considering each and every point along a retail demand relationship, a corresponding derived demand function can

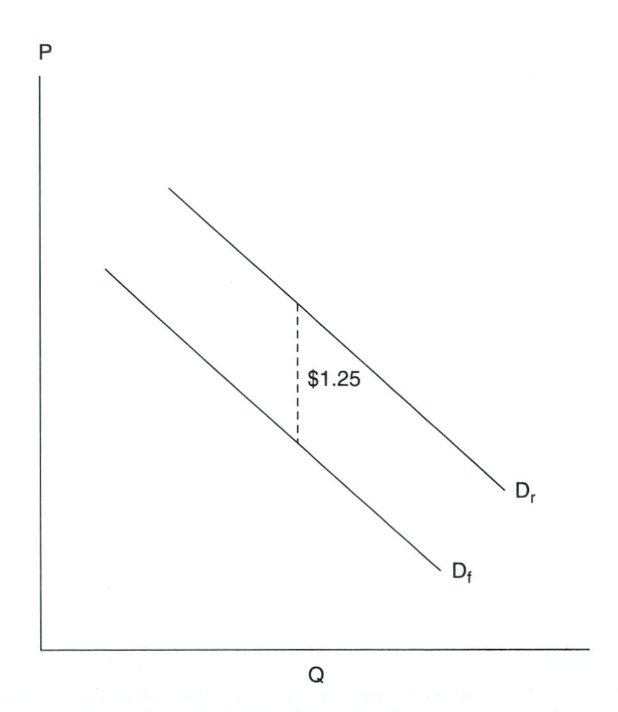

FIGURE 6–2 Retail and derived demands.

be obtained by subtracting the magnitude of the marketing cost. The farm level demand schedule reflects the effect of subtracting $1.25 from each and every price associated with the retail demand schedule. This provides a schedule of appropriate prices retailers could afford to offer for respective quantities at the farm level and still cover marketing costs involved in getting the products into the hands of the ultimate consumer.

The demand for raw products at the farm level will be to the left or below the retail level demand. The vertical distance between retail and farm demand relationships depends on how much it costs to move products from the farm to the retail level. The vertical difference between the primary and derived demand schedules would be the same regardless of what quantity of watermelons was considered assuming marketing costs per unit are constant. Because the demand for watermelons at the farm level is derived from the retail demand, it is referred to as a *derived* demand. Retail demand is a **primary demand** because that is where the demand originates, namely, with the consumer. The demand for products at all other levels in the marketing system is derived from the retail level.

If the retail demand schedule should change, then the demand schedule at the farm level would also change by the same amount as long as there is no change in the per unit cost of marketing. For example, if an increase in income occurs, the retail demand for watermelons would be expected to increase. This kind of change is represented by a new retail demand function D_r' in Figure 6–3. There is no reason to expect a change in income to have any effect on what it costs to move watermelons from the farm to the retail level. Thus, if the **marketing margin** remains the same, the same vertical distance used earlier would need to be subtracted from the new demand curve to obtain the new derived demand curve at the farm level (D_f'). This means that any change in primary demand causes a parallel shift in the derived demands.[3]

Similarly a change in the marketing margin can cause a change in derived demand even though primary demand remains the same. If the vertical distance reflecting marketing costs increases because of increases in the prices of certain inputs used in the marketing process, derived demand would decrease even though there was no change in primary demand. Thus, two forces influence the level of demand at the farm level: (1) the level of demand at the retail level (the primary demand curve) and (2) the size of the marketing costs (i.e., marketing margin). Both of these items influence the derived demand curve.

This means that one more item can be added to the list of possible shifters of derived demand curves. In addition to all of the potential shifters of retail demand functions, changes in marketing costs are another potential shifter of derived demand even though they do not shift primary demands for products. If marketing costs increased at the same time that retail demand changed then derived demand might remain unchanged if the two effects offset each other. For example, if retail demand increased vertically by exactly the same amount as the increase in marketing costs there would not be any change in the demand at the farm level. As far as producers were concerned in this case, the demand for the product they

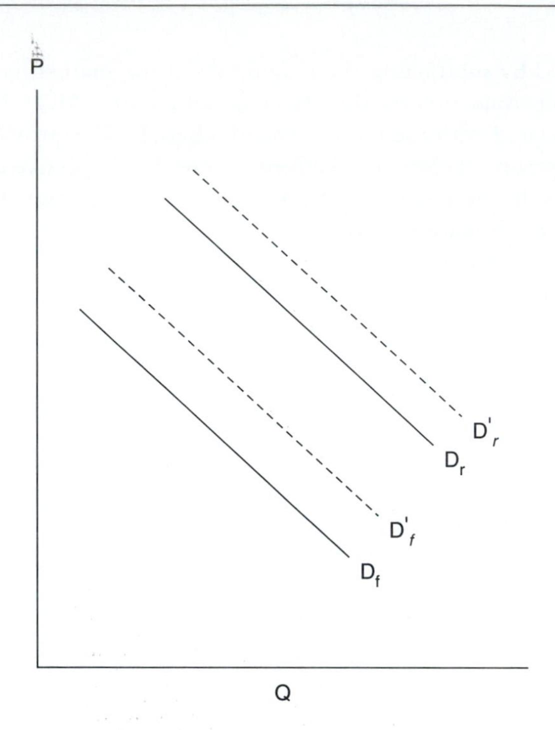

FIGURE 6–3 Original and new retail and derived demand relationships assuming no change in marketing margin.

faced would be exactly the same as it was despite an increase in retail level demand. In order for D_f to be the relevant derived demand curve associated with the new retail demand curve in Figure 6–3, the marketing margin would simultaneously need to increase by exactly the same amount as the vertical shift in retail demand.

DERIVED SUPPLY

The same concepts used in obtaining derived demand relationships are applicable for the supply side of markets because the supply of products at the farm level is not the supply that consumers face at the retail level. The difference in value of products between the two levels of supply in the marketing system is the same marketing cost per unit discussed earlier. On the supply side, the marketing margin must be vertically added to the farm supply relationship, however, in order to obtain the derived supply of the product at the retail level. For example, returning to the example of watermelons the retail price required to get a given quantity of watermelons to consumers will be $1.25 per unit higher than the farm level price associated with respective quantities. Thus, adding $1.25 vertically to each point on the supply schedule at the farm level would produce a

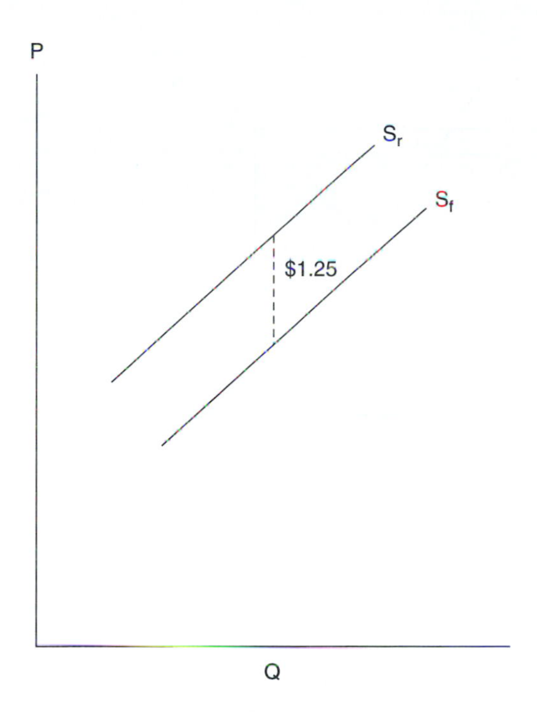

FIGURE 6–4 Primary and derived supply relationships.

schedule of quantities that would be available for sale at alternative retail prices (Figure 6–4). The farm and retail supply functions are parallel because the marketing margin is assumed to be constant.

On the supply side, the farm level relationship is referred to as the primary curve because that is where supply decisions originate. The supply curve at the retail level is the derived supply because it depends on the location of the farm level supply and the magnitude of the marketing margin. Thus, identification of primary and derived relationships is reversed between the demand and supply side of the market.

USING THE EXPANDED FRAMEWORK

When the demand and the supply relationships at the farm and retail level are put together in the same diagram, market clearing or equilibrium conditions at the farm and retail level can be examined simultaneously (Figure 6–5). The intersection of the supply and demand relationship at the retail level determines the retail price and market clearing quantity. Similarly, the intersection of the two farm level relationships determines the market clearing or equilibrium price and quantity at the farm level. Points A and B in Figure 6–5 involve intersections of retail and farm level relationships that do not have any particular

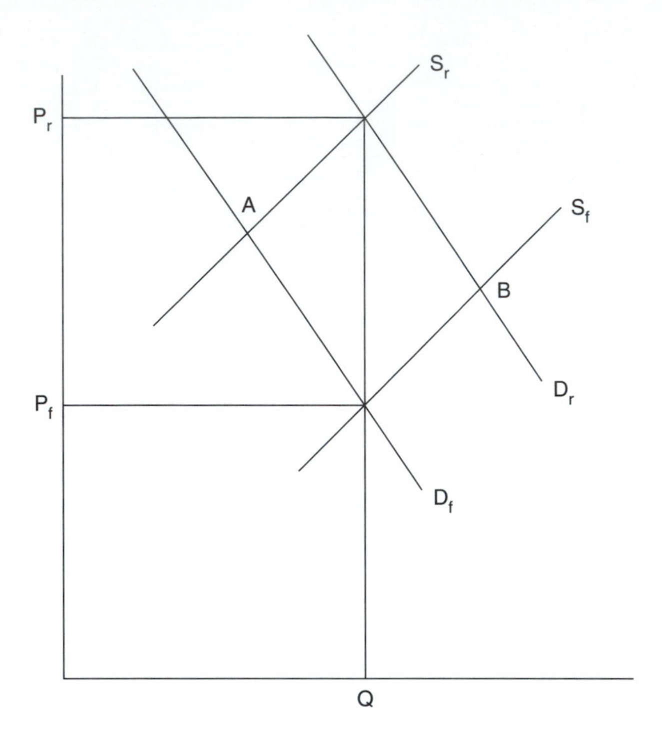

FIGURE 6–5 Combining demand and supply relationships for expanded framework of market.

economic interpretation or meaning because they represent behavior at different levels in the marketing system.

The difference in the market clearing prices at the retail and farm level in Figure 6–5 is exactly equal to the marketing costs. This occurs because the vertical distance between the two demand and supply relationships is the marketing margin per unit of the product. Also, the equilibrium quantity sold at the farm level is identical to the quantity being sold at the retail level. This is because it is assumed that everything that was produced was immediately available for sale at the retail level and there was no demand or supply for storing the commodity. This is especially reasonable for perishable commodities and is why watermelons provide a nice example for illustrating some of the basic ideas of an expanded framework for market analysis. Obviously most products are not as perishable as watermelons and consequently the effects of demand for storage and/or a supply from inventories need to be incorporated into the framework to make it more useful especially for shorter run analysis of nonperishable commodities. The effects of storage will be considered in more detail in Chapter 13.

A framework including derived demand and supply relationships even in its simplest form provides a very powerful tool of analysis. Graphically repre-

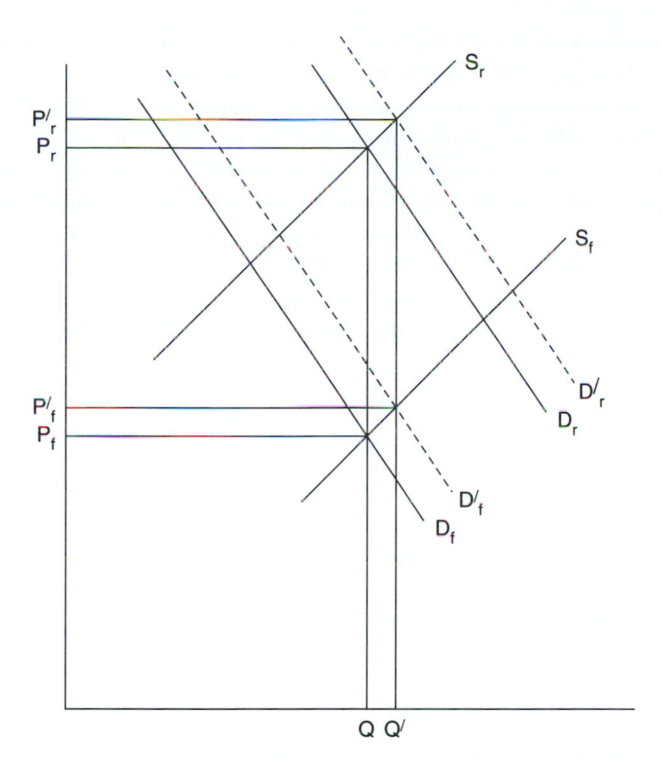

FIGURE 6–6 Effects of an increase in demand.

senting and manipulating these concepts provide a useful tool for qualitative analysis of market conditions. The ability to analyze directional effects on equilibrium quantity and prices at the retail and farm level for different directional changes in retail demand, farm supply, or marketing margin is available with such a framework. For example, suppose the retail demand function in Figure 6–5 increases. Assuming the marketing margin does not change, the derived demand curve would also increase as illustrated in Figure 6–3. Adding dashed lines to represent the new levels of primary and derived demands, indicates that the farm and retail prices and the equilibrium quantity would be expected to increase, Figure 6–6. As long as demand is the only thing changing, there is no reason that the primary and derived supply relationships would change. The quantity supplied at the farm and retail levels would increase as a result of the change in primary and derived demands, but there would be no change in the supply relationships at the farm or retail levels. The actual increase in prices at both levels resulting from a given change in demand would depend on the slopes of the supply curve.

An increase in marketing costs would be reflected by an increased distance between primary and derived relationships for both demand and supply sides of the market. This would result in prices at the retail and farm level moving in opposite directions. The relative magnitude of changes in prices at the two levels depends on the relative slopes or price elasticities of demand and supply.

It is useful to practice using alternative graphical scenarios of Figure 6–5 to demonstrate how changes in demand, **supply,** or the marketing margin cause changes in the equilibrium prices and quantity. After gaining practice considering the effects of changes in each of the major elements, it is useful to consider various combinations of changes. For example, a simultaneous increase in retail demand and a decrease in marketing margin or an increase in demand and a decrease in supply could be considered to analyze the resulting implications.[3]

Qualifications

Although the preceding framework is very powerful and useful, some qualifications need to be emphasized. First, the model represents the general tendency of farm and retail prices and market clearing quantity for a given period of time. Most markets will reflect fluctuating prices and quantities because of continuing changes in market conditions. Thus, what the framework represents is the tendency of market clearing quantity and prices for an interval of time. Different prices will exist for different qualities of the product and prices are likely to vary among locations for similar quality products. Quality and spatial characteristics are two important reasons why prices of commodities at a given time are likely to deviate from single values suggested by the above framework.

The framework also ignores any shortcuts in the marketing system. For example, consumers are not assumed to purchase any products directly from producers. All products are assumed to go through usual market channels. If direct purchases occur, it is assumed that either the producer or the consumer bears the marketing costs. For example, someone has to pay the costs of transportation by either going to get the product or having the product brought to them. Similarly, if there is any processing involved someone has to bear the costs. Thus, there is no way to avoid marketing costs.

The above framework can be modified in many ways. For example, if marketing costs vary with the quantity being handled then primary and derived demand and supply relationships would no longer be parallel lines. If the marketing costs decreased with larger quantities, the lines would be closer together as quantities increased. Similarly, if costs increased with volume the primary and derived relationships would be farther apart as volume increased.

Also, the framework is especially applicable when basic agricultural products are used for only one kind of retail product. Adjustments can be made, however, to consider the case of multiple retail products that originate from the same basic agricultural commodity with different marketing costs. The case of two retail products using the same agricultural product is considered in Appendix B. The same ideas can be used to consider the case of alternative sources of supply (e.g., domestic and imports) of a basic agricultural commodity.

Adjustment Time

The basic framework indicates that when an increase in demand occurs with no change in the marketing margin or farm supply, the farm price and the retail price will increase by an equal amount. Both prices have got to change by the same amount in order for the difference between the two prices to remain the same and be equivalent to the marketing margin. A question that might arise, however, is which price is likely to change first. The framework assumes that after allowing enough time for all adjustments to occur both prices are going to change by the same amount. It is going to take some time, however, for adjustments to occur because new market equilibria do not occur instantaneously.

If demand should increase, will retailers initially increase price? It is reasonable to assume that the first evidence of an increase in a retail demand would be an increase in sales and a more rapid depletion of inventories at the retail level. Retailers are likely to respond to depleted inventories by increasing their orders of products. Thus, the first effect of a change in demand in the retail level is likely to affect ordering decisions. As increases in orders get passed back through the marketing system, the available supply must be rationed among buyers. As orders for a raw product increase, its price will have to rise to allocate the limited quantity available for sale and encourage more to be produced and/or offered for sale. As retailers begin to pay higher prices for the products they purchase, they would be expected to incorporate this into their pricing decisions and increase retail prices. Therefore, this sequence of events suggests that prices at the farm level are likely to be affected first and then retail prices would adjust. This is consistent with monthly changes in basic farm commodities and wholesale prices being used as indicators of subsequent retail price changes.

In the case of a decline in retail demand, a decrease in orders by retailers is also likely to impact farm prices first. As orders decrease, producers will have to accept lower prices to move quantities available for sale as production is reduced. Retail prices would subsequently adjust downward as retailers experience a decrease in the cost of the products they purchase.

One exception to the above lag between retail prices and farm prices is in the case of perishable products. For these products, it may be necessary to adjust retail prices to sell an existing quantity at the retail level to avoid spoilage irrespective of farm prices at a point in time. For storable products, changes in retail demand are most likely to initially affect the volume of sales rather than price. As retailers adjust subsequent orders to reflect current sales rates, the first indication that total retail demand has changed for the entire market will be reflected at the farm level. As changes get reflected back to the farm level they will likely cause an adjustment in the farm level price unless there is an unlimited supply of the commodity (i.e., a perfectly elastic supply) at the farm level. After all the adjustments get worked out in the marketing system, prices at both levels would change by the same magnitude assuming a constant marketing margin.

SUMMARY

The chapter illustrates how simple demand and supply concepts can be used to analyze changes in market conditions. Different combinations of changes in price and quantity provide evidence of whether a change in demand or supply has occurred. The process can also be used to indicate directional changes in prices and quantities that might be expected from anticipated changes in demand and/or supply. Although this kind of analysis is extremely useful for considering a given market, a more general framework is required to illustrate the linkage between prices of commodities at the retail and farm level.

The broader framework uses derived demand and supply concepts to incorporate marketing activities explicitly into the analysis. Derived demand and supply relationships are determined by retail demand and farm supply functions and the cost of getting the products moved and transformed from producers to consumers. Vertical distances between primary and derived demand and supply schedules represent the value of marketing services. Primary and derived demand and supply functions can be depicted as parallel relationships in the same diagram assuming each unit of retail product requires a fixed quantity of basic agricultural product and marketing services and prices of inputs used by marketing firms do not vary with the quantity used.

The expanded framework is useful to illustrate anticipated effects of changes in retail demand, farm supply, and the cost of marketing services on retail and farm level prices as well as the quantity of products being produced and consumed under competitive **market structures.** Any change in primary demand or supply would be expected to alter farm and retail prices in the same direction and by the same amount under the simplifying assumptions used for developing the model. A change in marketing costs, however, causes retail and farm prices to change in opposite directions assuming no change in retail demand and farm supply relationships. The relative slopes, or elasticities, of the relationships determine how much of the change in marketing costs would be reflected in price adjustments at the retail level versus the farm level.

Changes in prices and quantities indicated by graphical manipulation of primary and derived demand and supply relationships in the expanded framework are indicative of how markets will change allowing sufficient time for adjustments to occur. The analytical framework should not be interpreted to imply that changes occur instantaneously. There is a tendency for initial adjustments in price to occur at the farm level and subsequent adjustments in price to occur later in the marketing channels. The implicit time period assumed by the model implies that everything that is produced is also consumed and consequently the effects of short-term changes in storage or inventories of nonperishable products are not considered in the simple model. The basic model can be modified to consider the effects of storage as well as multiple demands and/or sources of supply of a basic agricultural commodity. Modifications of the basic model involve the same kind of economic logic, but diagrams become a little more complex.

QUESTIONS

1. Over the last decade, total U.S. production of cranberries has increased a little while the average price (adjusted for inflation) per pound received by producers has decreased slightly. Based on this information what if anything can you conclude about relative changes in demand and supply of cranberries at the farm level? Explain your reasoning.

2. How would consumer expenditures on turkey be affected by an increase in the price of broilers assuming no change in the supply relationship for turkeys if the cross-price elasticity of retail demand for turkey and the price of broilers is positive? Briefly explain your reasoning.

3. Assuming no change in retail demand for dairy products, briefly explain how technological improvements resulting in a reduction in milk marketing costs would be expected to affect the demand for milk at the farm level.

4. Graphically illustrate the likely changes in the equilibrium quantity as well as retail and farm prices of poultry products if new environmental regulations result in an increase in water treatment costs at poultry processing plants.

5. Graphically illustrate the effects on farm and retail prices as well as the quantity of tomatoes that would be produced and consumed as a result of a reduction in the cost of production because of improved technology.

6. Assume that the income elasticity for strawberries is 1.5. Graphically illustrate the anticipated changes that would be expected to occur in the U.S. retail and farm prices as well as quantity of fresh strawberries next spring resulting from an increase in disposable income of U.S. consumers.

7. Assume that the retail demand for dairy products is relatively inelastic, but the farm supply of milk is relatively elastic (at least in the long run). If the marketing margin for milk increases because of higher costs of processing milk, would the change in retail price be expected to be smaller or larger than the change in price at the farm level? Briefly explain how you reached your conclusion.

REFERENCES

Schrimper, R. A. 1995. Subtleties associated with derived demand relationships. *Agricultural and Resource Economics Review* 24(2):241–246.

Wohlgenant, M. K., and R. C. Haidacher. 1989. *Retail to farm linkage for a complete system of food commodities.* Tech. Bull. 1775. Washington, DC: USDA, Economic Research Service.

APPENDIX A QUANTITATIVE ANALYSIS

The amount of change in quantity demanded and supplied resulting from a change in income and subsequent price adjustments can be expressed in the following form using price and income elasticities:

Demand: $\%\Delta Q^D = e_D\,(\%\Delta P) + e_I(\%\Delta I)$
Supply: $\%\Delta Q^S = \eta_S\,(\%\Delta P)$

where

$\%\Delta Q^D$ = % change in quantity demanded
$\%\Delta Q^S$ = % change in quantity supplied
$\%\Delta P$ = % change in price
$\%\Delta I$ = % change in income
e_D = price elasticity of demand
e_I = income elasticity
η_S = price elasticity of supply

A change in income implies that the price of a commodity would adjust by a sufficient amount so that the percentage change in quantity demanded was equal to the percentage change in quantity supplied. Setting the above two equations equal to each other and solving for the percentage change in price indicates that the change in price associated with a given change in income holding everything else constant would be expressed as follows:

$$\%\Delta P = \frac{e_I(\%\Delta I)}{\eta_S - e_D}$$

If the price elasticities of supply and demand were .2 and −.4 for a commodity then the denominator for the preceding expression would be .6. If the income elasticity was .3, a 10% increase in income would result in a 5% increase in price since $(.3 \times 10)/.6 = 5$.

APPENDIX B EXTENSIONS OF THE FRAMEWORK

In the case of multiple retail products obtained from a particular agricultural commodity, it may be necessary to aggregate the derived demands for all uses to determine total derived demand at the farm level. Thus, the total farm level demand for a commodity can be viewed as the horizontal summation of all derived demands for a particular product. For example, Figure 6–7 shows that total demand at the farm level (TD_f) is the horizontal summation of two derived demand curves (D_{f1} and D_{f2}) for the same basic commodity consistent with two alterna-

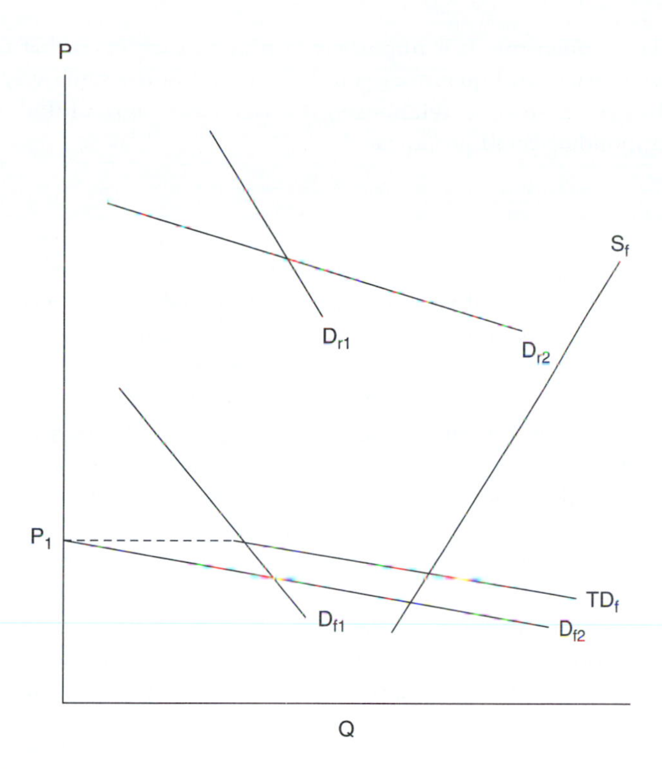

FIGURE 6–7 Total derived demand originating from two alternative uses at retail level.

tive retail demand functions (D_{r1} and D_{r2}).[1] In cases of multiple uses of raw products it is not realistic to consider obtaining retail or derived supply curve by vertically adding a marketing margin to the farm level supply. One reason is that the different end uses of the product may involve different marketing costs.

In such cases, the intersection of aggregate farm level demand with farm supply (S_f) determines the total production and equilibrium farm price. As far as producers of the commodity are concerned, they are interested in the price they receive for their product and not how potential buyers of the product are going to use the item. Identical raw products will have the same price regardless of how they are used to produce different retail products.

Once the equilibrium farm price is determined, each of the derived demand curves for different uses can be used to determine how much of the raw product would be allocated to alternative uses at a given equilibrium price. Adding the marketing margin for each of the retail products to the equilibrium farm price would produce a set of retail prices consistent with overall

[1]For all farm prices greater than P_1, the only relevant demand is D_{f1}. For prices less than P_1, some quantity of the raw product would be demanded for both of the alternative retail uses.

market equilibrium. It is important to note in such cases that the determination of retail prices and quantities is not illustrated in the same way as when there is a distinct one-to-one relationship between raw agricultural commodities and corresponding retail products.

NOTES

1. The alternative outcomes listed in Table 6–1 do not include situations if the new point happens to be on a boundary line between two quadrants. For example, this could occur if a change in quantity occurred without any change in price or vice versa. In situations such as these, the appropriate inference is that both demand and supply changed assuming the relationships have slopes that are not zero or infinite. For example, a new point on the boundary line between quadrants III and IV would imply that there had been a decrease in demand as well as an increase in supply of the commodity.

2. For further consideration of substitution possibilities see Wohlgenant and Haidacher (1989).

3. The ultimate test of mastery in using this framework is to consider simultaneous changes in all three elements to see if any directional inferences can be made about changes in prices and quantity depending on relative magnitude of the changes.

ALTERNATIVE MARKET STRUCTURES

The chapter discusses how different types of market structures affect buying and selling decisions. The competitive market environment implicitly assumed in developing the framework presented in Chapter 6 is considered in more detail and compared to various kinds of imperfect competition.

The major points of this chapter are:

1. Implications of a competitive market structure on individual buying and selling decisions.
2. Characteristics and nomenclature used for classifying alternative market structures.
3. Comparison of decision making under alternative market structures.
4. Necessary conditions for price discrimination.
5. Implications of different types of imperfect competition.

INTRODUCTION

The framework for linking agricultural producer and consumer behavior developed in Chapter 6 was based on several explicit and implicit assumptions. This chapter examines implications about individual buying and selling decisions for that kind of framework and contrasts the results with what is implied by alternative models of market structure.[1]

The first part of the chapter considers individual buying and selling behavior consistent with the kind of competitive market structure implicitly assumed in Chapter 6. The second part of the chapter describes the terminology used to identify different kinds of market structures and examines the characteristics used for classification purposes. Attention is then turned to showing how the principle of equating marginal cost to **marginal revenue** is useful for analyzing decision making in different types of markets. In particular, the way in which the existence of a **monopoly** (or **monopsony**) affects decision making is considered. The use of price discrimination to increase total revenue is also considered in some detail. The final parts of the chapter briefly describe some additional imperfect competitive market structures.

ASSUMPTIONS ABOUT INDIVIDUAL BEHAVIOR

The framework presented in the previous chapter implicitly assumes that when individual buyers and sellers make decisions about how much, if any, of particular products to purchase or sell they do not anticipate that their decision will have any bearing on the price they pay or receive. Essentially, they view their position as a "take it or leave it" situation as far as market price is concerned. It is assumed that each individual market participant operates within an environment in which their behavior has no effect on prices of various products. At first glance, this condition appears to conflict with earlier discussion about how the interests of each and every market participant affect economic activity, but in reality the assumption is reasonable once the structure of the overall market is considered. Also, several simplifying assumptions about the way in which marketing firms combine agricultural products with other kinds of economic resources to produce food products for consumers were made in discussing the nature of marketing margins. One of the assumptions was that individual marketing firms are able to purchase any quantity of agricultural commodities and marketing inputs they want at a fixed price.

The basic idea of individual producers, consumers, and marketing firms making decisions without considering their impact on market prices is a very pervasive and useful analytical concept. For example, it implies that the price of wheat, corn, hogs, or any other agricultural commodity remains essentially unaffected by how much of various commodities Farmer Jones, Farmer Brown, or marketing firms A to Z decide to produce and sell. Similarly, the retail prices of carrots, steaks, milk, and other retail food products are unaffected by purchasing decisions of John Smith, Mary Rose, or any other individual consumer. These situations can be represented graphically by perfectly elastic demand and supply curves. Thus each producer of a given commodity is assumed to face a perfectly elastic demand curve similar to that in Figure 7–1 when considering the potential sale of output. This means that a producer is assumed to be able to sell all of his or her production at the going market price. If he or she holds out for a slightly higher price, he or she will not be able to sell any amount because buyers have many alternative sources of obtaining the same product at a lower price. Also there is no incentive for a producer to offer to sell output at less than the market price because he or she would be overwhelmed by potential buyers looking for a lower price.

Similarly, each consumer is assumed to face a perfectly elastic supply similar to that in Figure 7–2 when deciding how much, if any, of various food products to purchase. The particular prices at which the perfectly elastic demand or supply relationships are located in Figure 7–1 and Figure 7–2 are assumed to be determined by the intersection of appropriate market demand and supply curves. For example, intersections of the aggregate derived demand and supply relationships at the farm level are assumed to represent the price that each and every producer would be able to get for their output of a particular product. Similarly, the intersection of the aggregate retail demand and derived retail supply relationships indicates the retail price that all consumers would face under a given

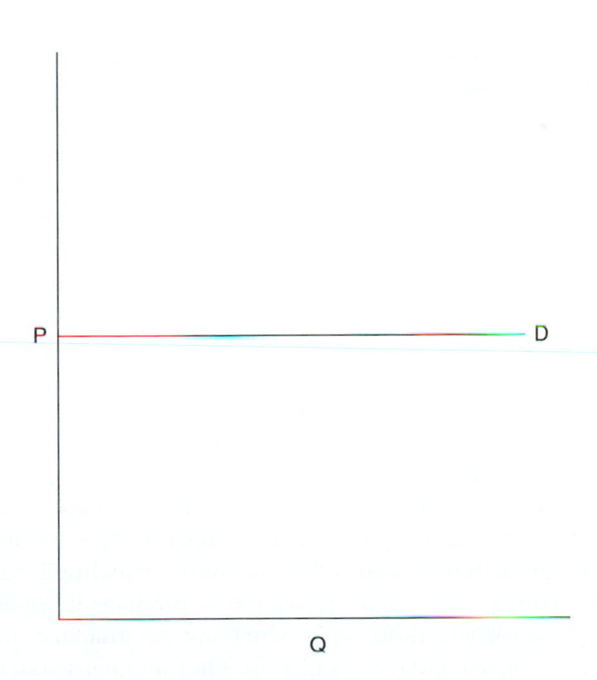

FIGURE 7–1 Demand faced by individual producer of agricultural products.

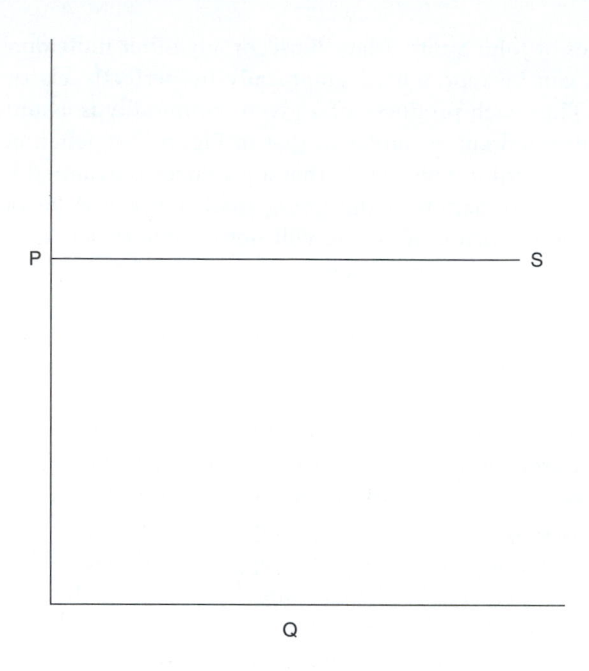

FIGURE 7–2 Supply faced by individual consumer of food products.

set of market conditions. The perfectly elastic supply and demand curves faced by individual producers and consumers are assumed to shift as aggregate market conditions change.

The reason that the above assumptions and implications are reasonable for many agricultural and food products is because of the kind of market environment or structure that characterizes the interactions that occur between buyers and sellers. It is reasonable to consider that prices of commodities are unaffected by the actions of any one individual as long as the aggregate market in which buying and selling decisions are made involves many participants and each participant's share of the total transactions or volume of business is small. The market clearing price and quantity are assumed to adjust consistently with the combined actions of all individuals indicated by aggregate market clearing mechanisms and not be influenced by the decisions of any single individual.

A competitive market structure framework is appropriate for analyzing many marketing issues related to agricultural and food products. It is especially applicable for considering world and national markets because of the large number of potential buyers and sellers of most agricultural and food products assuming a fairly generic characterization of products is applicable. The larger the scope of a market in terms of product and geographical dimensions, the more likely competitive conditions will exist. On the other hand, if market analysis involves a very narrowly defined product or a rather small geographical area, the existence of competitive market conditions may be more questionable. If a very narrow set of product or geographical attributes is an appropriate characterization for a market of interest, then it is less likely that market participants would ignore the influence of their decisions on the market clearing mechanisms. Even if there is some doubt about the existence of all the necessary conditions for a competitive market, the framework can be an extremely useful starting point for establishing a benchmark for comparing the implications of alternative models of market behavior.

The way in which individual buyers and sellers view the environment or market structure in which they make buying and selling decisions is an important element in understanding how markets operate. Buyers and sellers may interact in several different ways in determining market clearing prices and quantities depending on the characteristics of the market. In particular, the kind of environment in which individual sellers or buyers find themselves affects the way they operate in the market. The differences among market environments and how their characteristics affect individual buyer and seller behavior is a specialized subject body of knowledge related to market structure, conduct, and performance. Special interest exists in determining how different kinds of market structure lead to different prices and quantities of products being exchanged.

ALTERNATIVE STRUCTURES

The major categories used for describing different kinds of market structures are indicated in Table 7–1. The classification represents general categories along a continuum from the case of competitive markets to the case of only one seller (monopoly) or one buyer (monopsony). In the following sections, specific characteristics and implications of various market structures will be discussed in more detail and compared to competitive market structures.

The nomenclature in Table 7–1 indicates how suffixes are used to distinguish among alternative selling and buying environments. The *oly* suffix is used when referring to the selling side of the market and the *sony* suffix is used for describing the buying side of markets. For example, the major difference between monop*oly* and monop*sony* is whether a selling or buying side of a market is being considered. The various classifications in Table 7–1 mean that it is possible to

TABLE 7–1

DIFFERENT TYPES OF MARKET STRUCTURES

Selling Side of Market	Buying Side of Market
Competitive	Competitive
Monopolistic competition	Monopsonistic competition
Oligopoly	Oligopsony
Monopoly	Monopsony

have a number of different combinations of market structures. For example, a given market could be competitive on the selling side and **oligopsonistic** or **monopsonistic** on the buying side.

Classification Criteria

The major differences among market structures are basically reflected by the nature of demand and/or supply curves assumed to portray behavior of potential sellers and buyers of a product in a given situation. Given that these relationships are unobservable, other attributes of the market environment that presumably influence the nature of demand and supply behavior are used for classification purposes. The three characteristics that generally are considered are (1) the number and relative volume of business handled by individual buyers or sellers, (2) the ease or difficulty of market entry and exit, and (3) the nature of the product. Each of the three characteristics is somewhat subjective and depends largely on the context of the question or issue about market behavior being considered. Market characteristics that are relevant for one question or issue may be different from characteristics that are important for considering a different question. For example, one type of market structure might be appropriate for analyzing the general level of grain or livestock prices on the international or national level. At the same time a different set of assumptions about market structure might be appropriate if the relevant question was about the price of a very specific type of grain or livestock in a regional or local market. Thus, different market structures might be appropriate depending on the geographical or product orientation under consideration.

The number of market participants on the buying and selling sides of markets can vary from one to many depending on how a given market is defined. As noted earlier, competitive market conditions require many participants and consequently it is reasonable for each participant to make decisions assuming market prices are given. On the other hand, a monopoly and a monopsony are situations consisting of only one seller or buyer, respectively. Other types of market structure between competitive market conditions and a **monopoly** or monopsony involve an intermediate number of participants between one and many. Oligopoly or oligopsonistic market structures are assumed to exist when there are a few (more than one, but not many) market participants. There is no definitive rule as to what numerical count of participants constitutes a sufficient number to be considered many and there is no generally agreed upon demarcation point between few and many participants. The actual number of participants is not as critical as how they perceive the environment in which they make purchasing and selling decisions.

The general relationship between the number of participants and the extent to which they perceive their relative importance in affecting market outcomes depends on the particular context of the market being considered. In the case of agricultural and food products where every individual in the world is a potential consumer and numerous individuals produce rather similar, if not identical, agricultural products in different parts of the world, it is clear that many people are involved in determining market outcomes for agricultural and food products. If smaller geographical entities are selected as the market of interest, however, the number of participants on the buying or selling side of a market may be reduced considerably. There still could be a sufficient number however to result in sellers and buyers behaving essentially as price takers.

In addition to the number of participants, a key factor is how the total number of transactions or volume of business is distributed among participants. The existence of many buyers or sellers generally means there are so many competitors that the decision of any one participant has no perceptible impact on aggregate market conditions. In fact, the actions of every single individual no matter how little their share of total market purchases or sales may be, has some effect on the total market. However if the effect is too small to be perceptible it seems reasonable to assume that firms ignore this effect in making decisions.

The ease of entry and exit into a market is a second characteristic that is used to distinguish among different market structures. In some respects this characteristic is really a way of rationalizing differences in the number of potential participants in a given market. The ease of becoming a potential seller or buyer of a particular commodity directly influences the number of interested and active sellers and/or buyers in a given market at a particular point in time. It also indicates how rapidly higher than normal monetary returns associated with a particular activity could be dissipated by a change in number of participants and how rapidly an industry can adjust to new circumstances. In some cases, the number of potential competitors may be restricted by governmental policy decisions. For example, the U.S. government has restricted the production of various agricultural commodities from time to time as a way of raising the market price above what it otherwise would be if production had not been restricted. Another type of barrier on entry of firms results when governments issue a patent or exclusive right of production for a fixed number of years as an incentive for individuals to reap the benefits from creativity without worrying about the threat of immediate competition. If buyers or sellers have to be licensed or meet some other criteria, such as a minimum level of financial resources, to qualify as market participants the number of potential participants may be limited.

There have been circumstances also where a governmental agency was established to be the sole trader of agricultural and/or food products in international markets. For example, the Canadian and Australian Wheat Boards have controlled exports from their respective countries for several years (DeVos 1997). The existence of a few large firms that have an established market position to take advantage of economies of scale in production can be an intimidating barrier for potential new firms. If it is fairly easy for firms to change from production of one item to another as profitability varies, there will be mobility of resources among alternative uses and more potential competition. On the other hand if the technology of production requires some specialized resources and initial costs of production are high or it is difficult to get a foothold in a market, it is less likely that there will be as many potential market participants.

The third and final characteristic used for classifying market structures is the nature of the product being purchased and sold. The basic distinction is whether a given set of transactions involve exchanges of essentially nondistinguishable (i.e., homogeneous) items or **differentiated products.** In other words, to what extent does a given set of market transactions involve essentially the same product or a subset of products that have some similar characteristics, but are unique in some aspects considered by buyers and sellers as they make their decisions? Often the distinction between homogeneous and differentiated products again depends on the kind of marketing issue being considered. Basically the distinction between homogeneous and differentiated products is in terms of how market participants are assumed to view specific product characteristics. In particular, do potential buyers view the output of different sellers as perfect substitutes or are the outputs sufficiently unique to result in meaningful price differentials? In many cases, the distinction between homogeneous and differentiated products may appear to be clear but actually is quite subjective. If fairly broad product characteristics are used in defining a market of interest, a homogeneous product characterization might be appropriate. On the other hand if a very narrow and specific set of product attributes are taken into account by market participants in making decisions then it might be appropriate to consider the market consisting of differentiated products. For example, as a first cut for analyzing how the demand for soft drinks has changed relative to other kinds of beverages it might be appropriate to consider Coke, Pepsi, and other soft drinks as a group of homogeneous products. Nevertheless, there obviously are other circumstances in which it would be appropriate to assume each soft drink product is a very distinct product. Another example is the market for bacon. If bacon is viewed as basically a cured meat product derived from a particular segment of pork carcasses it could be viewed as a homogeneous product if one was trying to analyze changes in the general price of bacon relative to other pork or meat products. On the other hand if wholesale or retail prices of individual bacon transactions were examined closely it would become clear that various attributes of bacon (and almost any other product) are valued differently by market participants. For instance, if the price of bacon varies with how it is packaged or what kind of color or name is on the package it would be evident that bacon products are not entirely homogeneous. Thus it is possible for a given product to be considered a homogeneous or differentiated product depending on the kind of market analysis of interest.

There is no generally accepted reference source that can be used to determine if a product is homogeneous or differentiated. Rather it depends on which description seems to come closest to fitting the market inquiry of interest. As long as market participants act as if there is no difference between the wheat or anything else produced by Farmer Brown or Farmer Smith, the products are homogeneous. However if buyers' and sellers' actions depend on subtle differences within a class of products then the differentiated classification would be more appropriate.

Traditionally, most agricultural commodities have appeared to be fairly homogeneous as they entered marketing channels and became more differentiated as they got closer to retail outlets through the addition of marketing services.

Additional differentiation of crop and animal products through improved animal and plant breeding and biotechnology, however, is resulting in more differences in market values of characteristics of basic agricultural products. For example, several kinds of corn are being produced and handled separately in the market with different price premiums. High oil, white, waxy, hard endosperm, nutritionally dense, and high amylose value-enhanced corn products accounted for approximately 5% of total U.S. corn acreage in 1998 (Feedstuffs 1998).

In the future, there may be even more emphasis on recognizing and differentiating products at earlier stages of the marketing system. The real question however is how close to being substitutes the products are in the eyes of market participants and to what extent different prices emerge and exist for different characteristics of interest.

COMPETITIVE MARKETS

The framework for linking production and consumption of agricultural and food products presented in Chapter 6 represented the behavior of many buyers and sellers interested in making transactions involving a homogeneous product.[2] This meant that participants made decisions taking market prices as given or determined by aggregate market conditions. In this type of market environment the only decisions that participants have to make is about quantities. That is, how much do suppliers wish to produce and sell at the market-determined price? And how much, if any, quantity does a purchaser wish to buy at the market price?

Assuming each producer desires to maximize profit in this kind of market environment, the optimum amount of output is the quantity for which the increase in total costs associated with producing the last unit comes as close as possible (or equal) to the price that will be received. This is consistent with the principle of equating marginal cost (MC) to marginal revenue (MR) in order to maximize profit. Marginal cost is the increase in total costs associated with each incremental increase in quantity produced and marginal revenue is the change in total revenue associated with the sale of one more unit.

In a competitive market structure the marginal revenue associated with each additional unit of production is the price of the commodity because the price is expected to remain unaffected by how much is produced by a given individual. Thus firms only need to consider their production possibilities and the price of inputs to determine when the marginal (or extra) cost associated with producing the last unit will be as close as possible to the market price they expect to receive for the commodity. This implies that resources used in producing various goods and services will be allocated in such a manner that their opportunity cost equals the value of output. As noted in Chapter 5, when producers make production decisions they often do not have perfect knowledge of what the market price will be when the product is ready to be sold unless some kind of prior arrangement to guarantee a price has been established. Since production decisions take time it is not possible for market conditions to instantaneously adjust as quickly as theoretical models of perfect competition might imply. Nevertheless, the outcomes of the theoretical models provide a benchmark for predicting market behavior over time.

The only ways that producers in competitive market environments can influence their profits is to make sure their costs are as low as possible and possibly vary the timing of their sales. Given that all firms in this kind of market environment receive the same price for their output at a given point in time, profits can vary among firms because of differences in total costs of production even if their marginal costs of production are approximately the same.

Assuming that it is easy for firms to enter or leave the industry implies that changes in total production will occur until the least efficient firm in the industry is doing just as well as it could in any alternative activity. Thus the competitive environment creates a natural incentive for each firm to continually search for the cheapest way of producing a product and expanding production until the marginal cost of the last unit is equal to the market price.

Also there is no reason for firms in competitive markets to withhold information from others because each firm is assumed to be too small to have any effect on the price of the product. It is somewhat ironic that competitive markets are often where the most cooperation among groups of sellers or buyers occurs in terms of sharing information. Sellers do not necessarily view themselves competing with other sellers if they expect the price to be unaffected by their individual behavior. The same situation is applicable as far as buyers in competitive markets are concerned.

A competitive market implies that purchasers of a product can buy as much of the product as desired at a given price. This means that purchasers can obtain as much of the product up to the point where the perceived value of the last unit is equal to the price of the product. In other words, purchasers have the flexibility of buying as few or as many units of a product they want until the extra value obtained from the last unit purchased is exactly equal to the value of the money or resources that have to be sacrificed in terms of not being able to purchase other goods or services. The kind of evaluations potential buyers have to make in deciding how much, if any, of the product to purchase is greatly simplified when the market price of a commodity is assumed to remain the same.

MONOPOLY

A monopoly is defined as a market environment in which there is only one seller of a particular product. This is the polar extreme of a competitive market. A monopolist faces the entire market demand for a product because there is only one source of the product if anyone wants to buy it. The nature of demand faced by a monopolist depends on how much interest exists in a given product and the availability of possible substitute products from the perspective of potential buyers. Products may possess unique characteristics, but be sufficiently similar to substitutes that the demand may be very elastic and closely resemble the kind of demand faced by sellers in competitive market situations. Consequently, the nature of a demand curve faced by a monopolist depends critically on how products are defined.

To a certain degree every seller of anything is a monopolist in the sense that no one else can occupy or offer products at exactly the same location in a market. This type of geographical uniqueness may not be of much value or provide much market power if there are other sellers nearby from which similar items can be obtained. However, if a seller is geographically isolated from competitors there may be an opportunity to take advantage of local demands for products by charging slightly higher prices than similar products at different locations. An example of this might be a grocery store in an isolated rural area or a provider of some specialized marketing service to a group of geographically isolated producers. Later chapters will explore the spatial dimensions of markets in more detail.

The above discussion hopefully indicates that the consideration of a monopoly, like most other kinds of market environments, depends on how particular products are defined and the availability of potential substitutes. Nevertheless, this type of alternative market structure provides a useful way of considering the type of decisions that firms would make under a vastly different environment than competitive conditions. To the extent that a seller faces any kind of sloping demand curve for a product, decisions made by the seller have some bearing on the price and quantity traded in a market. In this sense a firm is assumed to have a degree of market power through influencing the market price and/or how much product is made available to prospective purchasers through its production decisions. In these circumstances, a seller could choose to charge a high price and sell a small quantity or charge a lower price and sell a larger quantity depending on the characteristics of the demand curve faced by the firm.[3] This provides an opportunity for the seller to unilaterally set whatever price the seller desires instead of having to accept a market price as in the case of a competitive market. An alternative strategy is to decide on a desired volume of business and let market conditions determine the price that would clear the market.

Even though a monopolist has more flexibility in making decisions than a competitive firm, there is no guarantee of profitability. For example, potential demand for a product may be so limited that there is no competition simply because it is not profitable for even one firm to provide the product. As noted above, a monopolist must decide how much to produce or what price to charge for the product and cannot arbitrarily pick just any combination of price and quantity, but is restricted to the set of opportunities represented by the demand for its product. It is clear that a monopolist has to operate under the constraint of what prospective buyers are willing to pay for alternative quantities.

One of the things that a monopolist must consider is really how unique is the product and whether there are any close substitutes that buyers might be able to turn to. Also, the firm would need to consider if it was really the only firm able to produce and sell this product. Is their unique market position the result of some natural condition or governmental regulation? Is there a possibility that other firms might begin to produce the product and attract some of their customers if they set too high a price or make too much profit?

In thinking about the kind of decisions that a monopoly faces it is customary to assume that the potential demand is known with certainty and all that has to be done is to select the ideal price and quantity combination consistent with market conditions. Actually firms never know the characteristics of the demand for products as clearly as assumed by economic textbooks. Many firms spend large sums of money on market research to try to find out more information about the demand they face, but they never have perfect and complete information. Firms have to make decisions based on what they think the demand for various products is and assess the consequences of their decisions and make readjustments in continually searching for an optimum market strategy as they observe consumer responses. Thus monopolists (and many other kinds of business firms) can be viewed as continually searching for the best pricing strategy rather than necessarily setting a price and never changing it.

How can a monopolist decide if it would be more profitable to sell a smaller quantity for a higher price or a larger quantity at a lower price? This is similar to the decisions that university athletic officials make when setting ticket prices to various events. Would they be better off with higher prices that might not result in sellouts or should they try to search for a price that will produce capacity crowds? Clearly the decision about price affects the number of customers and an unlimited quantity of tickets cannot be sold at a given price as firms operating in competitive markets do. If the demand for the product were known with certainty, total revenue for different combinations of price and quantity consistent with the demand curve could be compared. As noted previously, the demand curve for most products is never known with sufficient accuracy to permit this kind of calculation and furthermore the demand is subject to change in response to new market conditions. Conceptually, what monopolists would like to accomplish to maximize their profits is to find a price and quantity combination consistent with market demand so that the marginal cost of producing the last unit sold is as close as possible to the marginal revenue associated with that unit. This is the same principle (i.e., MC 5 MR) used by firms in competitive market structures to decide how much to produce, but the calculation of marginal revenue is more complicated in the case of a monopolist.

One of the things that makes calculations of marginal revenue complicated for monopolists is that they have to estimate how total revenue will change as price is varied to sell different quantities of a product. If a firm faces a downward sloping demand curve rather than a perfectly elastic one, the only way to sell a greater quantity is to lower the price to all customers.[4] There is a gain in revenue as more units are sold (the quantity effect), but there is also some decrease in revenue resulting from charging a lower price to all customers. There will be a loss in revenue (the price effect) associated with potential consumers who would have been willing to pay a higher price if fewer units had been available. For example, suppose that when the price of a product is $5.00 per unit, a firm is able to sell

10,000 units per year, but if the price were reduced to $4.95 per unit, sales would increase to 10,500 units. In this situation, total revenue would change from $50,000 (at a price of $5.00) to $51,975 (at a price of $4.95) or an increase of $1,975. The change in revenue resulting from selling an extra 500 units at a price of $4.95 produces an additional $2,475 of revenue but $500 is given up because of the loss of $.05 on each of the 10,000 units that could have been sold at a price of $5.00. Dividing the change in total revenue ($1,975) by the additional 500 units sold means that each additional unit contributed $3.95 to total revenue.[5] This value could be compared to what it cost the firm to produce each of the additional units to determine if it would be profitable to consider reducing the price. If the marginal cost was less than the marginal revenue, the firm probably would want to consider expanding production and possibly evaluate even further price reductions. On the other hand, if marginal cost was greater than the marginal revenue obtained from increasing sales the firm would want to consider higher prices to get closer to the point where MC 5 MR.

Whether total revenue for a monopolist increases, decreases, or stays unchanged as price is varied along its demand curve depends on price elasticity of demand. If the price elasticity of demand faced by a monopolist happened to be unity, total revenue would remain unchanged if the price was reduced or increased. If there is no change in total revenue regardless of how much is sold, then the marginal revenue associated with each additional unit sold would be zero.

If the price elasticity of demand is inelastic, total revenue would decrease as the quantity sold increased by reducing price, implying a negative value for marginal revenue. Clearly the firm would not want to produce and sell a quantity for which the price elasticity of demand was inelastic because each additional unit being sold results in a reduction in total revenue. Neither would a monopolist want to select a price and quantity combination on its demand curve with unitary price elasticity unless the marginal cost of producing each additional unit was zero.

Finally, if the price elasticity of demand is elastic, total revenue would increase as quantity of sales increased by reducing price. In this case the marginal revenue of each additional unit sold would be positive. If the firm continued to reduce price and thereby increase sales until the marginal cost of producing the last unit was equal to the marginal revenue the firm would be maximizing profit. By selecting such a point the firm will have squeezed as much profit potential out of the market as possible. If a positive difference existed between the marginal revenue and marginal cost of the last unit produced and sold, it would indicate some potential profit was not being realized. This also indicates why a monopolist would always want to operate at some point on a demand curve where the price elasticity was elastic.

The preceding relationships between the price elasticity of demand and marginal revenue can be summarized using the following formula: $MR = P(1 + 1/e)$ where P is market price and e is the price elasticity of demand. Substituting alternative values of e into the formula indicates whether the marginal revenue is zero, negative, or positive. For example if the price elasticity of demand is unitary (e.g., -1) it is clear that $1/e$ would be -1 and consequently MR would be equal to zero regardless of the value of P. Similarly if the price elasticity of demand is inelastic (e.g., $-.4$), $(1 - 1/e)$ would be a negative value and MR would also be negative. Finally if the price elasticity of demand is elastic (e.g., -2), $(1 - 1/e)$ and MR would be positive. Additional familiarity with the above relationship can be obtained by selecting different numerical values for the price elasticity of demand to verify that if the demand is inelastic marginal revenue would be negative and if the demand is elastic marginal revenue is positive.

The above ideas are graphically represented using a linear demand curve in Figure 7–3. According to the discussion in Chapter 4 the point elasticity of demand varies along a linear demand curve and total revenue would be greatest at the midpoint of the demand curve. Thus, if total revenue associated with alterna-

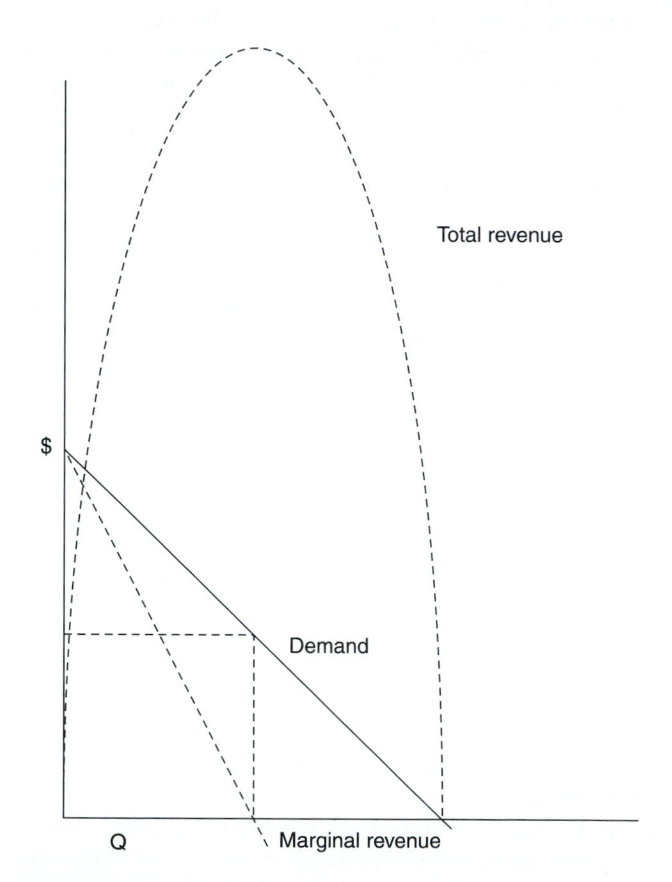

FIGURE 7–3 Relationship between total revenue and marginal revenue for a linear demand relationship.

tive quantities is included on the same diagram as the demand curve, it is clear that if price or quantity was equal to zero then total revenue would be zero. Thus the total revenue curve would be zero at the origin of the diagram, reach a maximum at the midpoint of the demand curve, and be zero at the quantity where the demand curve intersected the horizontal axis. The slope (or the first derivative) of the total revenue curve at various quantities represents the change in total revenue (marginal revenue) associated with each additional unit sold. If the values of marginal revenue associated with each quantity of a linear demand curve are plotted on the same diagram they constitute a linear line with twice the slope of the demand curve. The marginal revenue relationship intersects the horizontal axis at the midpoint of the demand curve consistent with the point where total revenue is maximized (and the price elasticity of demand is unity). For all quantities to the right of the midpoint of the demand curve, marginal revenue is negative consistent with total revenue declining. For all quantities to the left of the midpoint of the demand curve, marginal revenue is positive. If the monopolist had at least some idea of how the marginal costs varied with quantities, it would be possible to determine the quantity where the marginal revenue relationship intersected the marginal cost curve as illustrated in Figure 7–4. Once the optimum quantity (Q') of production was determined (where MC = MR), the firm could then decide what price (P') to charge based on how much consumers are willing to pay for that quantity.

The particular price and quantity outcome of a monopolistic situation is quite different from what would occur in the case of a competitive market if there were many firms with the same kind of marginal cost structure as illustrated in Figure 7–4. Under competitive market conditions firms would continue to expand production beyond Q¢ because the market price, which is also a competitive firm's marginal revenue, would be greater than the marginal cost of production. In this case, each firm in the industry would have an incentive to expand production assuming their individual production decisions had no effect on market price, but as they did the market price would adjust downward because of the increased production. This process would continue until the market price became equal to the marginal cost of producing the last unit. Thus a competitive market results in a greater quantity of production and lower prices than would occur under a monopoly with the same demand and cost of production schedules. A major qualification for this result depends on whether it is possible for many small firms to achieve the same economies of production as one large firm. This condition is a critical factor possibly explaining the existence of different kinds of market structure.

A monopolist might decide to produce more (and charge a lower price) than the point at which profits would be maximized for a couple of reasons. One reason could be to discourage other firms from entering production and making the market more competitive. Another reason might be to avoid governmental regulation of price and allocation of output that might occur because of public dissatisfaction if output was restricted too severely.

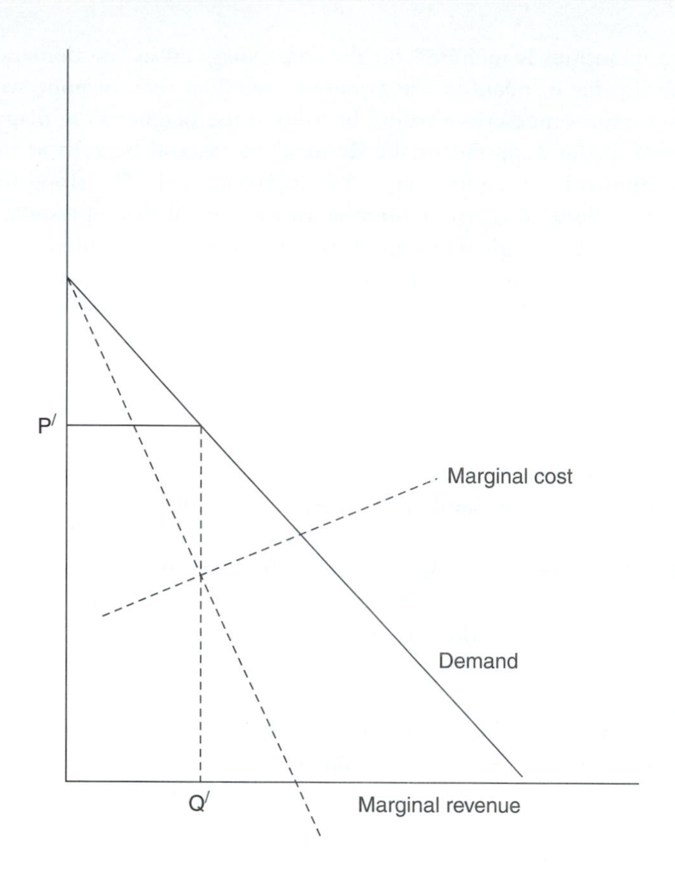

FIGURE 7–4 Determination of output and price that would maximize profits to a monopolist.

Price Discrimination

The basic principles of how monopolists can use information about potential market demand to set market output and price can be extended if a monopolist is able to segment customers into separate markets.[6] If the price elasticity of demand varies among markets and the monopolist has some way of preventing resale of the product it may be advantageous to charge different prices for the same product in different markets (i.e., price discrimination). Examples of this type of price discrimination are fairly common. For instance, restaurants often charge different prices for the same food at different times of the day or senior citizens often are charged a different price than other customers for the same items. Ticket prices for movies and airline travel may vary depending on the particular time that customers want to go to a theater or travel, respectively.

It is clear from the previous discussion that if a monopolist charged the same price in two or more markets with different price elasticities of demand, the marginal revenues associated with the last unit sold in each of the markets would differ. If the marginal cost of producing a commodity is the same regardless of who purchases it, a monopolist would prefer that the last unit sold in each market have the same marginal revenue. The only way to make sure that marginal revenues are identical if price elasticities of demand differ is to charge different prices by controlling how much of the product is available in each market. For example, the only way for MR_A (the marginal revenue in market A) to equal MR_B (the marginal revenue in market B) if the price elasticities of demands for market A and B are not identical is for P_A to be different from P_B. To consider a specific example, assume that when a monopolist charges the same price to all customers, the marginal revenue from the last unit being sold in market A is $2 and the marginal revenue associated with the last unit being sold in market B is $10. Consequently, if one less unit were allocated to market A and shipped to market B, an additional $8 (the difference between $10 and $2) of revenue would be realized. Reducing the quantities shipped to market A would permit a higher price to be charged in that market. However as extra shipments were sent to market B, a lower price in that market would be required in order to sell more. These kinds of adjustments could be continued until the marginal revenues realized from the last unit sold in each of the two markets became equal. In order to maintain different prices for the same product being sold in different markets, however, there must be an effective way of preventing entrepreneurs from buying the product in the market with the lowest price and reselling it in markets with the higher price, which would defeat the purpose of discriminatory pricing.

A graphical illustration of these ideas is presented in Figure 7–5. The demand curves for markets A and B are represented in the left-hand and middle panels. A horizontal aggregation of the demand curves for the two markets representing the total demand faced by a monopolist is represented in the right-hand panel. The marginal revenue curve associated with the total demand curve could be used along with the marginal costs of production to determine how much the monopolist would produce and what price to charge in order to maximize profits if a uniform price was being determined for customers in both markets, similar to what was illustrated in Figure 7–4. Using the relationships contained in Figure 7–5, the monopolist would decide to produce Q_1 and charge P_1. At that price market A would buy Q_A units and market B would buy Q_B units. The sum of Q_A and Q_B of course equals Q_1 because of the horizontal aggregation process used to determine the total demand.

The monopolist may be able to increase total revenue from the same total quantity (Q_1) of product by establishing different prices through reallocating sales between markets if the price elasticities of demand differ and there is a way of making sure no resale of the product occurred between markets. This can be seen by examining the marginal revenues associated with the original quantities (Q_A and Q_B) sold in each market. Assuming different price elasticities of demand in

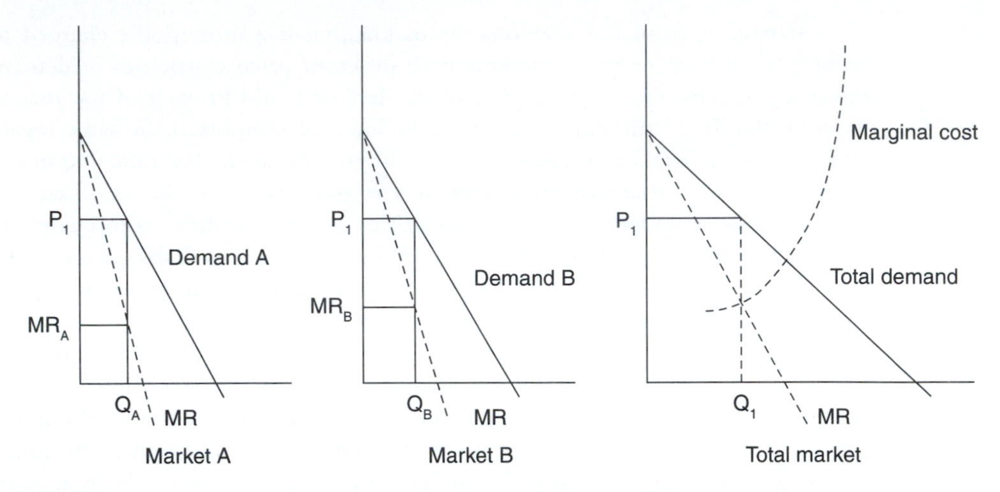

FIGURE 7–5 Case of a monopolist who faces separate markets with different price elasticities of demand.

each market means that MR_A associated with Q_A and MR_B associated with Q_B are not the same. If more of the product were sold in market B (at a lower price than P_1) and less of it sold in market A (at a higher price than P_1), it would result in the marginal revenues associated with each market becoming more similar and thereby increasing total revenue for a given quantity of product. The process of reallocation and adjusting the prices in the two markets could continue until the last unit being sold in each market resulted in exactly the same marginal revenue.

Some of the more common types of price discrimination associated with agricultural commodities concern dairy products and international trade. In the case of dairy products, a governmentally sanctioned classified pricing system has operated for a number of years in the United States. This system results in producers receiving a higher price for milk that is used for fluid milk products and a lower price for milk used for the manufactured dairy products like cheese, butter, and so on. As long as there is a difference in the price elasticities of demand for different uses of milk, a greater amount of total revenue is obtained from a given quantity of milk by setting different prices depending on how it is used than what would occur if a single price was paid for all milk regardless of use.[7]

Another example of discriminatory pricing techniques that has been used to enhance producers' revenue is the restriction of imports (or subsidization of exports) by governments in order to maintain a difference between domestic and world market prices. If the demand for domestic use is more inelastic than the demand in foreign markets, a government may be able to increase total revenue by restricting available quantity in domestic markets to produce a higher price and sell the extra quantity at a lower price on the international market. Selling a smaller quantity at a higher price in a market with an inelastic demand results in a greater revenue. Also selling a greater quantity at a lower price in a market with an elastic demand also produces a higher revenue. Thus greater revenue is generated in both markets by establishing different prices and effectively controlling how much is sold in each market.

MONOPSONY

A monopsony is very similar to that of a monopoly except the situation is that of having a single buyer rather than a single seller of a product. In this case the purchaser is assumed to be facing an upward sloping supply curve. Thus a monopsonist would need to take into account how much of an adjustment in market price would be required to obtain various quantities of the product. This kind of a decision-making environment again is quite different from the case of buyers in a competitive environment who only need to be concerned about how much to purchase at the price established by market conditions.

A common way to think about the existence of monopsonists is in a spatial dimension. For example, in a given location there may be one processing or packing plant that is a major buyer of locally produced commodities. If the firm desired to expand its operations it would need to increase the price it is willing to pay to encourage producers to expand production. The marginal cost of purchasing each additional unit, however, is more than the price of the last unit because of having to pay a higher price to all producers if they are providing a homogeneous product. If no perceptible quality differences exist among the products being purchased, all producers would expect to receive identical prices. This is similar to the case of a major manufacturing firm that is a dominant employer of labor in a given area that essentially sets the wage rate for the area depending on how many workers it hires. The marginal cost of the last worker is more than the wage paid to that worker if it takes an increase in the overall wage rate paid to all workers to be able to employ the desired number of workers. This is a general result for any potential purchaser facing an upward sloping supply of any resource or product in a given market.

The effect of a monopsonist's decision on the marginal cost can be illustrated using the following formula: $MC = P(1 + 1/n)$ where P is the price being paid and n is the price elasticity of supply. If the price elasticity of supply was perfectly elastic as in the case of competitive markets, the MC would be equal to P because 1/infinity is essentially zero. To the extent that the price elasticity of supply faced by a firm is other than perfectly elastic, the marginal cost of purchasing the last unit will be somewhat greater than the price.

OLIGOPOLY AND OLIGOPSONY

In addition to competitive market situations and those consisting of a single seller or buyer, there are several other intermediate structures of imperfect competition between the two polar extremes. These include cases of two or three (duopoly, duopsony, triopoly, etc.), or a few firms on either the buying or selling side of a market. The few firms might be dealing with homogeneous products or differentiated products within a given product class, but clearly the firms are interdependent in terms of the actions of one having a potential impact on others. An empirical analysis of 40 U.S. food and tobacco manufacturing industries indicated 37 of them exercised statistically significant oligopoly power in setting output price (Bhuyan and Lopez 1997).

Frequently, some barrier to entry may be the reason why a market consists of a relatively small number of competitors. For example, there might be some government regulation regulating the number of firms. Or the technology of production within a given line of business in conjunction with potential demand may tend to support only a few relatively large firms. A few large firms may be able to achieve **economies of scale** that are not attainable if the industry consisted of many small firms. Entry of new firms may also be difficult if a large volume of business is required to compete effectively with existing firms.

Various quantitative measures of market activity among the largest firms in an industry often are used to consider the degree of market power. Two of the most commonly used measures are the four- and eight-firm concentration ratios.[8] These measures represent the percent of a total industry's value of output or market transactions accounted for by the four or eight largest firms in a given industry or market. For example, in 1997 the four largest beef packing firms accounted for 68% of the total cattle slaughtered in the United States (Nelson and Hahn 1998). This ratio was 28% for cows and bulls and 80% for steers and heifers. Similarly, the four largest hog slaughtering firms accounted for 54% of that industry's output in 1997. The four-firm concentration ratio for sheep and lambs was 70% in 1997. Although most concentration ratios are calculated at a national level because of the availability of data, the same concepts are applicable for other kinds of markets. In the case of some lines of business activities the existence of four or eight firms might be considered a small number of firms, but there is no magic number that would automatically indicate the existence of an oligopoly or an oligopsony.

For market environments with a relatively few firms, individual firm demand and/or supply schedules may have some slope and consequently firms would have an incentive to use some of the same principles that are applicable for monopolists and monopsonists. One difference however is that individual firms would not be facing the entire market demand or supply relationship for a product and would need to be conscious of actual and potential actions of rivals. Many theories have been developed to explain alternative pricing strategies in these market environments. Two of the more frequently discussed kinds of pricing are **price leadership** and **predatory pricing.** Price leadership can occur in situations where one firm may have a cost advantage relative to other firms, or some dominant role, and takes the lead in establishing a price that is followed by other firms in the industry. Smaller firms may be reluctant to change prices until it is clear that everyone is likely to follow the leader of an industry. An alternative strategy would be to deliberately reduce price and even experience temporary losses in hopes of gaining a larger market share and eventually greater profits by forcing one or more competitors out of business. Predatory pricing is often described as a deliberate attempt by one firm to sell products at less than "cost" in hopes of gaining a long-term advantage. One of the difficulties with the notion of predatory pricing however is defining the relevant costs decision-makers use in making short-run pricing decisions.

Whenever an industry or given kind of business activity involves only a few decision-makers, it is tempting for them to try to avoid competitive pressures by forming a **cartel** and attempting to operate somewhat as a single entity in controlling price and/or total volume of business. For example, in recent years major oil producing nations have attempted to control the price of oil by regulating output. Invariably cartels have difficulty maintaining their unity because of conflicts about regulating the volume of output or dividing up the market. Often it is easier to agree on a price objective than agreeing how to allocate production among members to achieve the desired market price. Individually each member of a cartel has an incentive to increase its market share if it can do so without inducing others to follow suit. Unless everyone's actions in a market can be monitored easily or the threats of retaliation or penalties for violations of agreements are severe, it is difficult for cartels to effectively control every firm's behavior. Also for a cartel to survive for any length of time, there must be some barrier to entry of new firms.

Recently in the United States, several states in the Northeast have formed an Interstate Dairy Compact attempting to increase the price milk producers in the states received above the values established by federal **marketing orders.** The compact, however, does not regulate production. The extent to which a higher price of milk results in greater production and less consumption in a given region creates potential adverse effects on other regions. Consequently, it is not surprising that other regions have attempted to take steps to avoid adverse consequences.

Different policies exist in various countries pertaining to promotion or restriction of competition. In the United States it is illegal for firms to collude in any way to establish a uniform price or reach agreement to avoid competing. The privilege is reserved for use by federal and state governments. Firms can cooperate in terms of general business practices and sharing of information as long as it does not involve agreements to fix prices or overt agreements to operate in ways to avoid competing in certain markets. This has been part of an explicit national policy to encourage and promote competitive markets. It is a delicate balancing act for policymakers to decide how to permit efficient firms in an industry to expand and be able to reap the benefits of their ingenuity, but at the same time keeping firms from gaining too much market power. Most of the attention of regulators is directed at restricting firms from expanding by merger or acquisitions simply to reduce competition in an industry. Mergers in industries with relatively few firms are scrutinized carefully by the Federal Trade Commission and Department of Justice to determine the extent to which competition may be affected. An example of this was Cargill's acquisition of Continental grain merchandising business that was approved by the U.S. Department of Justice in

July 1999. The merger was approved provided certain grain facilities in various states be divested (Muirhead 1999). This merger involved the two largest grain exporting firms in the United States and may enhance the efficiency of U.S. grain exporting. At the same time, the proposed merger was controversial because of a possible reduction in competition within the United States. The combined business of the two firms was projected to handle about 10% to 13% of the U.S. grain production (Feedstuffs 1999).

Some mergers are not permitted if the government decides that the decrease in competition would be worse than potential gains in economic efficiency. There are many ways, however, that firms can develop close marketing relationships other than through ownership. For example, the formation of strategic alliances for obtaining inputs and distribution of products by marketing firms is an alternative to ownership. One example of this was the sale of several of Pillsbury's food manufacturing facilities and the development of alliances with new owners to obtain products that meet particular specifications (Kuhn 1999).

MONOPOLISTIC AND MONOPSONISTIC COMPETITION

The last category of market structures to be briefly considered is monopolistic and monopsonistic competition. It is much more common to consider this kind of structure on the selling rather than the buying side of markets. Actually the use of the *mono* prefix is a little misleading because each of these structures consists of many participants dealing with differentiated products. Thus monopolistic competition is the existence of an environment with many sellers of differentiated products. Each firm is assumed to have many competitors offering similar, but slightly different products or services. Consequently, a major marketing strategy of firms in these environments is to emphasize why their products or services are different from those of other firms and thereby perhaps justify moderately different prices. The ability to charge different prices however is limited by the nature of the demand curves that exist for their products. Some minor price differences may be able to be maintained, but market power is limited by the fact that customers have many alternatives to turn to if they become dissatisfied with a particular firm. These firms have a special incentive to establish brands and do a lot of promotional activity to develop, cultivate, and maintain a customer base. This kind of environment is often used as a description of retail food markets in which there seems to be a proliferation of different brands with relatively minor differences in characteristics. Brands are used to try to convey information about a certain set of attributes of products to potential buyers. For brands to be successful, it is important that products maintain a uniform quality to fulfill continued expectations of consumers. Increased abilities to identify and control particular attributes of agricultural and food products have created additional incentives to coordinate production and marketing activities more closely to ensure uniform quality of retail food products.

Branding and special identification of products by firms have been much more prominent on the selling than the buying side of markets. However buyers may differentiate themselves in terms of reliability of payment, helpfulness and friendliness of employees, or adequacy of facilities for handling different volumes of merchandise. As the results of biotechnology lead to agricultural products acquiring more special characteristics at earlier stages of production, the emergence of more differentiation in terms of procurement may develop. Buyers may specialize more in the kinds of products they purchase for specialized uses.

SUMMARY

The chapter describes the way in which different market structures affect how buyers and sellers make decisions. In particular, implications of the competitive market conditions implicitly assumed in developing the framework linking producer and consumer behavior are compared to several alternative market structures. Different behavioral implications depend basically on the nature of demand and supply relationships that characterize perceived market situations for individual sellers and buyers. In the case of competitive market conditions these relationships are perfectly elastic.

Different market structures can be viewed as variations along a continuum from competitive markets to the case of one seller (monopoly) facing the entire market demand, or one buyer (monopsony) facing the entire market supply. Intermediate structures between these extremes are oligopoly, oligopsony, monopolistic competition, and monopsonistic competition. The oly suffix is used to describe alternative structures from the selling side of markets and the sony suffix is used to classify alternative buying structures. The three characteristics used for classification purposes are (1) the number and volume of business of individual buyers and sellers, (2) the ease or difficulty of market entry and exit, and (3) whether the product is homogeneous or differentiated. The fact that operationally each of these characteristics depends on the geographical and product context of a given market means that the appropriate structure for analysis depends on how narrowly or broadly a particular market is defined. The competitive model is a useful starting point for considering international and national environments within which all other markets operate.

Participants in competitive markets make only quantity decisions. Producers maximize profits by determining the quantity for which the marginal cost of the last unit comes as close as possible (or equal) to price. Producers improve their profit position by making sure their costs are as low as possible and/or varying the timing of their sales if prices are expected to fluctuate. Consumers are assumed to buy a product up to the point where the perceived value of the last unit purchased equals the price of the product.

The marginal revenue of the last unit sold by a monopoly is $P(1 + 1/e)$ where P is the price of the product and e is the price elasticity of demand. Monopolists have to evaluate differences in total revenue from selling a larger quantity at a lower price or charging a higher price for a smaller quantity. It may be possible for a monopolist to increase total revenue by price discrimination if two or more separately identified markets with different price elasticities of demand for a product exist and resale of the product can be prevented.

Oligopoly and oligopsony market structures consist of a few firms that deal in homogeneous or differentiated products. The number of firms may be the result of barriers to entry, or the economies of production relative to the total size of the industry may support the existence of only a few firms. The degree of market power at the national or regional level for these types of markets is frequently inferred from four- or eight-firm concentration ratios. Different types of pricing behavior may occur in these markets depending on how the firms view their interconnectiveness. Whenever a market involves only a few firms there is a temptation to avoid competitive pressures by forming a cartel to operate more as a single entity in controlling price or total volume of business. In the United States any collusion among firms to establish a uniform price or limit competition is illegal. Also, other government regulations pertain to mergers and business practices that limit or restrict competition.

Two other major types of imperfect competitive market structures are monopolistic and monopsonistic competition. These structures consist of many firms handling differentiated products. A major marketing strategy in these environments is for firms to emphasize why their products or services differ from what other firms offer and thereby justify moderately different prices. Consequently, establishing brands and a lot of promotional activity are used to develop, cultivate, and maintain loyal clientele. The market power of these firms is limited by the existence of closely related substitutes.

QUESTIONS

1. What are the three basic characteristics used to classify different market structures?
2. Briefly describe the basic characteristics of a purely competitive market and the implications of this kind of market structure as far as marketing decisions by individual firms is concerned.
3. Briefly explain the difference between a monopoly and an oligopsony market structure.
4. What is the major distinguishing characteristic between monopoly and monopsony as alternative market structures?
5. If each of the 25 firms in an industry have exactly the same share of total market sales, what would be the value of the four-firm concentration ratio for this industry?
6. Suppose the following algebraic expression represents the demand for apples in a given market:

$$Q = 48 - 4P$$

where Q = thousand bushels of apples sold
P = price/bu

What price of apples would maximize total revenue in this market? Briefly indicate how you arrived at your answer.

7. Over what range of quantities would marginal revenue be positive if you were a monopolist facing the following demand function:

$$Q = 80 - 4P$$

where Q and P represent quantity and price respectively? Briefly explain your reasoning.

8. If the price elasticity of demand is −.5 at a particular point on the demand relationship, what information does this provide about marginal revenue at that point?

9. Briefly explain what market conditions must exist in order for discriminatory pricing to be successfully established.

REFERENCES

Bhuyan, S., and R. A. Lopez. 1997. Oligopoly power in the food and tobacco industries. *American Journal of Agricultural Economics* 79(3): 1035–1043.

DeVos, G. W. 1997. The elimination of the Canadian and Australian Wheat Boards: A move from triopoly to perfect competition in the world wheat market. *American Journal of Agricultural Economics* 79(5): 1742–1748.

Feedstuffs. 1998. Report shows growth for value-enhanced corn, 70(21). May 25, Minnetonka, Minn, p. 27.

Feedstuffs. 1999. EU clears Cargill grain acquisition, 71(6). February 8, Minnetonka, Minn, p. 5.

Kuhn, M. E. 1999. Teams that work. *Food Processing* 60(2) (February): Chicago, Il. Putnam, p. 19–26.

Muirhead, S. 1999. DOJ approves Cargill-Continental deal. *Feedstuffs* 71(28): 1.

Nelson, K., and W. Hahn. 1998. Concentration in the U.S. beef packing industry. *Livestock, dairy and poultry situation and outlook,* LDP-M-53. USDA, Economic Research Service. Washington D.C.

NOTES

1. Alternative models of market structure are used to analyze and contrast differences in how individuals are expected to react to different sets of circumstances. The models are simplifications of the real world designed to capture important elements of behavior, but not be overly complicated.

2. Specialized terminology is sometimes used to denote a subtle difference between different types of competitive market environments depending on the

availability of information. For example, some writers prefer using "perfect" competition to describe hypothetical competitive environments in which all market participants are assumed to have complete information about market opportunities and be able to make instantaneous adjustments. It provides a theoretical model for examining general tendencies of market behavior even though real-world situations fall short of satisfying the perfect information and instantaneous adjustment criteria.

3. Initially it will be assumed that all customers are charged one price. Later consideration will be given to situations where a seller has the potential to segment the market and establish different prices for different groups of customers.

4. This also means that it is technically not possible to define a supply schedule for a monopolist because the quantity that maximizes profits for particular costs of production turns out to be conditional on the nature of the demand curve. For example, the supply of tickets to a specific event may appear to be perfectly elastic (up to maximum capacity) once officials decide on the price to charge, but the price at which the maximum number of tickets is offered might differ if a different level of demand were assumed. Thus the willingness to offer or supply tickets is not independent of expectations of demand.

5. As alternative discrete changes in prices are considered, changes in quantities sold are likely to occur in multiple units, which means that any change in total revenue would need to be averaged over the change in quantity to estimate the marginal revenue per unit of quantity.

6. This assumes that monopolists are not able to charge individual customers different prices depending on willingness to pay a varying amount for the same item.

7. Part of this process is invisible to producers because they receive a blend price for all milk sold during a specified period of time. The price they receive is a weighted average of the prices for the different uses of milk in a given market monitored by governmental officials.

8. An alternative measure of concentration that can be calculated if data is available for each firm in a given market is the Herfindal-Hirschman index. A mathematical representation of this index is as follows:

$$I = \sum_{i=1}^{n} S_i^2$$

where S_i is the market share for the ith firm. If the industry consists of a single firm, $S_i = 1.00$ (or 100) and the index attains its maximum value of 1 or 10,000 depending on whether market share is expressed in percentage or decimal equivalent terms. The value of the index decreases as the number of firms in an industry increases or the share of output of each firm becomes more uniform.

CHAPTER

8

PRICE DISCOVERY

The purpose of this chapter is to discuss the variety of processes used by buyers and sellers to discover a mutually satisfactory price associated with ownership transfers of agricultural and food products.

The major points of the chapter are:

1. The meaning of ***price discovery*** and how it differs from ***price determination,*** which was the focus of previous chapters.
2. The nature of *agricultural industrialization* and how it has affected the marketing of agricultural and food products.
3. The major differences in individual and group behavior associated with alternative price discovery mechanisms.
4. The factors responsible for changing the type of price discovery mechanisms used for different commodities.

INTRODUCTION

Chapter 7 indicated how behavior of individual market participants could be affected by different market structures. However, it did not address the variety of processes used by buyers and sellers for reaching agreements about transferring ownership of agricultural and food commodities. The previous chapter assumed that in the case of competitive market structures, the invisible forces of demand and supply operated in some fashion to establish a market-clearing price all market participants somehow knew.[1] Also, nothing was indicated about how changes in underlying demand and/or supply conditions implying possible changes in values of products are discovered or market-clearing prices become known to market participants. Similarly, it was assumed that in noncompetitive market structures, prices were established by specified processes or somehow discovered based on perceptions of demand and/or supply conditions. Not much was said, however, about the actual processes by which buyers and sellers come to an agreement about a satisfactory price at which they are willing to transfer ownership rights.

The ways in which buyers and sellers interact and make decisions about mutually satisfactory terms of exchange for agricultural and food products is an interesting topic known as price discovery. This topic differs from the material in previous chapters that was intended to develop a conceptual framework illustrating how economic forces influence the price of commodities. Basically, the theory of price determination under alternative market structures, as presented in previous chapters, is concerned with how the relative strength of buyer and seller interests affect prices. The topic of price discovery deals with the characteristics of the actual processes used by market participants for arranging acceptable terms of trade for business transactions.

This chapter discusses some of the major elements of price discovery mechanisms used for food and agricultural products throughout the world. After a brief introduction, the most common systems of price discovery used by buyers and sellers are discussed noting some of the advantages and disadvantages of each. Then some key concepts are presented for thinking about the dynamic nature and evolution of price discovery mechanisms.

HISTORICAL PERSPECTIVE

The topic of price discovery is an especially timely topic in the United States because of the increasingly complex set of value-adding marketing activities associated with agricultural products moving through multiple transformation processes between initial production and ultimate consumption. An important issue is how various value-adding activities are coordinated with or without the use of explicit price signals to reflect values at each stage of the marketing process. Closely associated with this issue are how many, or how few, times ownership of commodities gets transferred as products proceed through various marketing channels.

Obviously, each time ownership of an item is transferred there is an element of price discovery involved and additional costs associated with making a business transaction. In earlier times when fewer value-adding activities occurred, the marketing of agricultural and food products often consisted of a single transfer or at least relatively few transfers of ownership between producers and consumers. In today's modern economies consisting of many specialized value-adding marketing activities there are many more opportunities for ownership transfers. As increased amounts of processing, transporting, and storage activities have been introduced into the marketing process and performed by specialized firms, additional transfers of ownership are often associated with moving products through the marketing process.

The number of times an item changes ownership between production and consumption is influenced by many factors. One is the economies of scale associated with specialization of activities. Another is the desire for greater control over production and/or marketing activities to guarantee quality and desired quantities at specified times. Although specialization of activities may lead to certain efficiencies and reductions in some costs, additional costs associated with more transactions can occur as more diverse activities contribute increased value to food products. One way of minimizing the costs of transferring ownership so many times, however, is for firms to continue ownership of a product for a longer period of time as it makes its way from initial production to ultimate consumption.

The term ***agricultural industrialization*** has emerged to characterize the evolving set of arrangements used for coordinating production and marketing activities of many agricultural products in the United States.[2] This terminology indicates that the production and marketing activities of many agricultural products in the United States have become increasingly similar to production and marketing processes used in the manufacturing and distribution of industrial products. One example is the decrease in movement of agricultural products through central and auction markets, and the increased reliance on alternative methods of coordination including various methods of forward-contracting delivery of products in the United States. Changes in the nature of production and marketing linkages associated with fewer exchanges of ownership of a given commodity have important implications on the way buyers and sellers interact and how price discovery systems operate.

DIFFERENT PRICE DISCOVERY SYSTEMS

The way in which ownership of products gets transferred in the United States varies widely among products. For example, houses and automobiles are bought and sold in quite different ways from grocery items, soybeans, or insurance policies. Even among agricultural products, entirely different processes and mechanisms are used for arranging transactions for vegetables, grain, milk, and other agricultural products. The diversity in price discovery processes used for different commodities (and even for the same commodity) within a given economy at a given point in time is often perplexing. Particular systems of price discovery appear to have evolved and work reasonably well for some products while entirely different price discovery mechanisms exist for other products even with identical or very similar market structures.

Considerable variation in price discovery mechanisms also exists in different parts of the world for the same commodity. For example, poultry products are bought and sold in the United States quite differently from other parts of the world. In many countries, cultural differences influence the way business transactions are negotiated and consummated. In some cultures, the practice of haggling or negotiating price seemingly appears to have more intrinsic value or is more customary than in other cultures. A basic question, however, is how much value is realized by market participants spending time haggling over price if the market is essentially competitive? Variation in the behavior of buyers and sellers under different market institutions is an interesting aspect of price discovery. For example, different kinds of behavior or conduct may be observed at auctions, roadside markets, or centralized buying stations for the same commodity.

Different mechanisms by which buyers and sellers discover each other's interest in conducting business require varying commitments of resources including time. Consequently, the total cost of purchasing an item is more than its monetary price if the value of resources used in making a transaction is taken into account. Similarly, if the value of resources used by sellers to finalize transactions is considered, the net return from a sale is less than the price or monetary value received. Consequently, the costs and returns associated with different mechanisms used by participants for arranging market transactions are important characteristics of alternative methods of price discovery. As circumstances change, individuals often find it useful to adjust the procedures used for price discovery and making transactions.

Most products are bought and sold quite differently now than in earlier years because new price discovery mechanisms have emerged to replace earlier methods of buying and selling products. Many of the changes in price discovery mechanisms for most agricultural products over the last 30 or 40 years have been facilitated by changes in transportation and communication technologies. Recently, satellite communication and the Internet have opened up all kinds of new opportunities for market transactions. If history provides any lesson at all, it appears reasonable to expect continuing changes in the manner in which exchanges of ownership are arranged and the accompanying prices of commodities are discovered or determined by market participants. One can only speculate on how price discovery systems are likely to change in the future in response to new opportunities for buyers and sellers to make arrangements for exchanging ownership of commodities.

INDIVIDUAL VERSUS GROUP ACTIVITIES IN PRICE DISCOVERY SYSTEMS

A major distinction among different systems of price discovery is the extent to which market transactions and/or negotiations consist primarily of individual activities versus more explicit group behavior. For example, many transactions occur as a result of scheduled or incidental individual encounters between prospective buyers and sellers whereas other kinds of market transactions are finalized in more of a public arena or environment with explicit group dynamics.

The distinction between private encounters and group dynamics is a useful starting point for considering certain aspects of price discovery, but it is important to recognize the limitations of this classification. Even the most private encounters or business interactions do not occur without some awareness of group behavior because everyone has some perception about what is happening elsewhere that might have a bearing on the current and prospective value of agricultural and food commodities. Private decisions to buy or sell products may be based partly on what other individuals are currently doing or expected to do in the future. Another aspect of the way in which group behavior influences individual negotiations is the manner in which the legal and overall environment for business transactions is affected by political processes. Governments influence price discovery mechanisms by many direct and subtle actions. In order to focus attention on the more customary mechanisms used by buyers and/or sellers to discover appropriate prices for market transactions, any direct governmental influence will be ignored initially. Some of the more explicit ways in which governments influence the price discovery processes will be considered later.

Individualized Agreements

Individual agreements involving a transfer of ownership at a mutually agreeable price can be characterized in many ways. One way is whether an exchange of ownership occurs immediately or whether the agreement establishes the terms of trade for exchange of ownership at a future period of time. A second dimension of individual agreements is whether they involve an explicit price or whether a mutually acceptable process for determining final terms of payment is established. A third way of characterizing individualized agreements is the manner in which interactions or communications between individual buyers and sellers occur. Each of these dimensions of individualized agreements will be briefly discussed.

Timing of Transfer of Ownership. An immediate transfer of ownership may occur as soon as there is a mutual agreement about an acceptable price and payment or financing for the purchase. This is often the case with many retail purchases in which customers pay a posted price and exchange of ownership occurs as soon as payment is made. Many transactions occur without any discussion about price or other details about transferring ownership. For example, if a buyer or seller agrees to a price proposed by another individual, an immediate exchange of ownership may happen as soon as payment is made.

Instead of an immediate exchange of ownership, individualized price discovery processes may involve determining appropriate terms of trade at which ownership will be exchanged at some time in the future. These agreements are usually referred to as ***forward contracts*** and involve a period of time between when a price is established and when transfer of ownership occurs. In some respects, forward contracts are similar to **futures contracts** that will be discussed in more detail in Chapters 14 and 15. Forward and futures contracts both involve a promise or commitment to carry out specific actions at a future point in time at a mutually acceptable price, but the nature of the promises are quite different. A forward contract involves the promise of a buyer to accept ownership of a specific lot of merchandise from a specific seller under specific conditions. Futures contracts are more standardized than forward contracts and there is more flexibility in how contractual obligations of futures contracts can be fulfilled. These distinctions will be made clearer in Chapters 14 and 15.

Contracting for the future delivery of agricultural and food commodities at a specified price is becoming increasingly common in U.S. agriculture. Contracts may be arranged before or after production begins. In some cases a distinction is made between production and marketing contracts depending on whether the agreement is reached before or after production is initiated and who owns the commodity during production, but terminology to characterize different types of production and marketing arrangements is not standardized. Data in Table 8–1 indicate that **production contracts** accounted for 10.7% of total farm output in 1993 to 1994. This compares to 8.3% of total farm output under production contracts in 1960.[3]

Firms that produce a large volume of commodities often find it desirable to negotiate contracts with specialized firms that provide an ensured outlet for products when they are ready to be sold. Forward contracting provides a certain degree of assurance to a seller that there will be a purchaser of the output when it is ready for sale. Thus some of the uncertainty of arranging for the purchase or sale of a product can be eliminated before the product is ready for sale. This is especially important if products are perishable implying a very short time period during which they can be moved into market channels.

Large retail outlets may also prefer to have some type of contractual arrangement that ensures them access to a given quantity and quality of products at specific times to avoid having to continually spend time locating suppliers, negotiating prices, and purchasing products. A forward contract provides some assurance about a forthcoming supply of a product to a firm dependent on having a certain volume of product for its operations. Being able to schedule the

TABLE 8-1

PRODUCTION CONTRACTS AND OWNERSHIP INTEGRATION

	Production Contracts[1]			Ownership Integration[2]		
Year	*1960*	*1980*	*1993–94*	*1960*	*1980*	*1993–94*
			(percent)			
Crops:						
Feed grains	0.1	1.2	1.2	0.4	0.5	0.5
Hay	.3	.5	.5	0	0	0
Food grains	1.0	1.0	.1	.3	.5	.5
Vegetables for fresh market	20.0	18.0	25.0	25.0	35.0	40.0
Vegetables for processing	67.0	88.1	87.9	8.0	10.0	6.0
Dry beans and peas	1.5	2.0	2.0	1.0	1.0	1.0
Potatoes	40.0	60.0	55.0	30.0	35.0	40.0
Citrus fruits	0	0	0	8.9	11.2	6.9
Other fruits and nuts	0	0	0	15.0	25.0	25.0
Sugar beets	99.0	99.0	99.0	1.0	1.0	1.0
Sugarcane	24.4	29.3	27.3	75.6	70.7	72.7
Cotton	5.0	1.0	.1	3.0	1.0	1.0
Tobacco	2.0	1.4	9.3	2.0	2.0	1.5
Soybeans	1.0	1.0	0	.4	.5	.4
Seed crops	80.0	80.0	80.0	.3	10.0	10.0
Livestock:						
Fed cattle[3]	—	—	—	6.7	3.6	4.5
Calves, slaughter[3]	—	—	—	1.5	1.8	10.0
Other cattle and calves	1.0	2.8	1.7	—	—	—
Sheep and lambs[3]	—	—	—	5.1	9.2	29.0
Market hogs	.7	1.5	10.8	.7	1.5	8.0
Fluid grade milk	.1	.3	.1	0	0	0
Manufacturing grade milk	0	0	0	2.0	1.0	1.0
Market eggs	7.0	43.0	25.0	5.5	45.0	70.0
Hatching eggs	65.0	70.0	70.0	30.0	30.0	30.0
Broilers	90.0	91.0	92.0	5.4	8.0	8.0
Market turkeys	30.0	52.0	60.0	4.0	28.0	28.0
Total farm output[4]	8.3	11.5	10.7	4.4	6.2	7.6

Sources: Table 2 of Food Marketing Review, 1994-95 USDA Agricultural Economic Report No. 743.

[1]Production contracts. Contracts entered into before production begins, excludes marketing contracts.

[2]Ownership integration. The same firm owns farms and other vertically related operations such as a hatchery, feed mill, processing plant, or packer-shipper. Excludes direct marketing to consumers such as producer-dealers of milk, roadside stands, or pick-your-own operations.

[3]Feeding of livestock by the meatpacker, some of which is under contract in feedlots owned by others.

[4]The percent of total farm output under production contracts and ownership integration includes only the products listed in the table and calculated using the same weights in each year so that changes in the share of, say, broilers do not affect the figure. The weights are the average share of cash receipts of each product in 1960, 1970, 1980, and 1990.

quantity of products that will be received facilitates the scheduling of subsequent marketing activities. Forward contracting can also help foster a relationship between parties encouraging continuing transactions without having to do a complete search of a market for prospective buyers or sellers.

One of the disadvantages of forward contracts is that the participants depend on each other's integrity to fulfill commitments. A contract may be worthless if one of the parties is unable or unwilling to fulfill its commitment. Even if there are legal remedies available in case the terms of a contract are not met, it may not be easy or inexpensive to recover damages. Also a forward contract that includes a specific price at which ownership rights will be transferred may prove to be unfavorable to one of the parties if the actual market value of the commodity at the time ownership is transferred differs significantly from the price specified in the contract. For example, producers might feel cheated if they agree to sell at a price that eventually turns out to be significantly less than the price at harvest time. Alternatively buyers may feel they paid too much if the price at harvest time is less than the contract price. This is why some contracts do not specify an exact price, but may include a process for determining a final settlement price.

Explicit arrangements for coordinating production of agricultural commodities with particular value-adding marketing activities is often referred to as ***vertical integration.*** A fairly restrictive interpretation of this term is to describe situations in which the same firm is responsible for multiple stages of production and/or marketing of a product totally using only its own resources. For example, if a firm grows, slaughters, and markets hogs, chickens, or other livestock products totally with its own resources, it would definitely be considered an integrated firm since more than one value-adding production and marketing activity is combined under one decision-making unit.

Other mechanisms for coordination of production and marketing activities vary depending on the ownership of resources and the basic commodity at different stages of production and marketing. For example, the term vertical integration is also frequently used to characterize situations in which some kind of contractual arrangement is used to coordinate two or more subsequent production and/or marketing activities for a commodity. The latter interpretation is often implied in describing situations where one firm contracts with several individuals to produce a commodity under a specific set of conditions. Some of these contracts are specific arrangements for producers to provide land, capital, and labor to grow crops or livestock. In many cases, the arrangement may not involve an explicit transfer of ownership upon completion of production because the contractor owns animals or crops throughout the process. These contracts are unique in terms of providing indirect remuneration for the explicit commitment of resources in terms of product produced rather than specifying a particular wage per hour or annual rent for land, buildings, or capital. The agreements usually specify a mechanism that will be used to establish the price or remuneration if specific production objectives are achieved.

Explicit Price or Formula. For most individualized market transactions that involve an immediate exchange of ownership, an explicit price is established by negotiation or one of the parties agrees to a price posted by the other party. Sometimes there may be considerable negotiation between buyers and sellers before a mutually acceptable price is determined regarding an immediate or future exchange of ownership of an item. At other times, exchanges of ownership occur very quickly by responding to posted prices. For example, individuals going to a large retail supermarket to purchase food products do not expect to negotiate price. However, if they purchase similar products in different environments like roadside stands or farmer's markets they may expect to engage in price negotiations.

Individual negotiations between buyers and sellers often consist of a series of bids and offer prices that eventually converge to a mutually satisfactory price. Often the amount of money involved in a given transaction will determine whether it is worthwhile for the two parties to spend time negotiating. For example, if one is buying a gallon of milk in the grocery store, it is not in the interests of customers or store management to spend time negotiating a price for each transaction. It is much easier for customers to make their decisions based on past observations (and possibly new information) about prices at different outlets. On the other hand, if one is purchasing an automobile or real estate, an asking price is often an open invitation to begin negotiations. Recently, some automobile dealers have begun advertising nonnegotiable prices for vehicles to appeal to buyers who do not like to negotiate. Even in these cases, however, some negotiation may be required if the transaction involves another vehicle as partial payment.

In cases where posted prices are used in competitive or noncompetitive markets, prices may be posted or announced either by sellers or buyers. As additional information about market conditions develops, adjustments in posted prices may occur. Sometimes posted prices are intended and understood to be nonnegotiable or said to be firm. Other times, posted prices are intended to be soft or interpreted as starting points for discussion or an invitation for counteroffers. For example, in the United States, retail food prices are generally viewed by consumers as firm or nonnegotiable. On the other hand, the use of posted prices at roadside stands or farmers' markets may be firm or soft. Market participants in a given environment generally know the customary way of interpreting posted prices, but sometimes the firmness or softness of a posted price may not be known until it is tested.

Some type of formula is frequently included in agreements for establishing a price associated with a future exchange of ownership especially when there may be considerable uncertainty about market conditions. Formulas can also be used to establish appropriate terms of remuneration after a product has been processed. In these cases a producer may not know exactly what price will be received for merchandise until it has been processed and/or ultimately sold. Nevertheless, the original sales agreement is likely to specify a particular procedure that will be used to establish the remuneration that producers will receive. In negotiating transactions involving a continuing flow of products whose value is likely to fluctuate because of changes in market conditions, formulae including automatic adjustments in the value of the commodity may be mutually satisfactory.

Forward contracts may specify an explicit formula or indicate how certain factors will be used to determine appropriate terms of payment for products that are subject to variation in value. The formula might be based on price information from some central market or an alternative source acceptable to the two parties. The formula might be as simple as agreeing on a fixed premium or discount relative to the price for similar quality merchandise reported by governmental or private sources. One problem in using this kind of mechanism these days is the decrease in publicly available price quotations because of the decrease in volume of products moving through central markets. In the absence of a reliable source of alternative price information, a mathematical formula may be devised that weights different types of economic information related to shifters of supply and/or demand relationships considered in earlier chapters. One example of this kind of formula is what has been used in the pricing of fluid milk in most U.S. markets based on the prices of processed dairy products.

Formulas are also used to adjust posted prices in response to rapidly changing regional, national, or world market conditions. For example, buyers and/or sellers frequently use predetermined differentials from the prices of selected futures contracts to set cash prices for grain, livestock, and many other commodities in the United States.

Type of Interaction. Buyers and sellers interact and/or communicate interest in exchanging ownership of products in a variety of ways. Transactions that involve negotiations about price and/or other details of the exchange may be handled directly through personal contact or handled indirectly through various forms of modern communication. Some market transactions are arranged with very limited interaction or communication between buyers and sellers. In some cases it is customary for buyers to initiate contact with sellers and in other cases sellers may approach prospective buyers. In other situations, specialized services of brokers provide the critical linkage between potential buyers and sellers.

Individual negotiations permit parties to feel that they benefit from superior negotiating abilities or individual knowledge of current or expected market conditions. To the extent that body language plays a role in negotiating strategies, some individuals prefer personalized contacts. A disadvantage of individual negotiations is that the process generally produces less public information about relative value of similar products and nature of market conditions.

Individual negotiations involving face-to-face contact may occur at a predetermined time at a buyer's or seller's location, at an alternative mutually agreed upon meeting place, or spontaneously whenever individuals encounter each other. For example, marketing of fruit and vegetable products may involve consumers or brokers traveling to production locations and buying directly from producers. An alternative is for producers to transport products to a centralized location more convenient to a larger group of potential buyers. Either way provides an opportunity for direct contact between prospective buyers and sellers, but there may be significant differences in costs and benefits associated with each system. From a buyer's perspective there may be a clear advantage of having access to more choices at a centralized location and thereby saving time, but the merchandise may not be as fresh as it would be if obtained closer to the site of production. From a seller's perspective, a centralized location may provide access to a larger pool of customers, but the extra cost of transporting products, renting space, and the time involved in operating or staffing a sales outlet must be weighed against potential benefits. This is analogous to decisions faced by households about whether it is better to take merchandise to a centralized flea market and pay a fee to rent some space that may have more customer traffic instead of inviting prospective buyers to come to their premises. Time involved in making appropriate meeting and traveling arrangements can also add to the costs of arranging transactions.

Variation in quality of items being traded may affect the manner in which transactions are arranged. In cases where considerable variation in quality of merchandise occurs, buyers and/or sellers have to decide if it is important to visually inspect merchandise or whether an independent third party's assessment of quality attributes is satisfactory.

Group Actions

An important aspect of group actions related to price discovery is how outcomes of individual negotiations affect subsequent buying and selling decisions. The dynamics of individual actions at a given location provide some of the emotional content of market behavior in terms of providing information about the intensity of buying and selling interests. The extent to which prices are observed or reported as public information is an important dimension of group activities. Price discovery mechanisms that involve even more explicit group behavior are auctions and **group bargaining.**

Auctions. The basic idea of an auction is an epitome of a competitive market environment in terms of an explicit mechanism for all prospective buyers and/or sellers to participate jointly in a bidding process. There are many ways in which auctions can be conducted to solicit and evaluate alternative ownership interests in homogeneous as well as unique products. For example, there are auctions at which single items (e.g., a farm or piece of machinery) or multiple units of the same item (e.g., pens of animals) are offered for sale. Auctions can be conducted to discover an appropriate price for an immediate exchange of ownership or a future exchange of ownership. The bidding process may consist of public expressions or submission of sealed bids of interest to buy or sell particular merchandise.

Auctions can be used by one or more buyers to solicit bids from prospective sellers. Another variation is a double auction where participants can be buyers or sellers depending on which side of the market they want to be on at a given moment of time. A good example of this kind of auction is the trading that occurs on futures markets by the open outcry of traders offering to buy or sell specific futures contracts. The difference between current and anticipated prices of commodities influences whether a given trader prefers to make a commitment to be a seller or a buyer of the commodity at a particular moment of time.

In the case of a public auction with ascending bids, information is produced for all market participants in that the successive bids essentially reveal the aggregate demand for what is being offered for sale at a given point in time at a given location. Before an auction begins, participants may have quite different perceptions of underlying market conditions, but competitive public bidding provides a process for a market-clearing price to be established reflecting relative interests of both sides of a market at a particular point in time.

The final price for eventual exchange of ownership determined by an auction process may be the winning bid (a **first-price auction**) or the bid of the closest nonwinner (a **second-price auction**) depending on the rules used for conducting the auction. In the latter case, the eventual buyer would be the person making the highest bid, but he or she would have to pay only the price offered by the next closest offer. There are several reasons why different ways of structuring an auction are likely to affect individual behavior. For example, some believe that it is best to structure an auction by asking individuals to indicate either publicly or secretly what they would pay if it turns out that they are selected as the winner. Others believe that a second-price auction may provide an incentive for individuals to communicate what they really think an item is worth rather than just making an offer that they hope will be the highest bid.[4]

Some auctions consist of soliciting secret bids from prospective participants and evaluating all the bids simultaneously at a designated time. The highest bid, or the highest set of bids if there are multiple homogeneous items, is selected in the case of a selling auction. The lowest (or lowest set) bid would be of interest in the case of an auction to select who to purchase merchandise from. When auctions involve multiple units of homogeneous items it is necessary to evaluate the entire roster of bids to determine a price that will be acceptable to a sufficient number of buyers depending on the amount of merchandise to be sold.

The usual **English** type of **auction** starts with a low bid and is successively raised until there is only one willing bidder remaining for a given lot of merchandise. An auctioneer may call out successively higher prices and get acceptances from members in the audience. An alternative is for an auctioneer to invite bids from an audience and solicit adjustments in offers until the auctioneer hears only one bid.

An entirely different kind of environment exists with a **Dutch auction.** This kind of auction is a descending price auction whereby the auctioneer starts with a very high price and successively lowers it until an individual agrees to pay the stated price. If more than one individual responds to a given price, the process can be reversed to see who will drop out.[5]

There can be several costs associated with the use of an auction as a method of price discovery. One example is the cost of organizing an auction and perhaps maintaining a facility where prospective participants can bring merchandise for interested parties to inspect. There may be costs of transporting merchandise and/or individuals to and from the site of an auction. Also the value of participants' time while engaging in the activity is another cost component of auctions.

Frequently items to be sold at an auction are displayed for inspection at the time of the sale to provide an opportunity for all prospective buyers to evaluate inherent characteristics and make comparisons with similar merchandise available elsewhere. The actual merchandise may not need to be displayed or located at the actual site of an auction if the characteristics of the item to be purchased or sold can be accurately described and/or communicated to all market participants. Indeed with modern means of communication, it is not even necessary for all auction participants to be at the same site to bid competitively. For example, electronic or televideo auctions can be conducted through simultaneous communication hookups. Remote video cameras can be used instead of moving the merchandise or participants to a given location if visual observation of the items to be auctioned is important. Various kinds of electronic auctions can reduce transportation and travel costs of participants, but also require other kinds of resources. The availability of computer technology to receive and sort information has added flexibility and speed to the way auctions can be used to buy and sell all kinds of merchandise.

Group Bargaining. Another form of price discovery involving group behavior is bargaining. This occurs when negotiations involving the transfer of ownership are conducted on behalf of a group of prospective sellers with a single buyer or on behalf of a group of buyers with a single seller. Cooperatives frequently have been used as vehicles for representing the interests of several producers in bargaining with one purchasing agent. Group bargaining may be an efficient manner for a single buyer to reduce the amount of time that would be required to negotiate with each individual cooperative member. In some cases, bargaining involves simply determining the kind of formula that will be appropriate to determine a reasonable settlement of the value for merchandise. This kind of agreement is especially important in reaching contractual terms between prospective sellers and buyers before production is undertaken.

Group bargaining as a method of price discovery requires a strong sense of unification among group members and a willingness to finance the costs of the bargaining effort. If individual members of a group believe they would be able to receive the same negotiated price without having to bear any of the overhead costs of maintaining a bargaining organization there is an economic incentive to break away from the group. One way in which some group unity problems have been overcome is with the help of explicit marketing orders sanctioned by the federal government for particular agricultural commodities. Marketing orders can be used to regulate the flow of output to particular markets and establish certain criteria such as quality standards and package sizes that all buyers and sellers must follow. Market orders cannot be used to regulate total production of commodities. Market orders have also been used as a vehicle for collecting revenue from all producers or handlers of specific commodities for the purpose of financing market development activities.

GOVERNMENTAL INTERVENTION

Different types of governmental actions can also influence the price discovery process in many ways. In some cases the actions may be very subtle such as facilitating the interaction between prospective buyers and sellers. This might consist of providing a central meeting facility, providing information about current and prospective market conditions, or the administration of marketing orders.

Other types of more direct governmental intervention are situations where governments may establish or attempt to control official prices at which ownership of products can be traded. Certain types of governmental actions may establish minimum or maximum prices at which particular goods can be exchanged. In these situations, the political process discovers relevant prices. Unless the prices that governments select are consistent with underlying market condition, some additional mechanisms will be required for handling excess production or rationing available quantities among potential buyers. For example, some means of production control or other means of intervening in the market are likely to be required if there is interest in maintaining a minimum price that is greater than what otherwise would occur.

EVOLUTION OF PRICE DISCOVERY MECHANISMS

As noted earlier, there are many methods that buyers and sellers have devised to arrange economic transactions. The diversity in price discovery mechanisms used within any economy at a given point in time as well as the emergence of new methods as others disappear is an interesting economic phenomenon. One way of thinking about this evolutionary process is to consider each kind of price discovery mechanism as a particular type of economic institution. The idea of an economic institution is that it represents a set of rules or guidelines that are codified or generally understood that influence the way individuals interact to execute economic transactions. Various methods for arranging business transactions are similar to other kinds of social and economic institutions that influence human behavior in a given society. For example, it is customary to think of families and communities as fundamental institutional units of any society. These institutional units have experienced changes over time and differ quite significantly around the world in terms of affecting individual behavior. Another example of institutional diversity and change is various religious or political organizations that affect certain aspects of human behavior.

Every society has many types of economic institutions that influence the way individuals perform economic activities. For example, economic concepts like money, credit, contracts, cooperatives, corporations, and property rights are important economic institutions that influence human behavior. One only needs to consider the introduction of credit cards, automatic teller machines, and computer technology in recent years to appreciate how individual and group behavior has been affected by particular changes traditionally handled by banking institutions.

The previous comments reflect some ideas contained in an article that T.W. Schultz (1968) wrote several years ago about institutional changes. In that article he used basic demand and supply concepts to consider why changes in economic institutions occur. The basic premise of his argument was that costs and benefits are associated with various economic institutions and as new circumstances alter the costs and/or benefits, it is likely that new institutions will emerge to displace alternative ways of performing similar functions.

Although Schultz (1968) did not directly comment on price discovery mechanisms, it seems that his ideas are very applicable for considering the evolution in price discovery systems. As noted earlier, it is clear that transaction costs are a critical element associated with various methods of price discovery. In view of current concern about the diminishing use of central markets associated with agricultural industrialization in the United States described earlier, it is useful to reflect on the situation that existed in the late 1920s and early 1930s when there was a lot of debate about alternative systems for planning and coordinating resource use in an economy. In thinking about these issues, Coase (1992) recognized an analogy between the controversy about centralized or decentralized decision making for an entire economy with the variation in how individual manufacturing firms depended on markets as a source of intermediate inputs. For example, he observed that some industrial firms apparently decided that purchasing intermediate products from other firms was more advantageous than producing everything they needed whereas other firms found it advantageous to internalize more production activities and rely less on market transactions. Coase postulated that a major factor affecting a firm's choice between the two alternative strategies was the perceived cost of arranging market transactions relative to the cost of coordinating more diverse activities within a firm.

Identifying transaction costs as an important component of any kind of buying and selling activity has had an impact on economic thinking about the way markets operate including those for agricultural and food products. A major implication is that the total cost of an item is more than the price or amount of money that is exchanged in a transaction, but also includes the value of time and other resources required to arrange a market transaction. If it is relatively easy to secure what is needed from other parties through market transactions that route might be more efficient than trying to manage and coordinate additional activities. On the other hand, if sources of supply were unreliable or unpredictable resulting in high costs of arranging market transactions, firms would want to consider alternative ways of making sure they had access to a continual source of products.

Changing transportation and communication technologies and an increasing value of time change the relative costs of using different methods of price discovery leading to changes in the way buyers and sellers do business. An increasing number of value-adding activities associated with food and agricultural products mean that there are more opportunities for specialized marketing firms. Also, individual marketing firms can expand the size of their business operations by becoming more integrated. As specialized marketing firms seek to handle an increasing volume of business, a continuous and certain source of product inputs and marketing outlets becomes critically important. This can lead to increased use of forward contracts for market transactions.

Other firms may decide to expand by performing more marketing services for the same product and thereby integrating more value-adding activities as part of the same business. By incorporating more activities as part of the same business operation, overall transaction costs may be less relative to what would be associated with multiple transfers of ownership between each marketing activity. Of course, offsetting the reduction in transaction costs are the additional costs of managing and coordinating more diverse activities noted by Coase (1992). Consequently, firms need to evaluate the costs of alternative ways of buying the desired quantity and quality of inputs as well as making sure there are customers willing to purchase the items they are producing.

As changes in communication, information processing, and transportation continue to produce differential changes in transaction costs associated with alternative price discovery mechanisms, it is likely that new methods of price discovery will emerge and replace some of the mechanisms currently used to buy and sell agricultural and food products. Alert entrepreneurs may be able to identify opportunities to reduce transaction costs and change price discovery and marketing mechanisms. Recently a whole new field of Internet marketing has begun to emerge. It is difficult to anticipate or forecast how this technology will alter the nature of food and agricultural markets, but it is clear that a new arena of opportunities exists to identify and negotiate with a broader spectrum of potential buyers and sellers than previously has been possible. Even within this technology, however, the same basic principles of price determination implied by the framework developed in earlier chapters are applicable.

SUMMARY

One of the major points of the chapter is highlighting the diversity of mechanisms used by buyers and sellers to discover a mutually satisfactory price associated with the transfer of ownership of agricultural and food products. The increasing number of value-adding activities linking production and consumption has been associated with a variety of mechanisms for coordinating subsequent activities. These have resulted in a decreased flow of products through central markets and more dependence on forward contracts and various forms of vertical integration for coordinating the exchange of ownership of commodities.

Various characteristics of individual and group behavior dynamics of **price discovery** were discussed in this chapter. The three dimensions of individual agreements highlighted are (1) differences in when transfer of ownership actually occurs, (2) whether an explicit price or some type of formula for determining final settlement price is specified, and (3) the way in which prospective buyers and sellers interact. Two particular types of price discovery involving more explicit group activities are auctions and group bargaining. In particular, several different ways in which auctions are structured and some of the potential benefits and problems of group bargaining activities were highlighted. Finally some of the ways in which governments intervene in the price discovery process are described.

The final section of this chapter includes a particular way of thinking about alternative methods of price discovery as economic institutions that evolve as relative costs and benefits of different ways of discovering prices are affected by changes in communication, information processing, and transportation technologies. In particular, the key role of transaction costs in transferring ownership of products is an important component of all transfers of ownership. Thinking about the role of transaction costs provides insight about the evolution of price discovery mechanisms.

QUESTIONS

1. Briefly describe the difference between the concepts of price discovery and price determination.
2. Discuss one of the reasons why agribusinesses have been motivated to enter into production contracts with farmers and one of the reasons farmers have been motivated to enter into production contracts with agribusinesses instead of both parties relying on alternative systems for arranging market transactions.
3. Describe what **formula pricing** means and why it can be an advantageous means of price discovery as well as some of the potential controversies associated with this method of price discovery.
4. Briefly describe the difference between an English and a Dutch auction.
5. Briefly describe some advantages and difficulties of group bargaining as a method of price discovery.
6. Identify some of the different ways in which governments can affect or influence price discovery.
7. Briefly describe two of the various components of transaction costs of alternative price discovery systems.

REFERENCES

Bernard, J. C., T. Mount, and W. Schulze. 1998. Alternative auction institutions for electric power markets. *Agricultural and Resource Economics Review* 27(2): 125–131.

Castle, E. N. 1998. *Agricultural industrialization in the American countryside.* Maryland: Greenbelt. Henry A. Wallace Center for Alternative Agriculture. Policy Studies Program Report #11. Also at http://www.hawiaa.org/pspr11.htm.

Coase, R. H. 1992. The institutional structure of production. *The American Economic Review* 82(4):713–719.

Gallo, A. E. 1996. *Food Marketing Review, 1994-95* Agric. Econ Rep 743, USDA, Economic Research Service. Washington, D.C.

Milgrom, P. 1989. Auctions and bidding: A primer. *Journal of Economic Perspectives* 3(3):3–22.

Schultz, T. W. 1968. Institutions and the rising economic value of man. *American Journal of Agricultural Economics* 50(5):1113–1122.

NOTES

1. The price of all transactions may not be identical even in competitive markets because of location or other specific transaction characteristics that affect values.

2. See Castle (1998) for additional discussions about agricultural industrialization.

3. The data in Table 8–1 also indicate an increase in the proportion of farm output produced under some type of vertical integrated ownership arrangement between 1960 and 1993–94.

4. Evaluating the expected outcomes of alternative ways of structuring auctions under various assumptions about how individuals behave in group settings has received considerable attention by economists. Bernard, Mount, and Schulze (1998) and Milgrom (1989) provide good summaries of some of the theoretical and empirical literature on this topic.

5. Alternative forms of the English and Dutch auction systems can also be structured by inviting private bids from buyers or sellers in cases where there is a specific item to be purchased or sold and there is interest in soliciting each potential participant's estimate of the underlying value. This process enables buyers to select the lowest offer price and sellers to obtain the highest price.

Part

III

Spatial Aspects of Agricultural Markets

Part III consists of four chapters that focus on particular spatial dimensions of agricultural markets. The objective of these chapters is to help students understand interregional and international movements of food and agricultural products and why prices of commodities generally vary among locations.

Chapter 9 indicates why the value or price at which ownership of a commodity is exchanged depends on where the commodity is located. Increasing distances between the location of agricultural production and where the bulk of consumption occurs is noted. Also there is a discussion of how changes in transportation technologies have influenced the geographical movement of goods over time. Interregional and international movements of commodities are value-adding activities because they require the use of resources and would not occur unless the expected change in value was sufficient to offset the cost of transportation services. The location of value-adding processing facilities is discussed in terms of the trade-off between the economies of scale of processing larger quantities of products at one location versus the increased cost of assembling larger quantities of a raw product given the geographical dispersion of agricultural production activities. Various ways in which governments influence transportation costs are also discussed. Finally, the concept of producer and consumer price surfaces is introduced to show how transportation costs affect the rate at which the value of a commodity changes with distance from a market. Price surfaces also can be used to illustrate how transportation costs affect competition among alternative markets and the amount of buying and selling that occur at particular locations.

Chapter 10 considers how price differences among geographically separated markets create incentives for interregional movement of products and increased regional specialization of production. A two-region trading model is presented to illustrate how interregional trading of a single commodity affects producers and consumers differently depending on whether they are located in an exporting or importing region. Excess demand and supply concepts are introduced to represent potential demand and supply of goods for movement between regions. The two-region model is developed initially ignoring transportation costs to highlight the incentives for interregional transfer of commodities. Interregional trade tends to increase price in the region where shipments originate and lowers the price in the region to which shipments are made. Interregional transportation costs are introduced into the model using concepts similar to those used to illustrate derived demand and supply. With this modification interregional shipments occur until the difference in regional prices becomes equal to the cost of transportation. No interregional shipments occur if the difference in price is less than the transportation cost.

Chapters 11 and 12 show how basic ideas of interregional trade are also applicable to international trade. The first part of Chapter 11 discusses the implications of recent trends in international trade of agricultural and foods on U.S. producers and consumers. Then some additional complexities of international trade relative to interregional trade are discussed. For example, language and cultural differences, specialized transportation services, multiple governmental policies, and different monetary systems are some of the items that make international trade a little more complicated than interregional movement of commodities within an economy. Since the value of a commodity can be expressed in terms of different monetary units most international transactions are influenced by two prices: the price of the commodity and the value of one country's currency expressed in terms of another country's currency (the exchange rate).

Chapter 12 shows how the interregional trade model can be modified to incorporate and illustrate the effects of different exchange rates on the amount of trade between two countries. Initially, in developing the model no transportation costs or barriers to trade are assumed. These assumptions are then relaxed to demonstrate how the model can be used to analyze the effects of changes in transportation costs, tariffs, subsidies, and other governmental policies on international trade. Chapter 12 summarizes some of the ways in which governments attempt to restrict and/or encourage international trade of particular commodities to accomplish alternative economic and political objectives. Recent changes in international trade policies associated with the Uruguay round of GATT negotiations, the establishment of the World Trade Organization, NAFTA, and other trade agreements are discussed in the final section of the chapter.

CHAPTER

SPATIAL CHARACTERISTICS OF MARKETS

This chapter discusses certain spatial aspects of agricultural markets that provide a foundation for understanding interregional and international trade, which will be considered in more detail in the following three chapters.

The major points of the chapter are:

1. Why the value of commodities varies among locations.
2. The factors influencing geographical movement of agricultural products.
3. The effects of changes in transportation technologies and governmental transportation policies on agricultural markets.
4. How transport costs influence the location of processing activities.
5. How producer and consumer **price surfaces** can be used for analyzing geographical price differences and potential competition among alternative market sites.

INTRODUCTION

Every market transaction involves an explicit or implicit agreement about when and where the transfer of ownership rights or provision of a service will occur. Many times an immediate transfer of ownership occurs at the time a price is established and/or payment is made. In these situations the relevant location for determining the value of the commodity is where the commodity is at the moment the transaction occurs. In other instances, there may be an interval of time between when an agreement is reached and the actual transfer of ownership rights and/or final payment. Sometimes relocation of a commodity may be required before the transfer of ownership rights are completed. It is important to know the exact location at which the ownership rights of a commodity are going to be transferred as part of any transaction because the value of commodities often varies among locations. If a transaction requires a change in location of an item before ownership rights are going to be transferred, it is important to know who is going to be responsible for transportation costs when agreeing upon a price for the commodity.

The first part of the chapter discusses the role of transportation as one of the value-adding activities linking production with consumption. Attention is then turned to considering how changes in transportation technologies and governmental policy have affected the cost of moving products and the evolution of marketing activities. In particular, the trade-off between increased costs of assembling raw products and the economies of scale in processing is noted. The final half of the chapter introduces the concepts of producer and consumer price surfaces and illustrates how they can be used for analyzing geographical price relationships and competition among markets.

GEOGRAPHICAL MOVEMENT OF AGRICULTURAL PRODUCTS

Considerable relocation of agricultural products occurs because of the distance between where most agricultural commodities (and other goods) are produced and where ultimate consumption occurs. The movement of agricultural commodities within (and among) countries definitely has increased over time. One indicator of this trend is the historical transformation from the **self-sufficiency** of individual household food and fiber production and consumption to more market-oriented economic behavior.

The geographical movement of agricultural and food products is a value-adding activity in the sense that items are moved to locations where their values are higher than where they originate, even though all other physical characteristics of an item may remain the same. The added value from relocating items is certainly true in an **ex ante** (expected) sense in that it would not be rational to consider moving an item unless the anticipated difference in value was sufficient to cover the costs of transportation. This should not be interpreted however to imply that all **ex post** (or realized) changes in value associated with moving products are always enough to cover transportation costs. Realized changes in value may differ from expected changes because individuals may have incorrect perceptions about values at alternative locations or market conditions may change while products are in transit. The difference in prices at two locations might be enough to suggest that movement of products would be profitable, but unless the delivered price is guaranteed or locked-in at the time the decision to move the product is made, the realized return could be less than the cost of moving the product because of a change in market conditions. On the other hand, the realized return could turn out to be greater than transportation costs if an unforeseen development results in a higher price than what originally was anticipated.

The potential ability to exchange or trade excess output with others has been an important factor that caused households, historically, to move away from producing all their own food and other items for household needs. In other words, the existence and viability of dependable markets is critical since people make decisions about how to use resources not only to survive, but to maximize their satisfaction. The potential gains from trade depend on the closeness of potential trading partners with different comparative advantages in production and the cost of transporting products. The ability to take advantage of the principle of comparative advantage by specialization in production depends on the availability of potential outlets for exchanging goods and services with others. Instead of individuals or collections of individuals (i.e., market, region, or for that matter entire economies) attempting to be totally self-sufficient in all goods and services, it may be advantageous to move products to different locations to exchange for goods or services that others have a comparative advantage in production.

Transportation of agricultural products requires several kinds of economic resources. Transportation costs associated with linking production to consumption is one of the components of the marketing margin for products. The number of people and other resources employed in providing transportation services for agricultural products comprises an important part of the agribusiness sector.

Effects of Alternative Modes of Transportation

Over time as new modes of transporting goods have developed and relative costs of transportation have declined, there has been a continuing increase in specialization of production and larger quantities of goods transported between production locations and points of ultimate consumption. Even farmers in modern economies depend on market channels to provide much of the food consumed in their own households rather than producing everything they eat. This behavior reflects gains from specialization and economies of scale that offset the extra costs of transportation of products. Even though it might not be possible to identify the exact production site of all food items you consumed yesterday, just thinking about likely locations of production and/or processing will probably suggest numerous geographical sources and a variety of transportation channels that were involved in getting food products into your hands.

Changes in the availability of different modes of transportation have affected the number and size of potential markets available to individuals and groups of households over time. Potential markets provide outlets for excess production and sources of desired goods and services. It is extremely difficult to imagine what the marketing of agricultural products was like in the United States or other parts of the world before motor vehicles (trucks as well as cars), airplanes, or even railroads existed. The kinds of transportation currently available for relocating agricultural products are a far cry from transporting goods on foot or by live horsepower. Perhaps the best illustration of the impact of new transportation possibilities on agricultural markets is to observe situations in certain parts of the world today where very little transportation infrastructure exists.

Similarly, it is difficult today to imagine earlier difficulties that U.S. consumers had in getting food supplies to their residences without the convenience of automobiles. It is little wonder that food retailing initially involved many small businesses located within reasonably short distances from consumers. The ease of being able to move products to markets and acquire other goods and services definitely affects the scope of household production and consumption decisions.

As recently as 160 years ago the extent of geographical movement of agricultural and other products in the United States and other parts of the world was basically limited by the availability of transportation possibilities provided by foot (human or animal) or water. It is not surprising that some of the earliest and continuing major centers of commerce in the United States and the world developed because of accessibility to water transportation. As technological improvements increased the efficiency of water transportation over time through better vessels and different forms of power, the extent of markets reflecting larger aggregates of potential demanders and supplier interests in the exchange process developed. This led to the emergence of a special set of entrepreneurs or merchants who were willing to purchase products for shipment and resale in other locations. In this way, the value-adding service of connecting producers in one location with consumers located elsewhere was provided as long as the expected difference in price at two locations was sufficient to cover the costs of transportation and other costs of doing business.

Water transportation continues to be a very low cost means of moving products. The development of steamboats in the early part of the nineteenth century opened up new realms of geographical trading possibilities via water transportation. Steam technology was also instrumental in the development of railroads, which created new opportunities for linking "landlocked" production areas producing bulky or lower valued commodities with consumption centers. One result of the introduction of railroads was that animal production moved farther away from urban centers. Improved transportation facilities affected where animals could be raised and slaughtered because of not having to drive them on hoof to market. The development of railroads meant that live animals could be shipped longer distances at lower costs. Improved transportation facilities also meant that salt and other preservatives could be shipped to production locations and cured products could be transported farther. Prior to the development of refrigeration, fresh meat was only available if animals were slaughtered near retail outlets. The development of refrigerated railcars altered the economics of transporting and processing livestock and other perishable food products.

Railroads and water-based transportation facilities continue to be used for moving large quantities of agricultural and food products. However, the flexibility and adaptability of motor vehicles for transporting products over land have resulted in trucks accounting for an increasing share of transportation services despite higher costs (Brown 1998). For example, between 1978 and 1987 railroads accounted for 48% of all U.S. grain shipments, while trucks moved 30%. Between 1991 and 1995, however, trucks accounted for 41% of all grain shipments, while the volume moved by train decreased to 39%. The share of U.S. grain shipments accounted for by barges also decreased somewhat (*Feedstuffs* 1998).

Governmental Transportation Policy

Governmental decisions and policies play a key role in determining the type of transportation infrastructure in any economy. Governments clearly play a key role in terms of public investment, subsidization (and/or taxation), and regulating various aspects of transportation activities. For example, President Eisenhower's decision to establish a system of interstate highways in the United States is similar to earlier public decisions to build the Erie Canal, help establish a national railroad system, open up the St. Lawrence Seaway, build sea and airport facilities, and so on. These kinds of public investments influence the kind of transportation infrastructure available for moving goods and services from alternative locations of production through different marketing channels.

With the development of new transportation possibilities that reduce the cost of moving products, increases in distance between producers and consumers can occur. On the other hand, increases in fuel, other operating costs, or taxes on transportation facilities can have the opposite effect. Historical improvements in transportation facilities and reduced costs of transportation provide some insight into the U.S. pattern of settlement as agricultural production tended to move away from the initial urban centers of the original colonies on the East Coast.

A considerable amount of government involvement has also occurred over the years in regulating transportation charges in the United States. Some of the original impetus for governmental involvement and control of transportation rates was to prevent railroads from exercising their market power in the absence of effective competition. This same philosophy was extended with similar kinds of regulation on the commercial trucking industry as it developed. Beginning in the 1980s, several steps were taken to reduce governmental regulation of the transportation industry in the United States. This resulted in many transportation charges fluctuating with market conditions. Thus the price of moving a product at a particular time depends on the relative demand for transportation service as well as the willingness of alternative providers to offer service, just like the prices of many other goods and services in a competitive market situation. There are still many governmental transportation regulations regarding licensing, safety and health measures, and quality of service. For example, the Surface Transportation Board replaced the Interstate Commerce Commission in 1996 as the federal agency responsible for overseeing rail mergers (Brown 1998). The size of motor vehicles and weight of loads are often set by individual states. In many cases agricultural commodities have been exempt from certain transportation regulations. This is especially true in the case of vehicles owned by producers hauling their own agricultural products.

LOCATION OF PROCESSING FACILITIES

Initial processing of most agricultural products usually results in a reduction of weight and consequently transporting less products through subsequent market channels. Examples are the reduction in weight from slaughtering animals or the removal of water in converting fluid milk into processed dairy products.[1] Similarly, processing of fruits and vegetables also eliminates some weight that does not have to be shipped through marketing channels. From the standpoint of reducing the amount of product weight to be transported and thereby reducing total transportation costs, it would appear desirable to have processing facilities located as close as possible to the site of production assuming the quality of the product can be maintained during the remaining time it is in the marketing channels.[2]

It is usually not economically feasible, however, to process each commodity exactly where it is produced because the density and volume of production at individual locations is often insufficient to take advantage of economies of scale in processing. Consequently, a certain initial amount of transporting raw products may be necessary to assemble a sufficient volume at one location before processing becomes economical. Thus the economic trade-off in locating processing facilities is between the increased costs of assembling a sufficient quantity of the raw product and lower processing costs per unit associated with a larger volume. At some point, however, it may not be economical to increase the volume per processing plant because of increased transport costs associated with assembling the raw product. In other words the extra cost of transporting a given amount of product to fewer processing locations may not offset the reduction in processing costs given the existing distribution and density of production. As the total amount and distribution of production change over time, the optimum number, size, and location of processing facilities may also change.

PRICE SURFACES

The effect of transportation costs on the values of commodities shipped to a given market or purchased at a single site can be illustrated using the concept of a price surface. Initially this topic will be discussed from the perspective of producers, or what is involved in assembling a product to an initial buying or processing station. Later the ideas will be extended to what consumers face at the retail level.

Producer Price Surfaces

Differences in the location of producers along a given transportation route at varying distances from a central market imply that the net price realized per unit of a homogeneous product decreases with the distance the product must be shipped to market. The net price is defined as the amount of money realized after subtracting the cost of transportation per unit of the commodity from the market price.[3] This assumes that all buyers at a central market pay the same price for homogeneous commodities regardless of where they are produced.[4]

The above concept is illustrated in Figure 9–1, which assumes that initial assembling of a commodity occurs at a market located at Point A. Alternative production sites at varying distances from the central market are represented by other points on the horizontal axis of Figure 9–1. The value of a commodity for any location is plotted on the vertical axis. The value at various locations along the horizontal axis can be determined based on a known value at Point A and the transportation cost per unit of commodity per unit of distance. For example, if the value of the commodity is $10 per unit at Point A and it costs $.40 per mile to transport the item to the market, the item's value at a distance of 10 miles from the market would be $6 (or $10 2($.40 times 10 miles)). The slopes of the two linear lines that connect at Point A reflect how fast the value of a commodity diminishes with distance from the market because of the cost of transporting it from various points of production. The steepness or flatness of the price surfaces emanating from Point A are determined by the magnitude of transportation costs per unit of the commodity. If there is a technological improvement or something else that reduces the cost of transportation services, the effects could be illustrated by a change in the slope of the price surface to reflect a less rapid reduction in values with increasing distance from the market. A change in market conditions that resulted in a change in the value of the commodity at Point A would be illustrated by vertically shifting the entire price surface up or down by the appropriate amount.

Price surfaces can be more complicated than what is illustrated in Figure 9–1 because transportation costs per unit of product may not necessarily be related to distance in a linear manner. This could occur especially in the case where different modes of transportation might be used depending on how far the product is moved. For example, it might not be feasible to use the same mode of transportation all the way from where it is produced to the initial buying station. The relevant price surface reflects the cost of the cheapest mode of transport at each point. Another modification in Figure 9–1 that might be necessary to reflect real-world conditions would be to include a discontinuity at

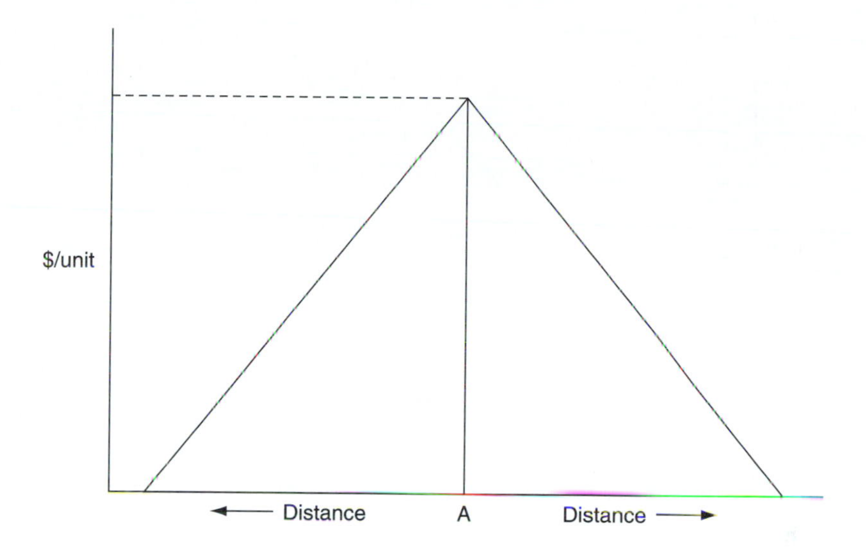

FIGURE 9–1 Price surface for producers located at varying distances from Market A.

Point A to represent loading and/or unloading costs or other fixed costs of transferring ownership.

The depiction of net values of a commodity at various distances from a market indicates that it would not be economical to produce a commodity at some distance from Point A because transport costs could exhaust the entire market value of the commodity. That point would occur where the price surface intersects the horizontal axis. Thus, the maximum potential distance from which products might be received at Point A can be determined assuming a given price at Point A and specified transportation cost possibilities. It is likely that the potential drawing power of a given market would be less than the distance at which net value becomes zero because of several reasons. First, at some point prior to the maximum distance, producers would likely decide it was not profitable to produce the commodity especially if they were going to realize a very low price because other products competing for use of the same resources might yield higher net values. Second, even if a product was already produced it might have more value for home consumption than taking it to a market if the market price was very low. Finally, other alternative locations might be better outlets for selling the product as will be illustrated.

Although Figure 9–1 illustrates the nature of price surfaces for one dimension in space, the concept can be readily extended to two-dimensional space. What is involved in developing a two-dimensional price surface is like spinning a top with the triangular shape illustrated in Figure 9–1 around Point A. Assuming that the same price surface is applicable for all directions emanating from

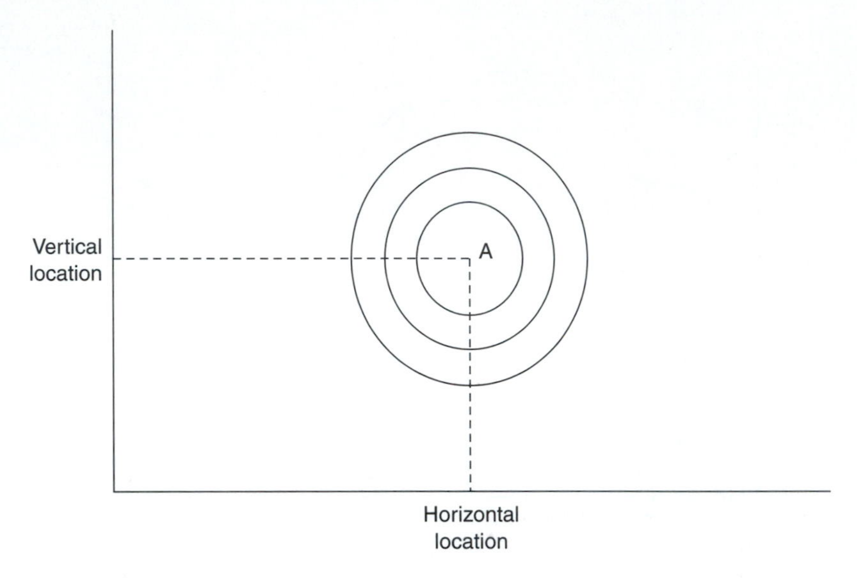

FIGURE 9–2 Isoprice lines for producers at varying locations relative to Market A.

Point A leads to a depiction of **isoprice** (or isovalue) **lines** around Point A as illustrated in Figure 9–2. Each of the isoprice lines in Figure 9–2 indicates the set of locations with the same net values of products sold at a fixed price at Point A. Taking into account geographical irregularities and the transportation possibilities of the real world, the actual isoprice lines might not be as symmetrical as those illustrated in Figure 9–2. In any case, these ideas are fundamental concepts for thinking about how prices vary with location and analyzing forces that influence the movement of products from where they are produced to where they are consumed.

Consumer Price Surfaces

The concepts illustrated previously are equally applicable for describing price surfaces relevant to consumers located at various places relative to a purchasing or distribution point. The cost of moving a product from where it is purchased to where it is ultimately used must be added to the price of a product purchased at a given market to reflect the total cost of products where consumers reside. In Figure 9–3, the slopes of the price surfaces relevant to consumers located at varying distances from a market at Point A reflect the extra costs of transporting a product to their residences. By rotating these price surfaces around Point A, isoprice lines of increasing amounts with increasing distances from Point A would be generated.

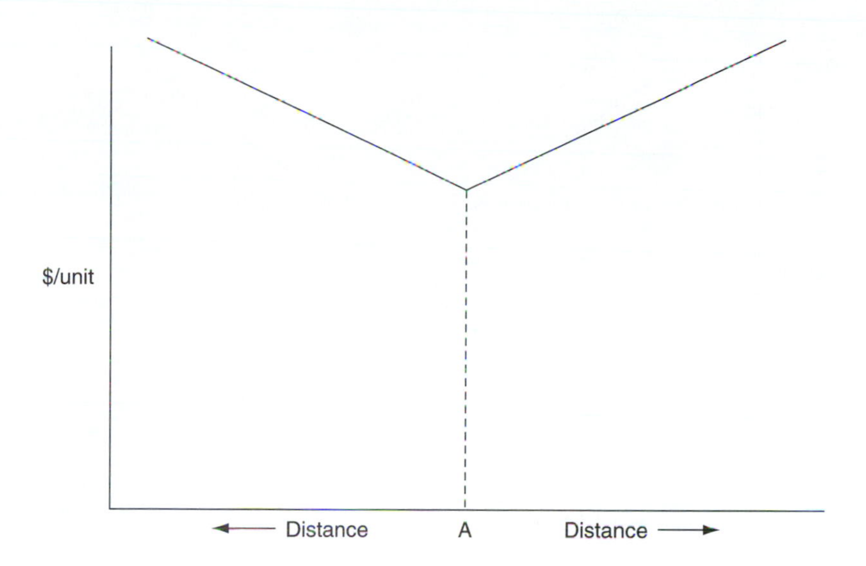

FIGURE 9–3 Price surface for consumers located at varying distances from Market A.

The concepts underlying isoprice lines and surfaces for assembling and distributing producer and consumer goods are quite similar, but result in different shapes and implications about how prices vary with distance from a given market. It is important to remember that the isoprice surfaces and lines for producer and consumer goods usually pertain to entirely different commodities in that other value-adding activities often occur between the time goods are assembled from producers and made available for consumers to purchase. If all other value-adding activities are assumed to take place at Point A, price surfaces for the initial and final retail product could be illustrated in one dimensional space as shown in Figure 9–4.[5]

Alternative Markets

The above concepts associated with isoprice lines and surfaces for a single market can be extended easily to illustrate how different values of a product and costs of transportation influence the size of market areas. For example, assume an alternative market located at Point B is introduced into the analysis. The price in Market B could be the same or different from that in Market A ignoring for the moment factors that determine the actual price at the two locations. Also transportation costs of getting products to Market B could be different from the costs of getting products to Market A.[6] Relevant price surfaces for each market in one-dimensional space could look like Figure 9–5 depending on appropriate assumptions about price and transport costs for each market.

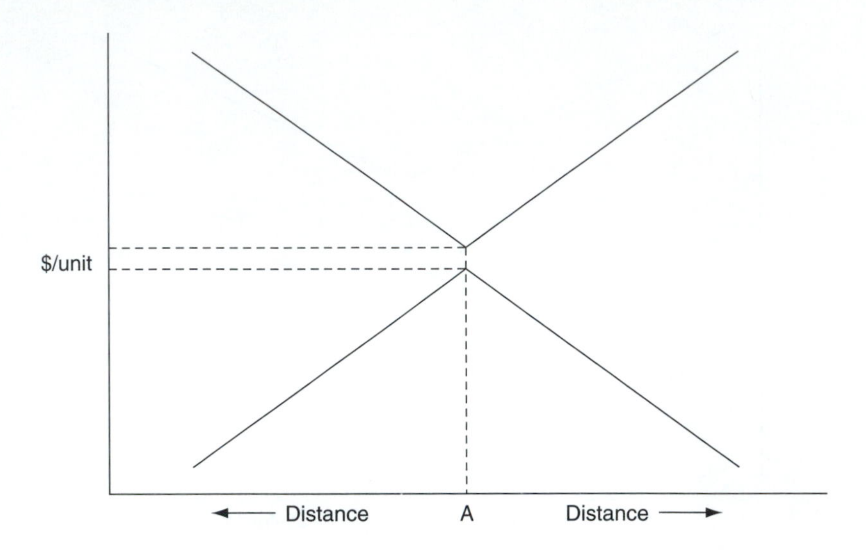

FIGURE 9–4 Price surfaces for assembly and distribution of commodity around Market A.

The point at which the price surface for Market A intersects the price surface for Market B in Figure 9–5 indicates where producers would be indifferent as to which market they shipped their product because they would realize the same net value. All producers to the right of the point of intersection of the two price surfaces would be better off shipping to Market B than Market A because they would realize a higher net return. Similarly all producers to the left of this point of intersection would find it advantageous to ship their product to Market A.

The location of the intersection of the two price surfaces could be altered by higher (or lower) transportation charges in one or both markets (i.e., differential changes in transportation infrastructure) or by changing the price in each market by different amounts. Although a distinct boundary between the two markets is illustrated in Figure 9–5, it is possible that transportation costs could be sufficiently high that each market's price surface intersected the horizontal axis before intersecting the price surface for the other market. In fact, there could be some intermediate area between the two markets where it was not feasible to produce and/or ship the commodity to either market in view of existing prices at the two markets and/or the costs of moving products to market. In this case each market for the commodity might be considered to be isolated from the competitive forces in the other market. This scenario is especially interesting for considering how a reduction in transportation costs makes production profitable in new areas farther from market centers and eventually leads to competitive forces from alternative market outlets. This kind of situation illustrates how agricultural production has evolved over time from self-sufficiency to depending on

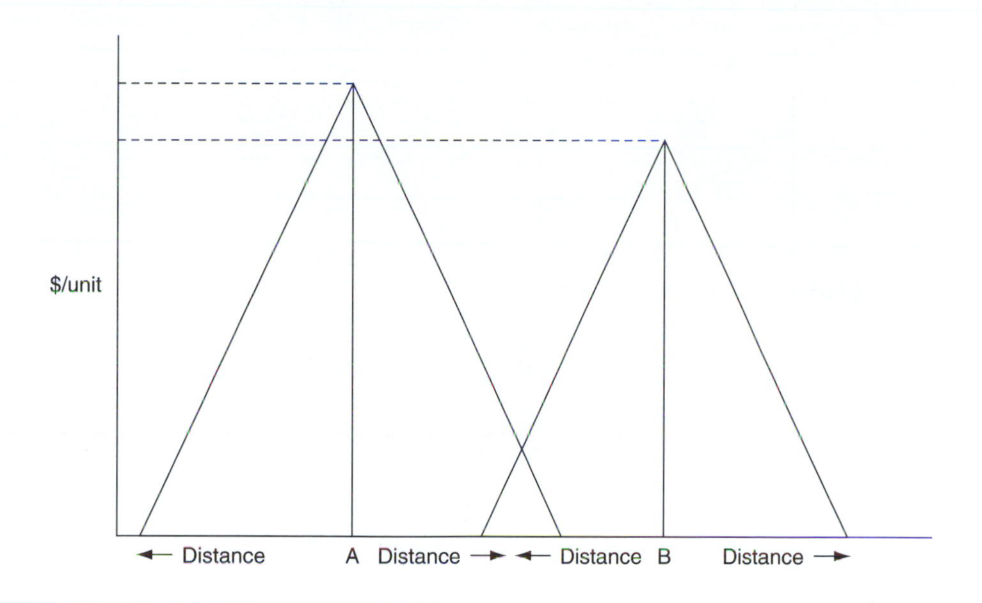

FIGURE 9–5 Price surfaces for producers located at varying distances from Markets A and B.

local markets and later on interregional and international trade. Price surfaces provide a framework for considering some of the historical evolution of agricultural markets.

The possibility of different prices at alternative markets illustrates why some markets disappear over time. To illustrate this point, a third market located at Point C is introduced in Figure 9–6. If something caused the price or transportation costs for Market A to change holding everything else constant, it is conceivable that the price surface for Market A could increase by a sufficient amount to entirely dominate that of Market B and move the competitive boundary to the intersection of the price surface of Market A with that of Market C. This would effectively eliminate the viability of Market B. The only way for Market B to continue to operate would be for it to increase its price to compete with the other markets.

Similar concepts to those just discussed can also be used to illustrate the evolution of retail markets for consumer goods. For example, as transportation costs have become lower for consumers, the size of retail market areas has expanded. An increase in a retail market area enables additional economies of scale in retailing to be realized, thereby lowering the entire price surface for a given location and starting the competitive process between markets all over again. A similar type of diagram as in Figure 9–6 can be used to gain insight about the historical evolution of retail food outlets over time.

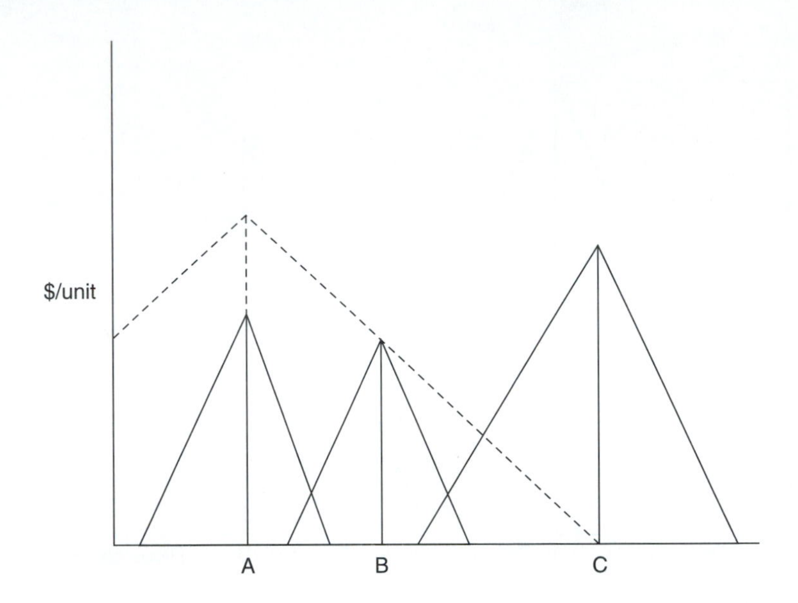

FIGURE 9–6 Price surfaces for producers located at varying distances from Markets A, B, and C and effects of a change in value at Market A.

SUMMARY

This chapter indicates why prices of identical items can vary depending on where they happen to be located at a given moment of time. This fact is especially important when considering the place where ownership of something is going to be transferred. Transporting items is a value-adding activity in that it would not be done unless the expected increase in value was sufficient to pay for the cost of transportation.

Potential gains from exchanging excess output of one good or service for other items depends on the cost of moving products and the relative closeness of potential trading partners with different comparative advantages. Reductions in transportation costs over time have permitted increased specialization of production as well as greater quantities of products moving longer distances. Transportation opportunities have expanded markedly over time as new modes of transporting products such as railroads, motor vehicles, and airplanes were introduced. However water transportation is still widely used when feasible because of its relatively low cost.

Governmental policies play an important role in influencing the quality of an economy's transportation infrastructure. Public investment decisions about transportation facilities, taxation and/or subsidization of transportation activities, and various types of governmental rules and regulations can affect the ease or difficulty of moving items within an economy or between countries. During the last 20 years there has been much less governmental regulation of transportation charges in the United States relative to earlier years. Consequently, transportation charges are more variable and are able to respond more quickly to supply and demand conditions.

Transportation costs play a key role in determining the location of processing activities. The closer processing can occur to where agricultural commodities are produced usually means a savings in terms of moving less weight. On the other hand, the economies of scale of processing usually require a sufficient volume of raw products be assembled before it is economical to process them.

Producer and consumer price surfaces illustrate how the value of commodities varies with distance to a market or central place where transfers of ownership occur. The fundamental principle that identical products at a given location have the same value means that products produced farther from a given market will have less value because of the additional costs of transporting products. Similarly, the actual or total cost of products purchased at a given location will increase with the distance consumers live from the market because of additional transportation costs. Isoprice or isocost lines in two-dimensional space are an alternative way of graphically indicating locations where values of an item are identical.

Intersections of price surfaces can be used to identify potential boundaries between alternative markets. This idea can be used to indicate how producers and/or consumers have incentives to change where they sell or buy items as differential changes occur in transport costs around alternative markets. A reduction in transport costs permit products to move greater distances and influence the amount of buying and selling that occurs at particular locations. Reductions in transport costs tend to expand the potential volume of transactions at fewer locations that have a comparative advantage for conducting transfers of ownership of goods, assuming a fixed geographical distribution of producers and consumers.

QUESTIONS

1. Briefly discuss why governmental regulation of transportation rates for agricultural products was initiated originally.
2. Suppose that the vertical dimension in the following diagram represents prices being offered for a commodity at two alternative locations (A and B) as well as the effect of transportation costs on net prices available to producers located at varying distances from each market.
   a. Graphically show how an improvement in transportation facilities surrounding Market B that reduces transportation costs would be expected to alter the net price relationships.
   b. What does your answer to part a imply about the volume of products shipped to Market A assuming no change in production patterns or transportation facilities in Market A?

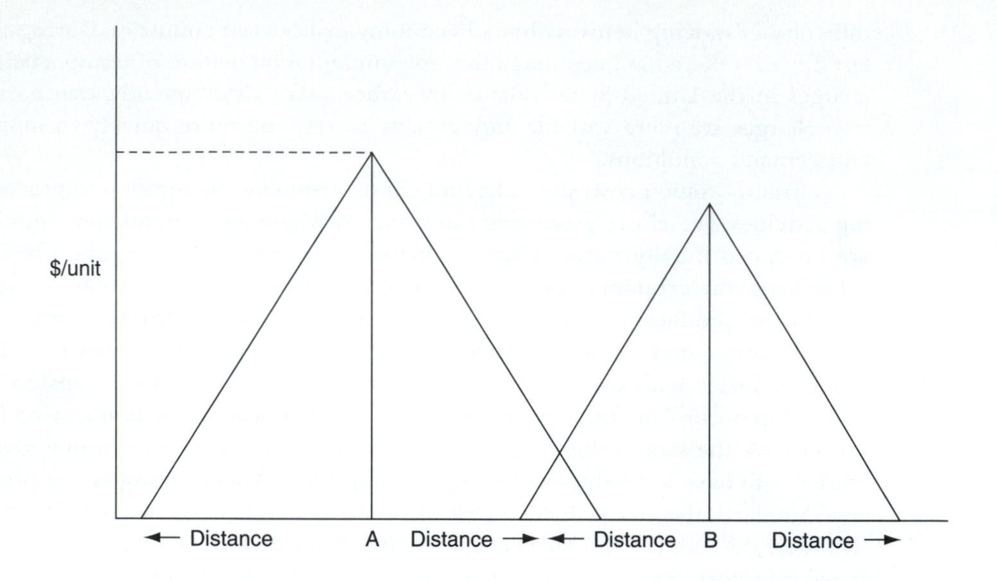

3. Suppose that the vertical axis in the following diagram represents the net cost of a homogeneous product that can be obtained from two alternative market sources, A or B, faced by consumers at various locations.
   a. Graphically show how you would modify the diagram to illustrate the effects of a governmental action that reduced the costs of transportation around Market A, but had no effect around Market B.

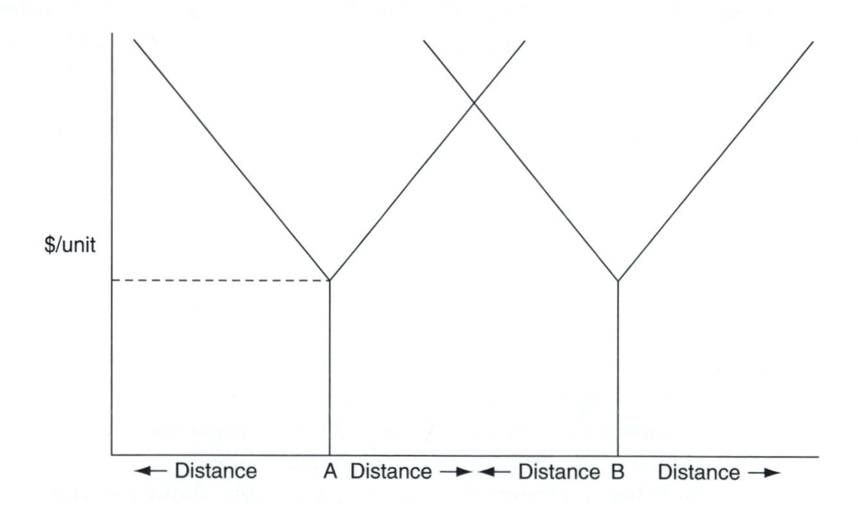

b. Based on your answer to part a, briefly interpret how the change would likely affect merchants in Market B.

c. What actions might merchants in Market B consider taking to increase their number of customers?

REFERENCES

Brown, D. M. 1998. Rail freight consolidation and rural America. *Rural Development Perspectives* 13(2): 19–23. USDA, Economic Research Service.

Feedstuffs. 1998. Trucks now nation's top grain haulers, 70(25): 23. June 22, Minnetonko, MN. p. 23.

Food Processing, 1997. If its Borden's it's got to be good, 58 (7): Chicago, IL: Putnam Publishing Company, p. 16.

NOTES

1. In 1852, Gail Borden discovered a way of producing condensed milk by removing water under low temperature and adding sugar as a preservative, which made it possible to distribute milk greater distances from farms (*Food Processing,* 1997).

2. Many kinds of processing actually extend storability enabling a product to be shipped greater distances than if the product remained in an unprocessed form.

3. Even if the producers do not transport the product to the market themselves, the most anyone would be willing to pay at a given location for a commodity that they plan to resell would reflect the necessary costs of moving it to the central market.

4. If a product has unique identifiable characteristics because of where it is produced, it can be considered as a separate product. Even in this situation, however, potential value at a point of production would still depend on the cost of transporting it to a market.

5. There is no easy way to simultaneously illustrate di fferent values for consumers and producers using isoprice lines in two-dimensional space. It is best to use producer or consumer isoprice lines separately in two-dimensional space depending on whether consumer or producer goods are being considered.

6. Transport costs could differ simply because of differences in availability of alternative modes of transportation.

CHAPTER

10

LOCAL MARKETS AND INTERREGIONAL TRADE

This chapter concentrates on economic factors that affect the movement of products between regions of an economy.

The major points of the chapter are:

1. The effects of interregional shipments of products on market prices.

2. The use of **excess demand** and **excess supply** concepts to analyze the underlying determinants of interregional trade.

3. The differential effects of interregional trade on producers and consumers of a particular commodity.

4. The use of a two-region trade model to analyze the effects of changes in transportation costs and regional demand and supply conditions.

INTRODUCTION

Many simplifying assumptions were involved in developing concepts about price surfaces and isoprice lines for individual markets in the previous chapter. For example, it was assumed that the price did not vary with quantity of the product being assembled or distributed at a given site. This assumption facilitated showing how transportation costs and distance from a particular site influence spatial variation in the value of a product. A fixed market price also facilitated illustrating how the volume of business at a specific assembly or distribution point could be influenced by changes in transportation costs affecting the number of producers and/or consumers attracted to a particular site.

Assuming that the price does not vary with quantity in a given market would be appropriate only if all demands at assembly points or the supply of products at all points of distribution were perfectly elastic. These characterizations could be appropriate if the price in a given market was determined entirely by national or world market conditions. A fixed price would also be applicable in the case of governmental actions to keep prices fixed regardless of market conditions.[1] In the absence of governmental intervention, however, prices of homogeneous commodities seldom are identical among all markets at a given point in time.

The major purpose of this chapter is to consider how price differences among markets create incentives for interregional movement of products and how interregional product movements affect geographical price patterns. The first part of the chapter discusses basic underlying geographical differences in product values that create incentives for interregional shipments and the simultaneous effect of interregional shipments on geographical price variation. The second part of the chapter presents a model using excess demand and supply concepts for analyzing trade between two regions of an economy. Graphical and algebraic examples are used to illustrate the operation of market-clearing forces and how interregional trade affects producers and consumers in each region. The final part of this chapter indicates how the analytical model can be used to illustrate the effects of changes in transfer costs and differential changes in regional demand and supply conditions.

INTERREGIONAL TRADE

Whenever the word *trade* is encountered, the first thing that individuals often think about these days is some kind of international connotation. This is not surprising in view of the increased importance of the international flow of commodities and increased awareness of the growing connectiveness of world markets. Actually the flow of commodities or services among countries is simply a geographical extension of the movements of products between producers and consumers within an economy. In this chapter, attention will be concentrated on the spatial movement of products within an economy linking production and consumption activities.[2] Chapters 11 and 12 will consider various aspects of international trade.

The extent of interregional (however broadly or narrowly a region is defined) movement or trading of commodities reflects the effects of different values that commodities have, depending on where they happen to be located (or desired), and the relative costs of transportation. The extent of variation in local market prices depends on the intermarket movement of commodities as well as how producers (or sellers) and consumers (or buyers) respond to prices at various locations. Under a given set of national and world conditions, the price of a given commodity at different locations is likely to be affected by the amount of product transferred between markets. At the same time the economic feasibility of moving products between markets depends on the cost of transportation and intermarket price differences. Consequently, it is necessary to consider how the volume of business at a given market site varies with price and how price differences among markets are affected by the amount of interregional trade.

Intermarket transport costs may differ from assembly or distribution costs that were considered in the previous chapter for a couple of reasons. First, different modes of transportation may exist between market centers that are not available for the assembly or distribution of products to or from a given market. For example, water, rail, or better land transportation facilities may exist between market centers; whereas more costly or less efficient modes of transportation may be all that individual producers and consumers have available to get products to or from particular initial assembly and distribution points. Second, transportation costs per unit of product per unit of distance may vary with the total quantity being shipped. The cost per unit might be considerably lower if there are economies of scale associated with larger shipments between market centers than is possible with usual quantities involved in assembly or distribution activities.

An Analytical Model

To develop a framework for considering interregional trade and price differences among markets, it is easiest to begin by considering two "isolated" markets. The isolated terminology implies that the two markets are far enough apart geographically, or local transportation costs are sufficiently high, that the markets do not compete as far as acquiring raw products from producers or distributing products to consumers at the retail level.[3] Under these circumstances, conventional demand and supply concepts can be used to represent aggregate interest in buying and selling a commodity in each market. These kinds of relationships are represented in the left- and right-hand panels of Figure 10–1. The relationships represent the amount of interest in buying and selling a commodity at alternative prices at a given location independent of what happens in the other market.

The use of a single demand and supply schedule for each market assumes that the relationships pertain to a particular form of the product at some stage of the marketing process. The model is equally applicable for a raw agricultural

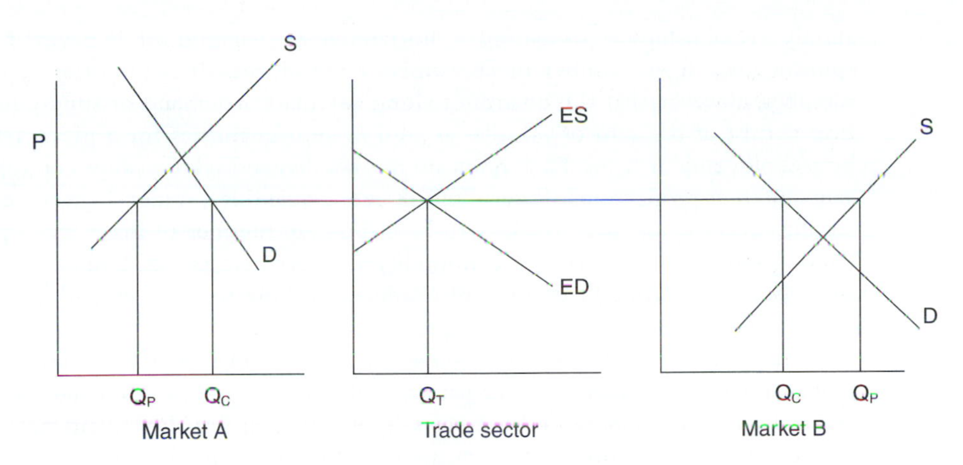

FIGURE 10–1 Two-region trade model.

product or a retail food product as long as the necessary adjustments are embodied in the demand and supply relationships to represent buying and selling interest at the appropriate level of the marketing system at different locations. In other words, the model is applicable for interregional trade of a given form of a product and does not incorporate the transformation process that also occurs to most agricultural products as they move through the marketing system.

Thinking of the supply and demand relationships as reflecting producer and consumer behavior is convenient regardless of what form of product is being considered at any particular level of the marketing system. In some cases the relevant demand function might be a primary demand relationship or a derived demand schedule in other cases. Similarly, the supply function might be a primary or derived supply relationship depending on what kind of product or what level in the marketing system interregional relationships are being considered.

Aggregation Issues

For purposes of considering demand and supply relationships for a given market site and at a given point in time, the total number of potential consumers (buyers) and producers (sellers) is assumed to be fixed. Thus the demand and supply relationships for each market involve the same type of aggregation of individual demand and supply responses considered in Chapters 4 and 5. As total population or the number of potential producers changes in a given region over time, the demand and supply functions for a given market would be expected to shift.

The actual number of consumers and/or producers actually participating in a particular market at a given point in time would be expected to vary with price. A changing number of market participants at different prices might at first glance seem to be in conflict with earlier discussions about market demand and supply relationships representing a horizontal aggregation of behavioral responses for a fixed number of consumers or producers. It is important to remember, however, that the quantities along any market demand or supply function consist of the sum of positive as well as zero quantities for a given set of households and/or firms. Each point on market demand relationships can represent different proportions of zero and nonzero quantities, under a given set of circumstances. Thus there is likely to be a different number of households purchasing different amounts of a commodity at alternative prices. Similarly, there is likely to be a different number of producers willing to produce and offer an item for sale at different prices in a particular market.[4]

In the case of isolated markets, changes in the number of active market participants associated with different prices could be interpreted as changes in the proportion of the population that are self-sufficient or not dependent on local markets. If the total number of consumers or producers in a given area changes over time, local market demand and supply curves would be expected to increase or decrease similar to the effects considered in Chapters 4 and 5. These kinds of changes would be represented by a shift in the quantities demanded or supplied at all prices.

Market-Clearing Forces

The left-hand and right-hand panels of Figure 10–1 represent the demand and supply forces that would determine market-clearing prices and quantities in the absence of any movement of product between markets. The implications of the model are as if the participants in each market were not aware of what was happening in the other market or even the existence of another market. Under these circumstances, the market-clearing prices in the two markets would not necessarily be expected to be identical. Prices in the two markets could differ because of differences in available technology, resource productivity, tastes and preferences, consumer incomes, and so on.

Once someone realized that more than one market exists for the same homogeneous product they might be surprised to discover that prices were not necessarily identical. If the difference in market-clearing prices between markets was small, there would be no incentive for anyone to ever consider wanting to move any of the product from one market to the other. This would certainly be the case if the difference in prices was less than the cost of transportation required to move the product.

If the difference in the market-clearing prices between two markets was large enough, however, an incentive to transfer some of the product from the market with the lowest price to the other market would exist if intermarket movement of the product was feasible. An incentive to transfer some of the product would continue to exist as long as the difference in prices in the two markets was greater than the per unit cost of relocating some of the product.

As some of the product was removed from the market with the lower price, the price in that market would tend to increase. Similarly, as more of the product entered the higher priced market, the price in that market would tend to decrease because of an increase in product availability. This process of quantity and price adjustments would be expected to continue until the resulting difference in price was equal to or just enough to pay for the cost of transportation.

Graphical Analysis

The way in which the preceding price and quantity adjustment process would operate can be graphically illustrated using the concepts of excess demand and excess supply. An excess demand schedule represents the schedule of quantities that consumers (or buyers) would like to purchase over and above the quantity that would be supplied (in the absence of any interregional trade) in a given market, at alternative prices. This concept can be illustrated initially by considering the left-hand panel of Figure 10–1, which indicates that Market A would have a higher price than Market B in the absence of any interregional movement of the product. At all prices below P_A in the left-hand panel of Figure 10–1, there is interest in purchasing and consuming a greater quantity than what would be available in Market A. The difference between the quantity demanded and supplied at each of the prices below the market-clearing price that would occur in the absence of trade determines the excess demand schedule. The excess demand schedule obviously is relevant only for prices less than what the market-clearing price would be in the absence of any interregional trade. In essence, the excess demand relationship represents a demand for importing a product at a lower price from another market.

Similarly, an excess supply schedule can be defined for Market B. An excess supply function represents the schedule of differences between the quantities producers (or sellers) would be willing to provide and the quantities consumers (or buyers) are willing to purchase at alternative prices. In the case of excess supply functions, the relevant range of prices is above the market-clearing level P_B that would exist in the absence of any intermarket movement of the product.[5]

The usefulness of excess demand and supply concepts to illustrate how much of a commodity would be traded or moved between two markets is illustrated by the middle panel of Figure 10–1. Initially it is assumed that there is no cost involved in transporting the product between the two markets. This assumption is obviously unrealistic, but can be modified as soon as the basic idea of how excess demand and supply functions determine the amount of interregional trade and intermarket price relationships is understood.

The middle panel of Figure 10–1 combines the excess demand and supply relationships from Markets A and B. For all prices lower than P_A there is a schedule of quantities that consumers (or buyers) in Market A would be willing to purchase over and above what is available in that market. Thus the excess demand function for Market A intersects the vertical axis of the middle panel at the same point as the market-clearing price in the absence of any interregional trade in Market A.

Similarly, for all prices above P_B there is a schedule of quantities that producers (or sellers) in Market B are willing to sell over and above what would be purchased or used in that region in the absence of any interregional trade. Thus the excess supply function for Market B intersects the vertical axis of the middle panel at the same price that would exist in Market B in the absence of any interregional trade.

The intersection of the excess demand and excess supply relationships is interpreted as representing market-clearing forces like intersections of usual demand and supply relationships. That is, the intersection of the excess demand and supply relationships represents the point where the market-clearing forces of excess demand and excess supply are equivalent. This means that the intersection in the middle panel of Figure 10–1 can be interpreted as indicating how much interregional movement of the product between the two markets would be required to equate prices in the two regions, in the absence of any transportation or **transaction costs.**

The intersection of the excess demand and supply functions implies that increasing quantities of interregional shipments of an item would occur until prices became the same in each market. This kind of equalization process is analogous to what happens to the flow of fluids between containers that are connected in some manner. For example, imagine two fish tanks or bowls with different depths of water on a table. Compare that picture to what would occur if the two containers were connected in some way that permitted the flow of water between the containers. In the latter situation, the water levels would obviously be the same because air pressure would cause water in one container to be transferred to the other until the water levels were equal.

In the case of interregional transfer of commodities, the economic incentive to equilibrate prices in the absence of transportation costs is accomplished by movement of products from lower priced markets to higher priced ones. The profit motive or incentive to take advantage of price differences between markets is like the unseen force of air pressure that causes water levels to equilibrate in containers that are connected.

Algebraic Representation

Algebraic expressions for the excess demand and supply functions represented in the middle panel of Figure 10–1 can be used to determine a numerical value for the amount of product moved between markets. The appropriate algebraic representations of the excess demand and supply functions can be easily obtained if the underlying demand and supply structure for each market is known. For example, assume the demand and supply functions for Market A are the following:

$$\text{Eq. 10.1} \quad \text{Demand: } Q = 100 - 4P$$
$$\text{Eq. 10.2} \quad \text{Supply: } Q = 20 + 4P$$

The above functions can be used to solve for the market-clearing price and quantity in the absence of any interregional trade. Setting the demand equation equal to the supply equation and solving for price and quantity would indicate a market-clearing price of 10. That is,

$$\begin{aligned} \text{Demand} &= \text{Supply} \\ 100 - 4P &= 20 + 4P \\ \text{or } 80 &= 8P \\ \text{or } 10 &= P \end{aligned}$$

Putting a value of 10 for P in either the demand or supply relationship indicates a value of 60 for the market-clearing quantity for this market.

The representation of the excess demand relationship for this market at all prices less than 10 can be obtained by simply subtracting the supply equation from the demand equation. That is, the excess demand for Market A is:

$$\text{Eq. 10.3} \quad ED_A = \text{Demand} - \text{Supply} = 100 - 4P - (20 + 4P) = 80 - 8P$$

Equation 10.3 indicates that there would be no excess demand at a price equal to 10. At any price less than 10, however, there would be a positive amount of excess demand in that the difference between the quantity demanded and the quantity supplied would be a nonzero amount.[6] For example, at a price of 4 the excess demand is 80 2 8(4)5 48. This value is consistent with the difference between 84 units being demanded and the 36 units that would be supplied at a price of 4 according to the underlying assumed demand and supply relationships.

Similarly, the excess supply function for Market B can be developed based on information about the underlying demand and supply structure in that market. For example, assume the following expressions represent the demand and supply conditions for Market B.

$$\text{Eq. 10.4} \quad \text{Demand: } Q = 46 - 2P$$
$$\text{Eq. 10.5} \quad \text{Supply: } Q = 30 + 2P$$

These relationships indicate that the market-clearing price and quantity in this market would be 4 and 38, respectively, in the absence of any interregional transfer of products between the two markets. Thus for any price above 4 there would be a greater quantity available than demanded. The actual expression for the **excess supply** could be obtained by subtracting the quantity demanded from the quantity supplied at each price above 4. Algebraically this would be equivalent to the following:

$$\text{Eq. 10.6} \quad \text{Excess supply} = \text{Supply} - \text{demand}$$

or

$$\begin{aligned} ES_B &= 30 + 2P - (46 - 2P) \\ ES_B &= -16 + 4P \end{aligned}$$

Again, Equation 10.6 indicates that at a price of 4 there is no excess supply in Market B, but at every price greater than 4, a larger quantity would be supplied than demanded.

Setting the equations for the excess demand and supply relationships equal to each other and solving for price is equivalent to finding the price for which excess demand and supply intersect in the middle panel of Figure 10–1. That is,

$$\text{Excess demand} = \text{Excess supply}$$
$$80 - 8P = -16 + 4P$$

or

$$96 = 12P$$

or

$$8 = P$$

Substituting a value of 8 for P in either the excess demand or supply equation indicates that the quantity of excess demand and supply in the trade sector is 16, or the market-clearing quantity. Analytically the intersection of the above excess demand and supply functions indicates that at a price of 8 there would be an excess demand of 16 units in Market A and exactly the same quantity of excess supply in Market B.

Effects of Trade on Producers and Consumers

Comparing the difference in prices in the two markets with and without trade indicates that trade would clearly lower the price in Market A and increase it in Market B. A change in price affects producers and consumers in each market as indicated by the underlying demand and supply schedules. Clearly, producers in Market A would reduce the quantity of production in response to the lower price resulting from interregional trade. On the other hand, consumers in Market A would be pleased to be facing a lower price and would increase their purchases of this product. Thus as far as Market A is concerned, producers of the product being considered would be adversely affected by trade and consumers would be better off.

The effects of trade in terms of Market B would benefit producers and make consumers worse off. These results occur because the price in Market B would tend to be higher with trade relative to what it would be in the absence of trade.

The different effects of trade on various groups within each market as well as between markets illustrates why particular individuals are likely to feel differently about whether the movement of a specific product between markets is favorable or unfavorable. The way trade affects a given individual not only depends on which region the individual is in but also whether their interest in a given product is predominantly from a consumer or producer perspective. Of course, everyone is a consumer of many products as well as a producer of something. Consequently, an individual's economic interest in the interregional movement of a given product could vary depending on which product is being considered.

Obviously, producers (or sellers) like higher prices whereas consumers (or buyers) prefer lower prices.

As prices in different markets adjust in response to interregional movement of all kinds of products, many adjustments in production and consumption activities would be expected to occur. Although the above analysis is in terms of a single product, the implications of adjustments for other products must also be considered. It is clear that the possibility of trade between markets results in some of the production of an item being transferred from a higher cost area to a lower cost area. Resources that are no longer needed for the production of an item in the higher cost area would be available for producing something else in which that market or region had a comparative advantage. This results in an overall net economic benefit in terms of lower costs of production through more specialization in production and why the unfettered movement of products among locations in an economy can be viewed as a value-adding component of the marketing system. Even though some individuals may be made worse off as a result of interregional trade, the aggregate benefits are greater than the costs. It is analogous to an increase in the size of a pie, although everyone's slice may not be larger relative to what they had before the possibility of trade was considered.

NONZERO TRANSPORTATION COSTS

The above graphical and algebraic illustrations of how excess demand and supply relationships can be used to determine the amount of product moving between markets assumed zero costs of transportation or essentially a frictionless world. The analysis can be modified to handle nonzero transportation costs once it is clear how excess demand and supply concepts represent economic forces affecting intermarket movement of products. Transportation costs are introduced by modifying the excess demand and supply relationships to reflect an additional cost component that separates markets. The separation of markets in a geographical sense is similar to the kind of separation between farm and retail markets considered in Chapter 6. In the latter case the marketing margin or the total cost of moving a product between the farm and retail levels in the marketing system was used as the linking mechanism. A similar situation exists in terms of considering the costs of transportation as an intervening cost element connecting Markets A and B.

The amount of transportation cost can be incorporated to ensure that intersections of excess demand and supply relationships are economically meaningful using the same type of reasoning introduced in Chapter 6. This means that the cost of transporting a unit of product should be subtracted vertically from each point on the excess demand schedule of Market A to obtain a realistic representation of the "effective" extra demand from Market A in terms of Market B's prices. In essence this is like obtaining a "derived excess demand" schedule from Market A that would be meaningful in terms of considering the excess supply conditions in Market B. This means that the appropriate offer prices for different quantities of **excess demand** from Market A would be adjusted downward

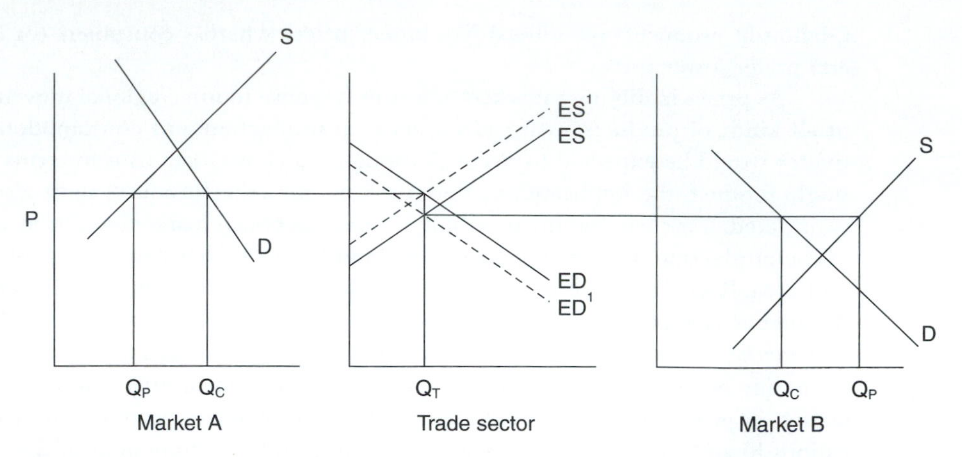

FIGURE 10–2 Two-region trade model with transportation costs.

by the amount that it costs to transport the item between the two markets. Assuming that the transportation cost per unit of product between the two markets does not change with the quantity being transported, the derived excess demand function is illustrated as a parallel line below the original excess demand relation in the middle panel of Figure 10–2. The vertical distance between the two excess demand curves represents the cost of transporting a unit of the product between the two markets.

Similarly, by vertically adding the fixed amount of transportation cost per unit of product to each point on the excess supply for Market B, a new or "derived excess supply" function would be obtained. The derived excess supply represents the schedule of extra quantities from Market B that would be available to Market A in terms of Market A's prices. This is indicated by the parallel line vertically above the original excess supply schedule in the middle panel of Figure 10–2. The vertical distance between the two excess supply curves is exactly the same as the vertical distance between the two excess demand curves because it represents the same cost of moving a product between the two markets.

After the derived excess demand and supply schedules are incorporated into Figure 10–2 to reflect the cost of transportation, the resulting diagram is very similar to what was presented in Chapter 6. The intersection of the derived excess demand curve for Market A with the excess supply function for Market B indicates the amount of product that would have to be moved between the two markets consistent with market-clearing forces in both markets. This intersection occurs at the same quantity as where the derived excess supply for Market B intersects the excess demand for Market A. The prices associated with the two intersections would be the appropriate market-clearing prices in each market that differ by the amount of transportation cost. This is similar to the results in Chapter 6 in which primary and derived demand and supply curves were used to illustrate the determination of retail and farm prices that differed by the magnitude of the marketing margin.

Introducing transportation costs into the analysis produces a general expression for price differences among markets. It is that the absolute difference in prices in two markets must be less than or equal to the cost of transportation, that is,

$$\text{Eq. 10.7} \quad |P_A - P_B| \leq T_{AB}$$

where T_{AB} represents the transportation costs of moving a product between the two markets. Considering the absolute value of the price difference handles the situation of moving a product from Market A to Market B or vice versa depending on which market has the highest price.

The intuitive meaning of the equilibrium condition on price differences is that there is no incentive to move a product between markets if the price difference is less than the costs of transportation. This means that if the inequality in the above relationship holds, one can be confident that there is no interregional trade. If an economic incentive to transfer any of the product between markets did exist, enough movement of the product would occur until prices in each market reached a point where any additional transfer of product would require further adjustments in price such that the resulting difference in price between the two markets would not be sufficient to cover transportation costs. Furthermore, as long as interregional movement of a product is not prohibited, price differences greater than the cost of transportation will never be observed except in the very short run with insufficient time for markets to adjust to a new situation. In fact, if at any time it appears that there is a greater difference in prices than the costs of transportation, it may be wise to recheck how the prices were measured or the way in which transportation costs have been calculated to make sure there is not a mistake or something preventing market forces from operating.

Even though the above relationship about intermarket price differences is derived from the simple framework of a two-market model, the result is generalizable to multiple markets in that the same condition would be expected to be valid for any pair of markets. Another way in which the above framework is more generalizable is to consider one of the two markets as an aggregation of all markets other than the one of special interest. In this way, the concepts become more applicable to focus on a given market's relationship to the rest of the world rather than just considering two isolated markets.

USES OF THE FRAMEWORK

Once the basic concepts of the above framework are understood, it can be used to analyze a variety of issues. For example, comparing the implications of the model with nonzero transportation costs to what is implied in the absence of any transportation costs indicates how transportation costs affect the resulting prices and amount of trade between the two markets. The amount of trade indicated in Figure 10–2 is clearly a smaller amount than what would occur in the case of no costs of transportation. This result is consistent with the generalization that the amount of an activity (e.g., movement of a product) would be expected to be negatively related to its cost (e.g., transportation costs). Also the price in Market A is higher and the price in Market B is lower after introducing transportation costs relative to the case when transportation costs were ignored. The effect of transportation costs on the prices in each market would depend on the slopes of the excess demand and supply relationships and consequently the changes in each market's price would not necessarily be the same magnitude.

Analyzing the effects of increases or decreases in transportation costs could be accommodated by increasing or decreasing the vertical adjustments to initial excess demand and supply relationships. The effect of alternative costs of transportation on the amount of trade and prices in both markets would be analyzed in the same manner as how changes in the marketing margin affect equilibrium quantity and prices at the farm and retail level.

A change in any cost of transferring a product between two markets could be analyzed in exactly the same way as a change in transportation costs. For example, if a tax was placed on a product leaving a given location or coming into another market the economic effects would be just like an increase in transportation costs regardless of where the tax was imposed. On the other hand, if the shipment of goods was subsidized, the effects would be just like a reduction in transportation costs.

Finally, it is possible to use the framework to analyze the effects of changes in underlying demand or supply behavior in either market. For example, if the supply of a product should increase in Market B because of new technology, it would result in a change in that market's excess supply. This in turn would result in a different amount of product moving between the two markets and a different price in each market. The resulting set of prices, however, would still not differ by more than the costs of transporting a unit of the product between the two markets.

One of the advantages of the simple two-region trade model is that it provides a systematic way of identifying the implications of shifts in demand or supply relationships or transfer costs between regions. The model can be used for many different scenarios just like the basic model of linking farm and retail markets. It can be used to consider the effects of a change in a single element as well as the effects of more than one change occurring simultaneously.[7]

SUMMARY

This chapter has examined the relationship between interregional shipments of a homogeneous product and geographical variation in its price. A two-region trading model using excess demand and supply concepts was developed to illustrate how market-clearing forces determine the amount of interregional movement of a commodity. An excess demand function represents the difference between the quantity of a product demanded and supplied at each and every price less than what would occur if the region was completely isolated from other markets. Similarly, an excess supply relationship represents the schedule of quantities that would be produced in excess of what would be demanded at each and every price above the market-clearing level in the absence of any interregional trade.

Excess demand and supply schedules represent potential demand and supply of goods for movement between regions. The intersection of an excess demand schedule for one region with another region's excess supply determines the amount of interregional movement of a product that would equilibrate prices in the two regions if there were no transportation costs. Interregional trade tends to increase price in the region where shipments originate and to lower the price in the region to which shipments are made relative to what would occur in the absence of trade. Once prices in the two regions reach the same level, there is no incentive to geographically transfer any additional amount of a commodity.

Changes in prices associated with interregional trade clearly have differential effects on producers and consumers of a commodity depending on their location. Interregional trade benefits producers in the region where the commodity is shipped out, but adversely affects producers in an importing region relative to what would exist in the absence of any trade. Consumers in an importing region obviously benefit from having a greater quantity of product at a lower price to consume compared to what would occur in the absence of trade. On the other hand, consumers in an exporting region face a higher price and consume less than they would in the absence of trade in a given product. These implications pertain to production and consumption of a single product. The feasibility of interregional trade, however, permits greater regional specialization in production and more efficient utilization of resources as producers and consumers respond to market opportunities.

The two-region model as initially presented ignored transportation costs to simplify exposition and to highlight the incentives for interregional transfer of commodities. Interregional transportation costs can be incorporated into the two-region trade model similar to the way derived demand and supply relationships were handled in Chapter 6. Introducing transportation costs into the model means that interregional shipments occur until the difference in prices in the two regions becomes equal to the cost of transportation. Obviously, there is no incentive for interregional trade to occur if the difference in a commodity's price between two isolated regions is less than the cost of transporting a unit of the commodity.

The two-region trade model can be used to illustrate how changes in transportation costs between the two regions affect the amount of interregional trade and geographical price variation. The model is also useful for analyzing how changes in demand and/or supply conditions in any particular region affect the amount of interregional trade of a commodity.

QUESTIONS

1. Assume that the D and S lines in the following diagrams represent demand and supply relationships for a particular commodity in two regions of an economy in the absence of any interregional trade. Assuming zero transportation costs and no barriers to a flow of commodities between the two regions, graphically show the amount of consumption (Q_c) and production (Q_p) that will occur in each region with trade.

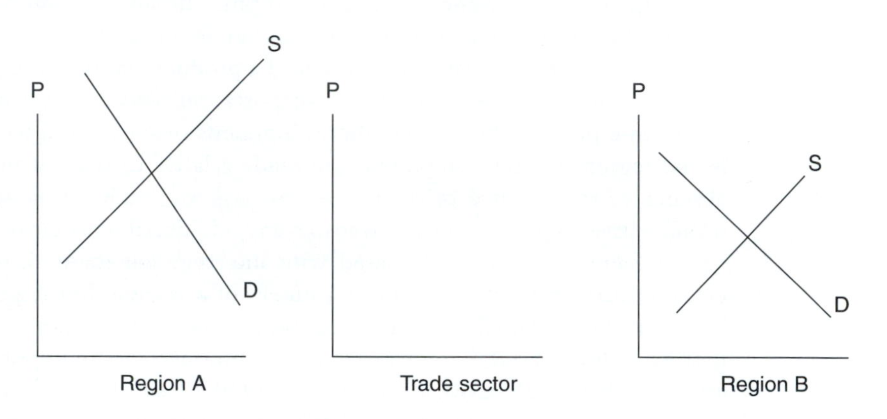

2. Assume that the following equations represent market demand and supply of a commodity at the retail level.

$$\text{Demand: } Q = 45 - 5P$$
$$\text{Supply: } Q = 25 + 3P$$

where Q and P represent quantity and price per unit, respectively, for this product.

a. What is the algebraic expression that would represent the excess supply schedule for this market?
b. What is the range of prices for which there is a positive excess supply in this market?

3. Assume that the following equations represent the demand and supply of a particular commodity for a given region of a country.

$$\text{Demand: } Q = 100 - 4P$$
$$\text{Supply: } Q = 20 + 4P$$

where Q = quantity and P = price per unit.

a. If this were an isolated region with no possibility of trade with any other region, how much of this commodity would be produced and consumed?
b. What is the algebraic expression that represents the excess demand for this region?

c. How much would be produced and consumed in Region A, if another region of the country was willing to offer unlimited quantities of the commodity to Region A for $6.00/unit?

d. How does the possibility of trade with another region affect producers and consumers of this commodity in Region A?

4. Assume that Region A has demand and supply conditions for milk that can be represented by the following two equations:

$$\text{Demand: } Q = 100 - 4P$$
$$\text{Supply: } Q = 10 + 2P$$

where Q and P represent quantity and price, respectively.

Assume that it is feasible for Region A to trade with Region B that has an excess supply equation of milk represented as follows:

$$Q = 20 + 4P$$

a. What will be the quantity of milk traded between the two regions ignoring the costs of transporting the product between the two regions?

b. Describe how interregional trade will affect milk producers in each of the two regions compared to their situation in the absence of interregional trade.

5. Over the last few weeks, an increase in federal fuel tax and other factors have resulted in an increase in the cost of transporting products in the United States. Briefly explain what changes you would expect these developments to have on regional differences in prices of agricultural commodities and the quantity of interregional shipments.

6. If you observe interregional shipments of veal from the northeastern part of the United States to the South and interregional shipments of pork from the South to the Northeast, what can you conclude about the difference in regional prices of each of these commodities? Explain your reasoning.

NOTES

1. This situation has occurred numerous times in the case of agricultural and food commodities by governments using various policy instruments to insulate consumers and/or producers from the effects of changes in market conditions.

2. This topic is closely related to a field of study known as logistics, which is applicable at the individual firm as well as at a broader aggregate level. Logistics in brief relates to the body of knowledge, techniques, and processes associated with acquiring and ensuring the appropriate flow of inputs to production sites and the distribution of output to potential customers; essentially, linking production with consumption activities.

3. If the markets were not isolated it would be necessary to consider a multiplicity of demand and supply relationships in each market depending on the

other market's price. The results would be similar to what is described later in this chapter, but the process involved in reaching an equilibrium solution for both markets would not be as straightforward because it would be necessary to take into account how initial demand and supply curves shift in response to price changes in other markets.

4. The attraction of additional consumers or producers from a greater distance to a given site in response to different prices is one reason that the relative price responsiveness (as measured by price elasticities) of demand and supply behavior may be relatively large for small markets.

5. Actually each market has both an excess demand and an excess supply function. The functions exist for different ranges of prices relative to the initial market-clearing price. It is possible to use a single concept of excess demand concepts over the entire range of prices by treating positive and negative quantities as exports and imports, respectively. For purposes of considering intermarket trade possibilities, however, it is easier to consider the excess demand function for the market with the higher initial price and the excess supply function for the market with the lower initial price.

6. For any price greater than 10, there would be a negative quantity of excess demand or a positive excess supply.

7. In cases of multiple changes, it may not be possible to determine directional effects on trade flows and/or prices without additional information about the magnitudes of the various shifts or slopes of demand and supply relationships.

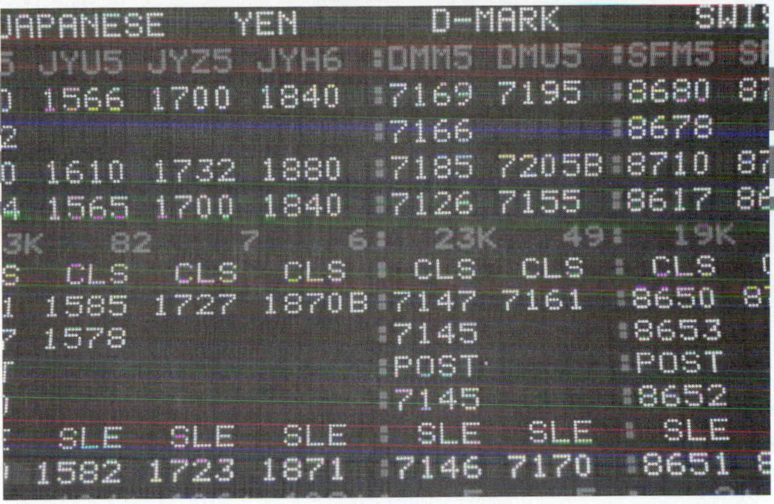

CHAPTER 11

INTERNATIONAL TRADE

This chapter discusses some recent trends in the international trade of U.S. agricultural and food products and some of the things that make international transactions a little more complex than interregional trade, which was discussed in the previous chapter.

The major points of the chapter are:

1. The important characteristics of international trade of U.S. agricultural and food products in recent years.
2. The difference between **complementary** and **competitive imports**.
3. The complexities of international transactions because of language and cultural differences, specialized transportation services, and multiple governmental regulations.
4. The dependence of international transactions on two kinds of prices: the per unit cost of commodities (or services) and **exchange rates** between currencies.

INTRODUCTION

Many of the concepts applicable to interregional trade discussed in the previous chapter are equally useful for analyzing trade between participants in different countries. As noted earlier, international trade is basically a geographical extension of the same activities associated with interregional trade, but some additional complexities exist. A brief review of some recent changes and trends in international trade of U.S. agricultural products will be presented before getting into some of the complexities associated with international trade. Chapter 12 will indicate how the interregional trade model presented in Chapter 10 can be adapted to analyze the effects of variations in exchange rates and other economic variables on international trade.

TRENDS IN INTERNATIONAL TRADE OF U.S. AGRICULTURAL PRODUCTS

Agricultural products have been an important component of international trade of the U.S. economy for many years. Tobacco and cotton were among some of the earliest major U.S. exports. Today, the variety of agricultural products that the United States exports as well as imports is much greater than ever.

Over the last few decades, U.S. exports of agricultural products have accounted for 20% to 30% of total U.S. farm output (Harris and Benson 1996). U.S. exports of agricultural products, however, have fluctuated over the last three decades (Table 11–1 and Figure 11–1). During the 1970s, the total value of U.S. exports of agricultural products increased very rapidly, but then decreased between 1981 and 1986. Between the mid-1980s and mid-1990s U.S. exports of agricultural products tended to increase, reaching a peak of almost $60 billion in 1996.[1] The value of U.S. exports of agricultural products began to decrease after 1996 and dropped by more than 16% by 1999 in response to a decrease in world demand associated with financial difficulties in several countries and increased competition from other exporting countries (Westcott and Landes 1999). The total value of U.S. agricultural exports is expected to increase by approximately 2% between 1999 and 2000 (Whitten 1999).

The prosperity of the agricultural sector in the United States has been affected by changes in the total value of agricultural exports. For example, some of the growth in total agricultural income during the 1970s, the late 1980s, and early 1990s was associated with an expanding export market. Similarly, some of the financial crisis in U.S. agriculture in the early and mid-1980s as well as during the latter part of the 1990s can be attributed to decreases in the value of exports.

Historically, U.S. export markets have been relatively more important for grains than livestock products, however increasing amounts of the latter items are now being sold abroad. The United States is the world's major supplier of wheat, corn, and soybeans. These commodities along with rice, feed grains, cotton, and

TABLE 11-1

VALUE OF U.S. AGRICULTURAL EXPORTS, IMPORTS, AND TRADE BALANCE, 1970–1999

YEAR	EXPORTS	IMPORTS (MILLION $)	TRADE BALANCE
1970	6,958	5,686	1,272
1971	7,955	6,128	1,827
1972	8,242	5,936	2,306
1973	14,984	7,737	7,247
1974	21,559	10,031	11,528
1975	21,817	9,435	12,382
1976	22,742	10,491	12,251
1977	23,974	13,361	10,613
1978	27,289	13,886	13,403
1979	31,979	16,185	15,793
1980	40,467	17,292	23,176
1981	43,783	17,338	26,445
1982	39,097	15,457	23,640
1983	34,769	16,276	18,493
1984	38,027	18,905	19,122
1985	31,201	19,740	11,461
1986	26,312	20,885	5,427
1987	27,876	20,650	7,226
1988	35,316	21,014	14,301
1989	39,674	21,571	18,103
1990	40,365	22,700	17,705
1991	37,780	22,722	15,064
1992	42,625	24,479	18,158
1993	42,879	24,624	18,255
1994	43,960	26,590	17,369
1995	54,725	29,852	24,873
1996	59,891	32,565	27,326
1997	57,365	35,788	21,577
1998	53,730	37,004	16,727
1999	49,000	37,500	11,500

Sources: 1970–1997 data from http://www.econ.ag.gov/briefing/AgTrade/data.htm. 1988 and preliminary 1999 data obtained from Agricultural Outlook, August 1999, at http://www.econ.ag.gov/epubs/pdf/agout/aug99/ao263j.pdf.

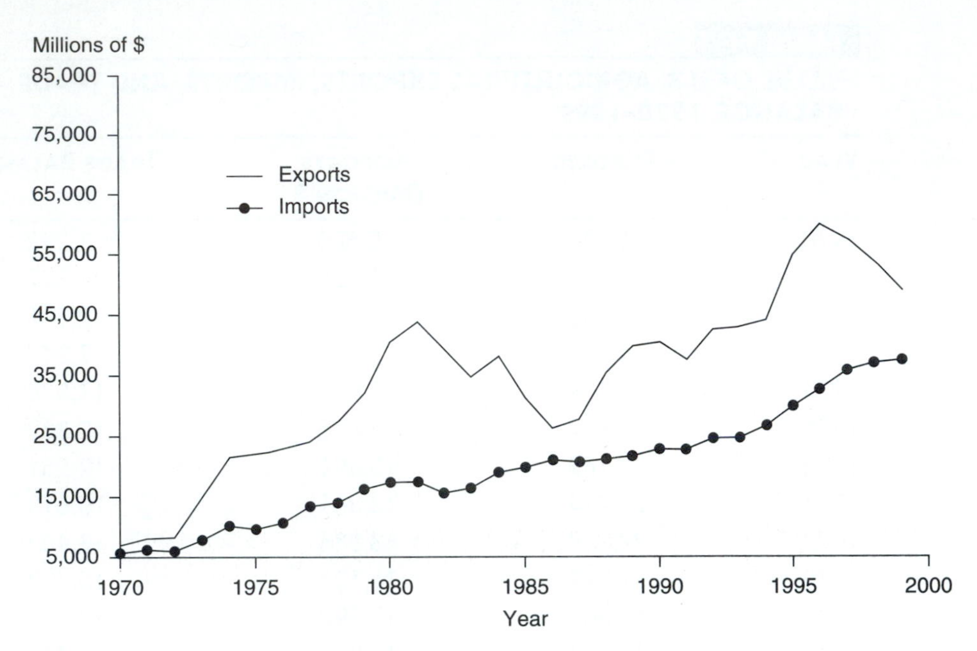

FIGURE 11–1 Value of U.S. agricultural exports and imports, 1970–1999.

tobacco are considered to be bulk commodities. Recently high-value products have clearly accounted for a larger share of total exports of U.S. agricultural products than bulk commodities, but both categories of exports have fluctuated during the last two decades (Figure 11–2). High-value products are defined as semi-processed and processed grains and oilseeds, animal products, horticultural products, sugar, and other tropical products. In essence the total value of high-value agricultural exports is calculated as the difference between the total value of agricultural exports and the export value of bulk commodities.

The three largest U.S. export markets for agricultural products currently are Japan, Canada, and Mexico. In fact, these three countries accounted for over 40% of the total value of U.S. agricultural exports in 1998. The relative importance of sales of U.S. agricultural products to European countries has decreased as exports to Japan and other countries in the Pacific Rim have increased.

The total value of imports of agricultural products into the United States has not fluctuated as much as exports (Figure 11–1). The value of U.S. agricultural imports has been rising steadily and nearly doubled over the last 15 years. Approximately one-third of the total value of agricultural products imported into the United States in 1998 were from Canada and Mexico. Other countries that accounted for more than a billion dollars worth of agricultural products imported into the United States in 1998 include Colombia, Indonesia, Italy, the Netherlands, France, Brazil, and Australia.

The total value of U.S. agricultural exports has consistently been greater than the value of agricultural imports (Table 11–1, and Figure 11–1). This is quite

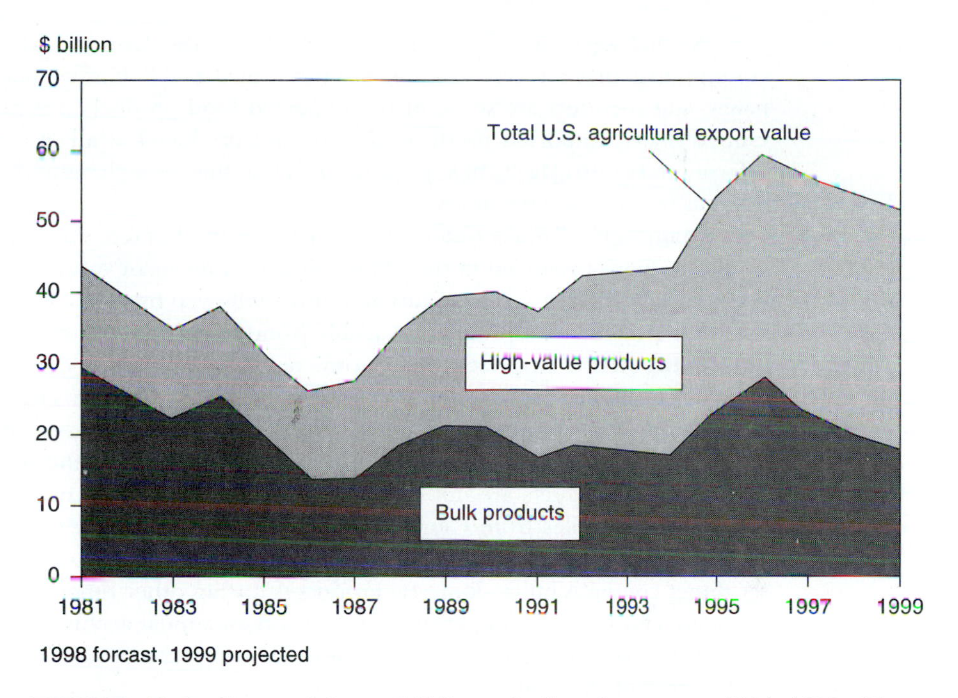

FIGURE 11–2 Composition of U.S. agricultural exports, 1981–1999. (From *Agricultural Outlook,* October 1998, P. 10.)

different from U.S. trade in nonagricultural products for which the value of imports has exceeded the value of exports for some time. The trade surplus from the agricultural sector has helped offset some of the trade deficit associated with other products. In one sense, U.S. exports of grain, poultry, and livestock products provide food for individuals in other countries who devote their time to producing electronic products, textiles, and other items desired by U.S. consumers. All too frequently, international markets are viewed only as potential outlets for increased domestic production of particular products. Trade is a two-way street, however, and one of the benefits of international trade is providing alternative sources of lower cost products. International trade increases the purchasing power of consumers allowing them to enjoy a larger bundle of goods and services. When products are obtained from other countries, it is natural to view imports initially as displacing domestic production and causing an associated decrease in jobs and income. The economic expansionary effects resulting from increased purchasing power of domestic consumers because of lower priced commodities, however, are diverse and difficult for people to observe and appreciate. The additional purchasing power acquired by other countries from international trade can also lead to additional economic expansionary effects associated with increased exports of certain products. The latter type of expansionary effects again may not be as evident to individuals who view imports as a competitive threat to their livelihood.

Several agricultural products imported into the United States consist of commodities that are typically not produced in the U.S. Coffee, tea, spices, bananas, and coconuts are some of the imported food products not grown in the United States. Imported products that are not produced at all in a country are referred to as **complementary imports** in that they complement domestic production to provide consumers a wider variety of choices.

Many agricultural products imported into the United States, however, are similar to foods produced in the United States. In terms of value, the largest category of U.S. food imports is fruits and juices followed by coffee. Imported products that are similar to products available from domestic production are referred to as **competitive imports** even if some or all of the imports occur during seasons when the importing country does not currently produce similar products. For example, the importation of fresh fruits and vegetables from the Southern Hemisphere provides more choices to consumers in the Northern Hemisphere when similar products are not available from local sources of production. Fresh fruits and vegetables from Central and South America do not compete directly with U.S. production during certain months of the year, but the imports may offer direct competition to domestic producers during other times of the year. Imports are classified as competitive products on an annual accounting basis, however, if they are similar to what is available from domestic sources of production at any time of the year.

The distinction between complementary and competitive imports also depends to some extent on how broadly or narrowly product groups are considered. For example, imports of pistachio nuts into the United States might be considered as a complementary import in the context of a fairly narrow product definition, but the same product might be considered as a competitive import under a broader consideration of all nut products.

DIFFERENCES FROM INTERREGIONAL TRADE

Although there are a lot of similarities between interregional and international trade, there are a few differences that make international transactions a little more complex. Some of the major differences consist of (1) language and cultural differences, (2) specialized modes of transportation services, (3) multiple sets of governmental regulations, and (4) monetary exchange rates. Brief comments about each of the first three topics will follow. A more detailed discussion about the role of exchange rates in international trade is included to help set the stage for material in the following chapter.

Language and Cultural Differences

For any kind of market transaction to occur, obviously potential buyers and sellers (or their representatives) must be able to communicate with each other in an effective and timely manner. One potential complication in conducting international business negotiations can be language differences. Unless buyers and sellers are fluent in more than one language, arranging international business transactions may require the assistance of an intermediary who can serve as an interpreter. An interpreter may not only be useful for initial negotiations involving individuals from countries that do not use a common language, but may also be valuable in making sure that there are no misunderstandings of technical terminology pertaining to the terms of a proposed transaction. Individuals who have multilingual skills and international business experience can be instrumental in making sure that all parties understand a proposed agreement. With the increasing globalization of agricultural markets these skills are more valuable than ever.

International transactions may occur by individuals agreeing to meet at a mutually agreed-upon place or may be conducted by some other means of communications. If the negotiations do not occur at a face-to-face meeting, differences in time zones can create a problem in finding a convenient time for conversation between parties located at different latitudes around the globe. This is not as much of a problem for individuals located in different countries having the same latitude, but normal business hours can vary depending on the culture of countries. International East–West communications are just an extension of the kind of complications that people on the East or West Coast of the United States encounter in conducting business with people in different time zones.

In addition to language and time zone differences, cultural and protocol differences influencing the way business is conducted in different countries are very important for the successful completion of international transactions.

Specialized Transportation

Another area in which additional specialization is usually required for international transactions is in the area of transportation services. It is important that appropriate transportation arrangements are available for delivering the quantity and quality of product at desired times. When a prospective price for a transaction is being discussed, it is important that the buyer and seller have a clear understanding as to what each party's responsibility is for making transportation arrangements and bearing some or all of the costs of relocating the commodity. It is also important to specify when ownership of an item is going to be officially transferred to know who will bear the risk of any loss that might occur in transit especially if third parties are involved in providing transportation services.

Some form of water or air transportation is required for all international shipments from or into the United States except for transactions with Canada and some other parts of Central and South America. Even though rail and truck modes of transportation are feasible for moving U.S. products to Canada and Mexico multiple transportation providers, operating under different sets of rules and regulations may be required to get products to their final destination.

The use of water or air transportation for international movement of agricultural products is often combined with some type of rail or truck transportation to get products to and/or from a port or air terminal. Once a product is ready to leave or come into a country, additional inspection and/or certification than if the product was just being sent to another region of the same country may be required. This is analogous to additional documentation and processing required for international travel relative to the ease with which individuals are able to travel within a country.

A set of **incoterms** has been developed and periodically revised by the International Chamber of Commerce to reduce uncertainty about interpreting different terms used by international traders. For example, FAS is an acronym representing Free Alongside Ship, which indicates a seller's obligation to deliver specified items to a carrier designated by the buyer. This is just one of several three- or four-letter combinations that have universal meaning among traders and are similar to specialized jargon in other lines of activity. The 1990 revision of these terms was undertaken in response to increased use of electronic data interchange of documents required for customs clearance or proof of delivery.

Governmental Regulations

International transactions have to conform to rules and regulations pertinent to both the country of origin as well as the country where the product ends up. Some of the philosophical reasons for differences in rules and regulations related to the international movement of products will be discussed in more detail in Chapter 12. At this point, however, it is sufficient to note that various countries have different policies that have a bearing on the amount of international trade.

Countries can restrict or encourage the amount of international trade in a variety of ways. Many countries permit any individual or firm to engage in international trade provided that products conform to established health and safety specifications and other governmental requirements. In some cases, a governmental agency may be designated to handle all exporting or importing of products into a particular country. In these cases all prospective exporters and buyers of products must deal with a designated government agency. In other cases, governments may designate or license a particular set of individuals or firms to be involved in exporting or importing products.

It is important to know what kind of rules and regulations apply to particular international transactions. Keeping up with changes in rules and regulations in two countries may be more than twice as difficult as keeping informed about rules and regulations pertaining to interregional trade within a country. Monitoring the changes in rules and regulations of many countries can be burdensome, and is one of the reasons why some international trading firms specialize in particular commodities or particular parts of the world. Specialized trading firms keep up with rules and regulations affecting a particular class of products as well as maintain contact with potential buyers or sellers in particular countries.

Another aspect of having to deal with multiple sets of governmental rules and regulations is the complications that can arise in the settlement of business disputes. In the case of interregional trade, it is usually quite clear what legal channels of recourse exist within any country for resolving business disputes. It is not always clear what legal remedies are readily available for settling disagreements about international business transactions. Some reciprocity agreements exist between countries, but every country's legal and court system has its limitations as to the applicability to citizens and firms of other countries.[2]

Monetary Systems

Even though significant complexities exist in each of the three areas just discussed, one of the biggest differences between international and interregional trade is dealing with different monetary systems. This means that international transactions generally involve two prices: the price of the physical commodity and the price or value of another currency. Exporting or importing products among countries generally involves at least one extra financial step. The extra complications are similar to what tourists face in making decisions based on prices expressed in different currency units. The way this problem is handled by most tourists is to acquire currency of the country where purchases are going to be made. This is accomplished by exchanging a given amount of currency of one country for a certain amount of the currency of another country before making purchases.

The ratio between the amount of one currency that must be exchanged to obtain one unit of another currency is known as the exchange rate. For example, if a U.S. citizen were planning on traveling to Japan and the current exchange rate was 100 yen per dollar, he could take \$1,000 to a bank or some other financial institution and receive 100,000 yen. He could then put the yen in his pocket and be able to use it to purchase what he wanted when he arrived in Japan and faced prices expressed in yen. Similarly, a Japanese visitor to the U.S. could exchange 100,000 yen for \$1,000 dollars and be able to purchase products in the United States where prices are expressed in dollars.

There are two ways that an exchange rate between two currencies can be expressed at any point in time. In the previous example, 100 yen per dollar implied the same thing as \$.01 per yen. The two ways of expressing an exchange rate depend on which currency is used as the base for comparison. The exchange rate between the currencies of any two countries may represent the number of currency units of Country A that one can get per unit of Country B's currency or vice versa.

Current values of the exchange rates of many of the world's major currencies are reported in the business sections of many newspapers and various sources of financial information on the Internet. For example, foreign exchange rates for selected countries relative to the U.S. dollar on October 5, 1999, are indicated in Table 11–2. Individuals engaged in international trade need to monitor the exchange rate market just as carefully as market conditions for physical commodities because many exchange rates continuously fluctuate.

TABLE 11-2

FOREIGN EXCHANGE RATES FOR SELECTED COUNTRIES AS OF OCTOBER 5, 1999, ON NEW YORK MARKET

	Foreign Currency in Dollars		Dollar in Foreign Currency	
	Current	Previous	Current	Previous
Australia	.6631	.6623	1.5081	1.5099
Great Britain	1.6565	1.6559	.6037	.6039
Canada	.6794	.6804	1.4718	1.4697
Euro	1.07430	1.07350	.9308	.9315
France	.1638	.1637	6.1059	6.1105
Germany	.5493	.5489	1.8206	1.8219
Greece	.003252	.003260	307.55	306.75
Hong Kong	.1287	.1287	7.7680	7.7677
India	.0229	.0230	43.580	43.570
Ireland	1.3637	1.3618	.7333	.7343
Israel	.2344	.2345	4.2665	4.2651
Italy	.000555	.000554	1802.36	1803.70
Japan	.009398	.009414	106.41	106.23
Mexico	.105854	.105630	9.4470	9.4670
New Zealand	.5230	.5223	1.9120	1.9146
Norway	.1300	.1301	7.6950	7.6890
Spain	.006453	.006444	154.96	155.18
Sweden	.1231	.1232	8.1202	8.1172
Switzerland	.6750	.6744	1.4815	1.4829
Taiwan	.0315	.0315	31.79	31.78
Venezuela	.0016	.0016	628.5000	627.0000

Source: News & Observer, October 6, 1999.

If the Japanese–U.S. exchange rate was such that 150 yen were received per dollar instead of 100 yen per dollar, it would indicate that the dollar had appreciated relative to the yen. An alternative way of describing the adjustment in the exchange rate would be to say that the value of the yen had depreciated relative to the dollar. On the other hand, if 50 yen were received in exchange for a dollar it would indicate that the dollar had become less valuable (or depreciated) relative to the yen compared to an exchange rate of 100 yen per dollar. Another way of describing the latter situation would be to say that the yen had appreciated relative to the dollar compared to when the two currencies were trading at 100 yen per dollar.

The exchange rate is just like a price in other transactions in terms of expressing the amount of purchasing power that is required to acquire one unit of something. In the case of exchange rates, it represents the amount of one currency that must be given up to get a unit of another currency. For instance, in the above example, 100 yen would be required to obtain 1 dollar or 1 dollar would have to be given up to obtain 100 yen.

International transactions involve the relative value of one currency in terms of another currency as well as the usual kind of prices representing values of commodities. Prospective international commodity buyers and sellers obviously have an idea about a commodity's current value expressed in their home country's currency. At least one of the parties, however, will need to be aware of the current (or prospective) relevant exchange rate in order for a given price to be meaningful in terms of another currency. For example, a U.S. meat exporter could negotiate an international sale of pork in Japan based on a price quoted either in dollars or yen. As long as the current exchange rate between the currencies of the two countries was known, it would not matter whether the price was stated in dollars or yen per lb. Furthermore, once a satisfactory price was established, it would not really matter if the U.S. firm was paid in yen or dollars, except for the extra transaction cost of converting yen into dollars and possible changes in the exchange rate between the time an agreement is reached and payment is received. Understanding specific details about how payment or transfer of funds is going to be handled as well as all the usual temporal and spatial characteristics of a transaction that have a bearing on price or value of a specific commodity is very important in international trade negotiations.

International financial markets are very specialized and a full treatment of this topic is beyond the scope of this book. In considering the effects of international trade on a single commodity, the exchange rate can be considered as just another price at which the ownership rights of two items are traded. The only difference is that the two items being exchanged are currencies rather than physical items or services that ultimately will be consumed or used by someone.

Individuals, banks, and other firms have interests in holding or using given currencies to accomplish various objectives. These objectives may include domestic or international purchasing as well as holding a particular amount of wealth in particular currencies. For someone who does not travel abroad or directly engage in international transactions there may be no interest in holding anything but one kind of currency required to make transactions in their own country. For international travel or trading, acquiring currencies of other countries may be necessary in order to conduct international transactions. Some individuals hold different currencies simply to speculate on changes in values of different currencies much like other individuals speculate on changing values of other kinds of assets.

In most cases, market forces representing the degree of interest in holding or exchanging different currencies determine the value of one currency relative to another. Sometimes governments try to control or fix exchange rates by being a provider or buyer of whatever quantity of their currency is required to stabilize

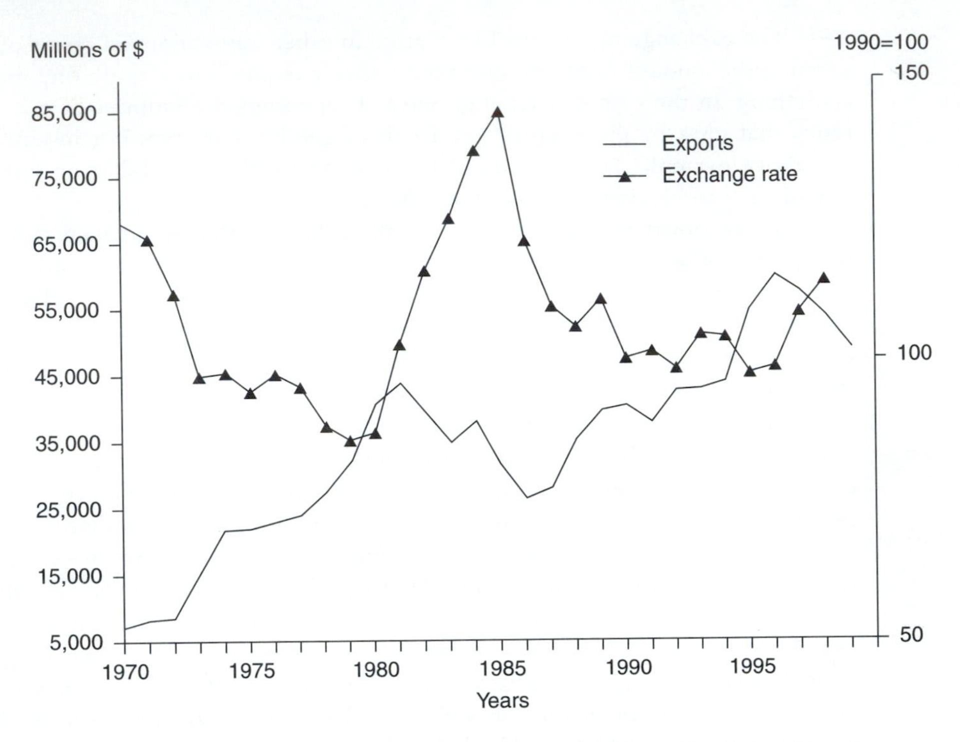

FIGURE 11–3 Value of U.S. agricultural exports and index of weighted exchange rates of U.S. dollar, 1970–1998.

its value at a particular level. A government may announce an official rate of exchange for their currency, but unless the government is willing to actively enter the market to maintain it, market forces may result in the actual exchange rate deviating from an official rate.

At one time, the United States attempted to maintain a value of the dollar tied to the value of gold. This restriction was eliminated several years ago and thereafter the value of the dollar was allowed to float relative to other currencies. The willingness of individuals and financial institutions to exchange dollars for other currencies determines the value of the dollar. This situation is referred to as a case of flexible or floating exchange rates that vary with financial market conditions. In recent years, many governments have allowed the value of their currencies to float rather than trying to maintain a fixed exchange rate.

Changes in the real value of exchange rates in terms of foreign currency per U.S. dollar for 1970–1998 are illustrated in Figure 11–3. The index values are based on exchange rates of the U.S. dollar with the currencies of several countries adjusted for changes in prices.[3] The information indicates that during most of the 1970s the value of the dollar decreased relative to other currencies and then increased during the early 1980s before drifting downward again. The index did not fluctuate much between 1987 and 1996, but moved up a little during 1997 and 1998.

Comparing the pattern of changes in the exchange rate with fluctuations in the value of agricultural exports discussed earlier suggests that some of the cycles in U.S. agricultural exports have coincided with swings in the exchange rate. As the value of the dollar decreased during the 1970s, the value of agricultural exports increased very rapidly. As the dollar gained value in the 1980s, agricultural exports declined for a while, but later began to increase as the value of the dollar decreased. During the last few years, as the value of the dollar decreased, the value of agricultural exports decreased again. The way in which the exchange rate can affect the volume of trade of a given commodity between two countries will be examined more closely in the next chapter.

SUMMARY

International trade of U.S. agricultural products historically has been an important component of agricultural marketing. Fluctuations in the quantity and value of agricultural exports have important implications on agricultural prosperity because exports have accounted for 20% to 30% of total U.S. farm output in recent years. During the 1970s the value of U.S. exports of agricultural products increased very rapidly, but then declined during the early 1980s. Later the total value of agricultural exports again increased reaching a peak of nearly $60 billion in 1996 before declining again. The United States has been a major provider of bulk commodities such as wheat, corn, and soybeans on the world market. Recently, however, an increasing proportion of world trade in agricultural commodities has been in high-value products consisting of semiprocessed and processed grains, oilseeds, animal products, horticultural products, sugar, and other tropical products. Japan, Mexico, and Canada are the three major buyers of U.S. products accounting for more than 40% of the total value of U.S. exports of agricultural products.

The value of agricultural products imported by the United States has been consistently less than the value of exports but has been increasing rather steadily. Imports consist of many agricultural-based products not grown in the United States (complementary imports) as well as some that are similar to those produced domestically (competitive imports). Some of the latter products, however, are imported during certain times of the year when the items are not available from U.S. producers. Canada and Mexico are the two major sources of agricultural products coming into the United States.

International trade can be viewed simply as a geographical extension of interregional trade, but arranging and consummating international transactions often entails some additional complications. These can arise from language and cultural differences, the need for specialized transportation services, complying with multiple governmental regulations, and dealing with different monetary systems. Language and cultural differences can be especially critical in negotiating details of business transactions so that all parties understand their rights and responsibilities. The international movement of commodities often requires multiple kinds of transportation services to get a product from one country to another besides complying with additional governmental rules pertaining to imports or exports. Some of these issues are sufficiently complex to create opportunities for many individuals with specialized skills and knowledge to facilitate certain aspects of international business.

One of the things that makes international transactions different from interregional transactions is the role of two prices; namely, an exchange rate as well as the price or value of a physical commodity or service. At least one participant in international transactions needs to be able to translate the price or value of a commodity at one location into an equivalent value in terms of another country's currency. International transactions involve the exchange of a given quantity of one currency for a given amount of purchasing power expressed in terms of another currency. Participants in international business transactions follow changes in the relative value of currencies in the same way that other traders monitor changes in the prices of agricultural and food commodities. Fluctuations in the value of U.S. agricultural exports since 1970 are related to variations in the value of the U.S. dollar relative to currencies of several other countries. Chapter 12 will provide more details about how changes in an exchange rate can affect the volume of trade of a commodity between two countries.

QUESTIONS

1. Briefly explain the difference between a competitive and a noncompetitive (or complementary) import.
2. Which countries have been the largest importers of U.S. agricultural products in recent years?
3. Describe the general pattern of changes in U.S. imports and exports of agricultural commodities over the last two decades.
4. Write a brief essay describing similarities as well as differences in market transactions involving interregional or international trade of agricultural commodities.
5. If the current exchange rate between the German mark and the dollar is 1.66 marks per dollar, what would be an alternative value that would indicate a decrease in the value of a mark relative to the dollar? Briefly explain your reasoning.

REFERENCES

Baxter, T. 1998. Taking measure of the dollar's value: New exchange rate indexes. *Agricultural Outlook* (June–July): 8. USDA, Economic Research Service. http://www.econ.ag.gov/epubs/pdf/agout/june98/ao252a.pdf

Harris, H. M. Jr., and G. A. Benson. 1996. *Agriculture in a world economy.* Leaflet No. 1 in Southern Rural Development Center No. 198 International Trade Leaflets 1–10, Mississippi State University. Starkville, Miss.

Westcott, P., and R. Landes. 1999. Long-term agricultural projections reflect weaker trade. *Agricultural Outlook* (April): 34–39. USDA, Economic Research Service.
http://www.econ.ag.gov/epubs/pdf/agout/apr99/ao260g.pdf
Whitten, C. 1999. U.S. ag exports to turn up in fiscal 2000. *Agricultural Outlook* (October): 6. USDA, Economic Research Service.
http://www.econ.ag.gov/epubs/pdf/agout/oct99/ao265e.pdf

NOTES

1. This value is nearly 40% greater than the nominal value of agricultural exports in 1981, but the increase reflects changes in quantities as well as prices. If the index of prices received by U.S. producers is used to reflect changes in the prices of agricultural exports, the change in the real value of exports is less than half the change in nominal value.
2. Arbitration of disputes is possible under the International Chamber of Commerce provided the initial agreement contains a statement stipulating that both parties accept this process.
3. Adjusted bilateral exchange rates for different countries are weighted by the share of U.S. exports of particular commodities for 1990–1994 in calculating the index (Baxter 1998).

CHAPTER

12

INTERNATIONAL TRADE MODEL AND POLICIES

This chapter illustrates how the interregional trade model introduced in Chapter 10 can be modified to account for prices in different countries being expressed in different monetary units. The model can be used to illustrate how modifications in exchange rates influence the amount of international trade between countries. The model can also be used to show the effects of changes in transportation costs, taxes, subsidies, or other policy instruments on the amount of trade. International trade policies that influence the international movement of agricultural products are also discussed in this chapter.

The major points of the chapter are:

1. The way in which a two-region interregional trade model can be adjusted to incorporate exchange rates.
2. The use of the two-country international trade model to illustrate how a change in the exchange rate can affect the amount of trade.
3. The role of transportation costs, taxes, subsidies, and other policy instruments that influence the amount of international trade.
4. Some alternative rationales for particular international trade policies.
5. The highlights of some recent international trade agreements.

INTRODUCTION

The first part of the chapter indicates how the two-region interregional trade model developed in Chapter 10 can be modified to analyze the international movement of commodities. The same set of simplifying assumptions that was used in developing the interregional trade model is used again to illustrate how a given exchange rate can be introduced into the graphical analysis. After developing the basic two-country model, it will be used to demonstrate how a change in the exchange rate could alter the amount of trade and implied prices of a commodity in each country. The way in which changes in transportation costs and other factors influence the cost of moving a commodity between countries also will be illustrated. Some of the ways and reasons that countries influence the amount of international trade in particular commodities are discussed toward the end of the chapter. Finally, the last part of this chapter presents some highlights of recent international trade agreements.

BASIC ANALYTICAL MODEL

A graphical representation of trade between two countries uses the same basic concepts that were incorporated in the model to analyze interregional trade. In this case, the regional entities for which market demand and supply relations for a single homogeneous product are defined represent two countries. Initially, to keep things as simple as possible, it will be assumed that there are no transportation costs or barriers to moving the product between the two countries. Later it will be shown how these simplifying assumptions can be relaxed to make the analytical model more realistic.

A five-panel diagram like that shown in Figure 12–1 can be used to illustrate behavioral relationships underlying international trade. Panel a is used to represent the domestic demand and supply conditions for a commodity in Country A and panel e represents similar information for the same commodity in Country B. Each country's demand and supply functions for a given product can be used to define excess demand and supply relationships similar to those of interregional trade. The horizontal difference between the quantity demanded and produced under alternative prices defines a country's excess demand for imports and/or a country's excess supply for exporting purposes. For example, the excess supply situation of Country A is represented in panel b in Figure 12–1. Similarly, panel d represents the excess demand situation for the same commodity for Country B. If a country is not able to produce any quantity of a commodity because of climatic or other reasons, its excess demand for imports will be its total demand.

Excess demand and supply relationships for international trade can not be immediately plotted on the same graph as easily as in the case of interregional trade. The difficulty arises because the units on the vertical axes of the different panels in Figure 12–1 are not identical if each country uses a different currency. For example, the information in panels a and b in Figure 12–1 pertaining to Country A will be in terms of prices measured in the monetary units that people in that country

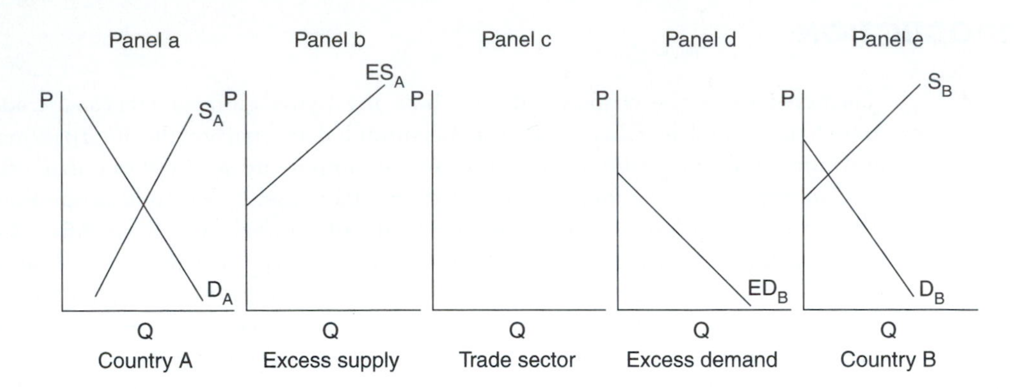

FIGURE 12–1 Two-country international trade model for a commodity.

use. Similarly, information pertaining to Country B will be in terms of the monetary units used in that country. If the example represented trade between the United States and Japan, the vertical axis for at least two of the panels would be in terms of yen and the vertical axis for at least two of the other panels would be in terms of dollars. Fortunately, there is more commonality in the measurement of quantities among countries. Thus the horizontal axes of different panels in Figure 12–1 are all assumed to have the same units of measurement.

Before an excess demand and supply schedule for two countries can be used to determine trade implications, a decision must be made as to which country's currency will be used for the vertical axis of the trade sector (panel c of Figure 12–1). Once this decision is made, it is rather straightforward as to how the excess demand and supply schedules can be placed on the same diagram so that the possibilities of a meaningful international market-clearing price can be examined. After a currency unit for the vertical axis of the trade sector is selected, it is clear whether the excess demand or supply relationship needs to be translated into equivalent terms of the other country's currency in order to be plotted in the middle panel representing the trade sector. This kind of transformation is analytically quite easy as long as the exchange rate between the two currencies is known. For example, if one of the points on the U.S. excess supply function indicated a willingness to export 2 billion lb of pork at a price of $.80 per pound, this could be translated into a willingness to export 2 billion lb of pork at a price of 80 yen per pound assuming an exchange rate of 100 yen per dollar. Using a similar conversion process for each and every point on the U.S. excess supply relationship would produce an equivalent expression for the U.S. willingness to export varying quantities of pork at alternative yen prices.

If the currency for Country B is used for the vertical axis of the trade sector in Figure 12–2, the ED_B relationship in panel c is exactly the same as that of panel d. In this case, the exchange rate would be used to convert the ES_A relationship in panel b to determine an equivalent price-quantity relationship in terms of Country B's currency.[1]

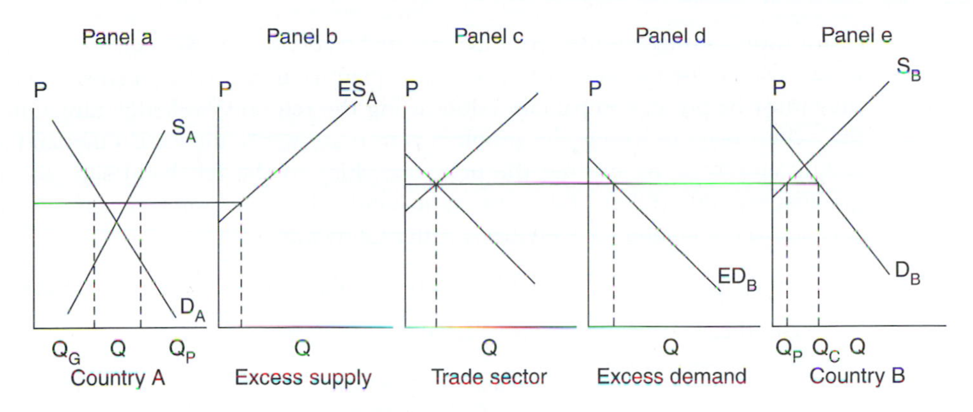

FIGURE 12–2 Market clearing in two-country international trade model.

Combining the transformed excess supply function with the excess demand function in panel c indicates an international market-clearing price and quantity consistent with both countries' internal demand and supply conditions for the commodity. Once a market-clearing price in the trade sector is determined in terms of one currency, the exchange rate can be used to calculate what that price represents in terms of the other country's currency.

It is not possible to extend a horizontal line from the intersection in panel c to both panels a and e to illustrate the international market-clearing price in both markets as was done in the case of interregional trade. The reason this is not possible is because the units on the vertical axes of some of the panels differ. A horizontal line from panel c representing the market-clearing price can be used for those panels that have vertical axes with the same units as panel c. For example, in Figure 12–2 a horizontal line can be extended from panel c to panels d and e to show the market-clearing price with international trade. The horizontal price line can be used to determine the amount of domestic production and consumption with international trade in the same way in which the implications of interregional trade were determined. International trade clearly produces a lower price, less production, and more consumption of this commodity in Country B than would occur in the absence of international trade.

Panels a and b in Figure 12–2 involve a different currency measure on the vertical axis from panel c and consequently the exchange rate would have to be used to convert the market-clearing price in panel c into equivalent terms in the other currency. An alternative way of specifying the international market-clearing price in panels a and b would be to find the point on the excess supply function in panel b that corresponds to the quantity of exports consistent with panel c. That point would indicate the prevailing price in Country A consistent with the quantity of exports demanded by Country B.

An equivalent algebraic representation of the graphical model would begin by obtaining the appropriate expressions for the excess demand and supply relationships similar to those of interregional trade. Before the equations can be used to solve for the unknown price and quantity in the trade sector, however, they must be put in comparable values using the relevant exchange rate. One of the easiest ways to handle this problem is to first rewrite the excess demand and supply equations by moving the price variables to the left-hand side and the quantity variable (ES or ED) to the right-hand side. For example, assume the excess demand equation for Country A is the following:

$$ED_A = 10 - 2P_A \qquad \textbf{(Eq 12-1)}$$

Another way of writing EQ 12.1 is

$$P_A = 5 - .5\ ED_A \qquad \textbf{(Eq 12-2)}$$

Equation 12-2 indicates how price varies with alternative values of excess demand whereas the former equation indicates how excess demand quantities vary with P_A. It is the same information expressed in two different ways.

Expressing the excess demand and/or supply equations with price as the dependent variables makes it easier to incorporate the appropriate exchange rate conversion to solve for the correct market-clearing price and quantity. For example, if 100 units of Country B's currency is equivalent to 1 unit of Country A's currency, each side of Eq 12-2, the excess demand equation for Country A, can be multiplied by 100 to convert it into the other country's currency (as follows:

$$100\ P_A = 100\ (.5 - .5\ ED_A) \qquad \textbf{(Eq 12-3)}$$

The left-hand side of Eq. 12-3 can then be replaced by P_B resulting in the excess demand having the same variable representing the market-clearing price as the excess supply equation for Country B.(

$$P_B = 500 - 50\ ED_A) \qquad \textbf{(Eq 12-4)}$$

Setting the latter form of the excess demand equation equal to the appropriate expression for the excess supply, with P_B as the dependent variable, and replacing ED_A and ES_B by a single quantity variable representing the amount of international trade enables a solution for the unknown quantity of trade to be determined. For example, assume the excess supply equation for Country B is ,

$$P_B = 100 + 50\ ES_B \qquad \textbf{(Eq 12-5)}$$

Setting Equation 12-4 (expressed in terms of P_B) equal to Equation 12-5 and solving for the unknown quantity variable would indicate 4 units would be traded. That is,

$$500 - 50\ ED_A = 100 + 50\ ES_B$$
$$500 - 50\ Q = 100 + 50\ Q$$
$$400 = 100\ Q$$
$$4 = Q = ED_A = ES_B$$

If the amount of trade is 4 units, P_B would be 300 according to either the excess demand or excess supply equation. This price is equivalent to a value of

3 for P_A because of the exchange rate assumed between the two currencies. It is also easy to verify that the solution is consistent with the original version of the excess demand (Eq. 12-1) expressed in terms of P_A and ED_A.[2]

Effects of a Different Exchange Rate

In the previous example, the mechanics of analyzing international trade were illustrated assuming a particular exchange rate. The model can also be used to show how a change in the exchange rate would affect the volume of international trade between two countries. To see how this works it will be assumed that the underlying demand and supply conditions for a given commodity in Countries A and B are exactly the same as in Figure 12–1. Consequently, the excess demand and supply relationships in the second and fourth panels of Figure 12–2 are exactly the same as before. To illustrate how a change in an exchange rate operates, the earlier example of the United States exporting pork to Japan can be used assuming the exchange rate adjusts from 100 yen per dollar to 150 yen per dollar. This is a rather large adjustment in an exchange rate, but it helps to make the directional changes in the diagram clearer than what would be the case if a smaller adjustment in the exchange rate were considered.

A different exchange rate means that each point on the original U.S. excess supply function implies a different point in the middle panel than previously. For example, a willingness of the United States to export 2 billion lb of pork at a price of $.80 per pound would now be equivalent to a willingness to export that quantity at a price of 120 yen per pound instead of 80 yen per pound (Figure 12–3). The new exchange rate results in an upward shift in the U. S. excess supply relationship in the trade sector assuming the vertical axis continues to be expressed in yen. The upward shift in the excess supply function associated with the new exchange rate is not a parallel shift because the exchange rate is translated into a proportional effect on prices at which alternative quantities would be made available on the world market. Thus a 50% adjustment (as assumed in this example) in the exchange rate will have a bigger effect on the location of points on the excess supply function corresponding to higher prices than for points associated with lower prices.

There is no reason why the excess demand function for Japan (or Country B) would be affected by the new exchange rate if that function were originally expressed in terms of yen. Perhaps the easiest way to think about this is to realize that most consumers and producers in a given country are not likely to even be aware of exchange rates as they make daily consumption and production decisions about a given commodity. They may realize that the price they face rises or falls as exports or imports change, but their willingness to buy or produce a commodity as represented by their demand and supply schedules are not altered by a change in exchange rates. As far as they are concerned, they only see prices of commodities in terms of their own currency. Individuals or firms who engage in international trade, however, have to be aware of

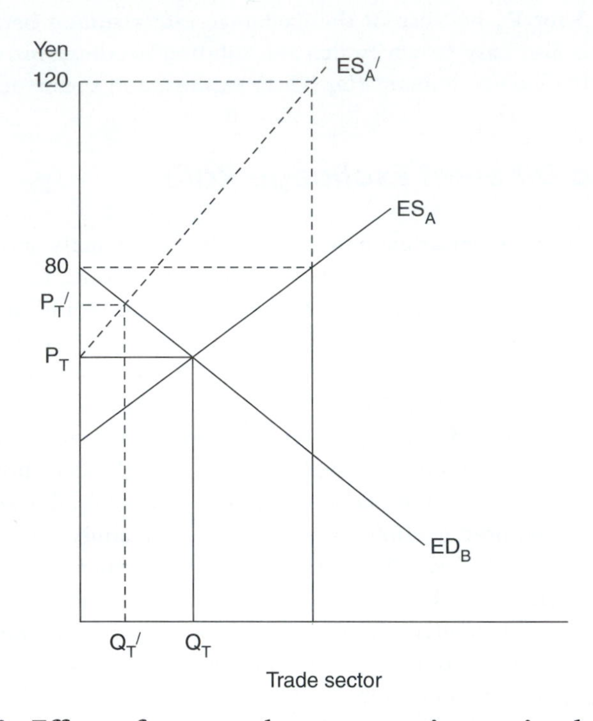

FIGURE 12–3 Effects of a new exchange rate on international trade.

exchange rates as they express interest in buying or supplying commodities in international markets at particular prices expressed in one currency or the other.

The net effect of an upward shift in an excess supply function associated with a new exchange rate is that a smaller quantity of the good (Q'_T rather than Q_T) will be traded between the two countries. As a result of the decrease in the quantity traded, the price in the importing country will be higher (P_T') than it would have been under the previous exchange rate (P_T). Similarly, the price in the exporting country will be lower than it would have been under the previous exchange rate as a result of a decrease in exports.

Identical results as far as changes in the amount of trade and prices in each country would be implied if the exporting country's currency had been used for the vertical axis of the trade sector panel. The only difference is the excess demand function in the middle panel would be shifted downward as a result of the new exchange rate while the excess supply schedule would not be altered (Figure 12–4). The new excess demand function would intersect the original excess supply function at a lower quantity (Q_T' rather than Q_T) and imply a lower price (P_T') in terms of the exporting country's currency.

The only difference between Figures 12–3 and 12–4 is whether the excess demand or supply function is shifted because of the new exchange rate. Which

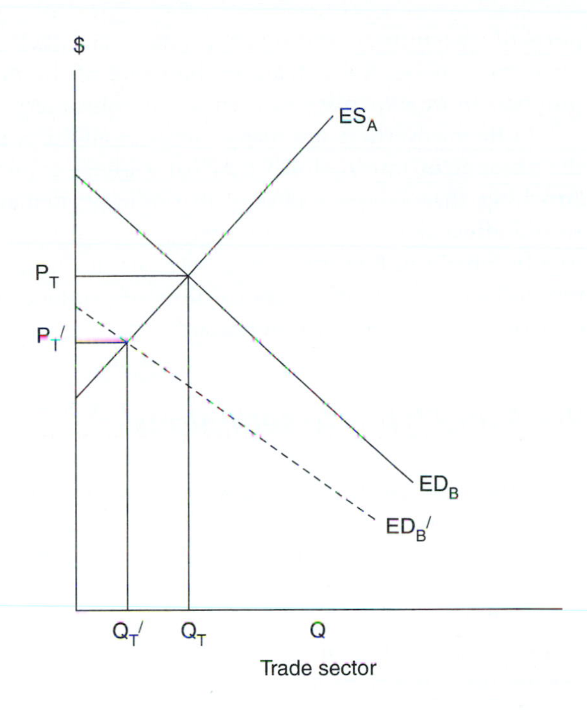

FIGURE 12–4 Alternative analysis of effects of a new exchange rate on international trade.

function is shifted by a given adjustment in the exchange rate depends on which function is converted when plotting in the trade sector diagram. Whether an excess demand or supply function is shifted upward or downward by a new exchange rate depends on how the new exchange rate compares to the previous one. In the previous example it was assumed that there was an increase in the value of the dollar relative to the yen (or the value of yen decreased) and the United States was the exporting country and Japan was an importing country. All kinds of different scenarios related to alternative adjustments in exchange rates and assumptions as to which country is importing or exporting a particular commodity can be analyzed using the same concepts.

The results of the previous analysis lead to the implication that increases or strengthening of the value of an exporting country's currency would be expected to cause the quantity of exports of a given commodity to decrease, **ceteris paribus.** Another way of stating this is that depreciation or weakening of the value of an importing country's currency is like a decrease in purchasing power on the world market which would be expected to cause a decrease in imports of most commodities. This conclusion is useful for predicting what is likely to happen to exports or imports of a given commodity as a country's currency appreciates or depreciates relative to other currencies in the world. For example, as the U.S. dollar strengthens or gains value relative to other currencies, U.S. exports of agricultural (and other) products are likely to decrease and imports are likely to increase. An increase in the value of the dollar makes U.S. agricultural products more expensive in the eyes of consumers in other countries.

Obviously, there are many things in addition to exchange rates that affect the amount of international trade of a given commodity at any point in time. Anything that causes a change in domestic demand and supply relationships would affect the location and intersections of international excess demand and supply functions. It is difficult to sort out the relative importance of all the factors influencing the international flow of products at a given point in time, but clearly exchange rates are important.

Relaxing the Simplifying Assumptions

Once the basic mechanics of how the two-country international trade model operates are clear, it is feasible to revert back to the basic three-panel diagram that was used for interregional trade. The only thing that must be remembered is that the units for the vertical axis for the trade sector will be the same as one country, but different from that of the other country. Transportation costs and certain kinds of trade policy instruments can be easily introduced into the analysis using the three-panel trade diagram.

Transportation costs can be introduced into the two-country international trade model in exactly the same way as shipping costs were handled in the case of interregional trade. Once the excess demand and supply relationships are expressed in the same monetary units, a corresponding set of parallel functions can be determined assuming the cost of moving the product between the two countries is known (Figure 12–5). A derived excess demand function (ED_B') can be specified by vertically subtracting the cost per unit of moving the product between the two markets from each point of the original excess demand function (ED_B). Similarly, a derived excess supply (ES_A') relationship can be obtained by vertically adding the transportation cost to each point on the original excess supply relationship (ES_A). The only thing about this process that is slightly different from how transportation costs were introduced into the interregional trade model is that the transportation costs must be expressed in the same monetary unit that is used for the vertical axis of the trade sector. This is true regardless of whether the buyer or seller pays the cost of moving the product.

Once the full set of original and derived excess demand and supply functions are placed in the same diagram, net shipping and receiving prices (P_A and P_B) of the commodity would be indicated by intersections of the appropriate relationships in the trade sector diagram. One of the intersections in this diagram that is economically meaningful is the one between the original (or primary) excess demand and the derived excess supply. The other economically meaningful intersection is between the primary excess supply and the derived excess demand functions. Interpretations of these two intersections are similar to those considered earlier in linking farm and retail markets in Chapter 6 or two spa-

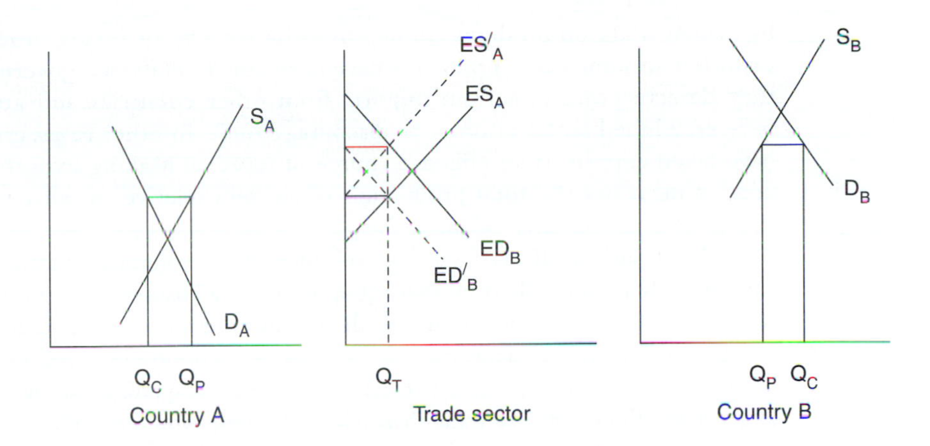

FIGURE 12–5 Introduction of transportation costs into two-country international trade model.

tially separated regional markets of a country in Chapter 10. The net shipping (or receiving) price in the middle panel of Figure 12–5 would need to be converted into an appropriate value for one of the countries using the current exchange rate.[3]

The above analysis can be used to show how changes in transportation or other transaction costs of moving the product between two countries would have a tendency to influence the amount of international trade. Improvements in transportation facilities or other changes that reduce the cost of moving the product among countries would tend to increase the amount of trade.[4]

ALTERNATIVE POLICY INSTRUMENTS

Additional costs besides transportation charges may be involved in moving commodities across country borders. Analytically these costs could be incorporated into the analysis just like another component of transportation cost. One type of extra cost that might be levied by either the exporting or importing country is a fixed fee per unit of commodity traded to raise revenue or as an explicit way of restricting the amount of product leaving or coming into a country. At the same time that one country may try to limit trade flows by charging a **tariff** another country may be encouraging trade by subsidizing or paying for part of the cost of transferring the product among countries. The effects of changes in tariffs or subsidies can be analyzed in exactly the same way as changes in transportation costs.

Countries have pursued a very mixed pattern of tariffs and subsidies in an attempt to alter international commodity movements (Fairchild et al. 1996). Many countries have viewed international trade as a lucrative source of revenue by placing a tax on products coming into the country or taxing products being exported. In some cases, producers have been able to convince governments that they deserve protection from imports from other countries and governments have established tariffs in order to discourage trade. In other cases, governments have taxed exports as an effective source of revenue. Making exports more expensive decreases the total production of an item relative to what would have occurred in the absence of such taxes.

Analyzing the effects of tariffs or subsidies on the volume of international trade can be easily accomplished by making appropriate adjustments in the vertical distance representing the cost of moving the products between two countries. The extent to which there are extra costs over and above transportation charges will reduce the volume of trade. The extent to which a subsidy per unit exists will result in a larger flow of international trade than what otherwise would occur, ceteris paribus.[5]

One way in which an importing country can attempt to protect itself to some extent from variation in world prices is to establish a **variable levy** system. The way in which a **variable levy** works is that the amount of extra charge levied on each unit of product imported varies inversely with world prices. If the price on the world market happens to decrease, the levy is increased to prevent the domestic market from being flooded with imports. As world prices increase, the levy is reduced in an attempt to maintain more stable domestic prices instead of having them fluctuate with world market conditions.

An alternative policy instrument that may be used to limit the amount of a product coming into or going out of a country is by establishing an explicit limit or quota on the quantity that can be traded. The quota may be administered by permitting a designated subset of potential traders to import or export a certain amount of product or simply announcing that no additional trade will be allowed once a particular target amount is reached. The establishment of a quota by an importing (or exporting) country can be incorporated analytically into the two-country international trade model by making the excess demand (or supply) function perfectly inelastic at a designated quantity (Figure 12–6). If a quota is established by an importing country it is analogous to the excess demand curve becoming perfectly inelastic regardless of how low the price might happen to go. Similarly, a quota on a country's exports could be represented as the excess supply becoming perfectly inelastic regardless of how high a price is considered. The effects of a trade **embargo** or a governmental edict banning trade with certain countries is similar to establishing a zero quota for international trade.

Imposing a high tax or tariff on the movement of products between two countries once the volume of trade reaches a certain level has essentially the same effect as the establishment of a quota. Recently, **tariff-rate quotas (TRQ)** have been introduced as part of the **tariffication** process around the world. A **TRQ** basically allows a certain quantity of a product to be imported at a low rate of tariff, but once imports reach a specified level, a higher tariff becomes applicable on any (additional) quantity above the quota.

Several kinds of nontariff trade barriers may also be used instead of quotas or tariffs to restrict the amount of trade. Such things as tighter quality spec-

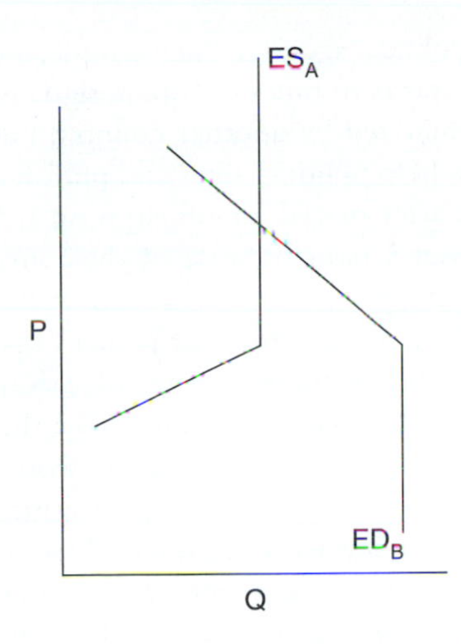

FIGURE 12–6 Representation of a quota on exports from Country A or on imports of a commodity into Country B.

ifications or stricter labeling requirements can reduce the volume of trade. Recent actions by European countries to limit or ban imports of biotechnologically developed products or livestock products produced with hormones are examples of these kind of restrictions.

The possibility of using the two-country international trade model to analyze the implications of various kinds of policy instruments indicates potential usefulness of the model to provide insight about possible consequences of alternative governmental actions. Although the preceding model is based on only two countries, its implications are much broader because one country could be considered as a composite aggregation of the rest of the world. This is similar to the way the simple interregional trade model can be generalized to consider more than two regions.

INTERNATIONAL TRADE POLICY

Implications from the analytical trade model indicate why it is natural for people to have different opinions about governmental actions to promote or restrict international trade of various products. This is nothing new, since U.S. independence arose because of the tax England placed on tea and other products. Later, sharp differences about trade policies between the more industrialized northern states and the more agriculturally oriented southern states existed at the time of the Civil War. Recently, sharp political debates have focused on the **North American Free Trade Agreement** (NAFTA), trading rules established as part of the **General Agreement on Tariffs and Trade (GATT),** and the creation of the **World Trade Organization (WTO).**

As previously noted, governments may attempt to influence international trade in many ways. Sometimes international trade policies may be motivated primarily by political agendas. For example, the sudden imposition of an embargo that prohibits exports to one or more designated countries may be a political response to actions initiated by another country. International trade policy has often been designed to help political allies and punish enemies. For example, several years ago President Carter placed an embargo on U.S. exports of particular products to the former Soviet Union because of their invasion of Afghanistan. Recently, several countries have placed limitations on trading with Cuba, Iraq, and selected other countries because of actions or policies that were deemed to be unacceptable.

International trade policies of a country also may be designed to protect particular producer groups. Sometimes the argument is made that some protection from the competitive forces of world markets is required in order for domestic producers to be able to get a critical industry established. This is referred to as the **infant industry** argument. One of the difficulties with this policy is that seldom is any timetable indicated as to how long it will take the infant to mature.

For a number of years, several countries believed that one way to promote economic growth was to increase production of items it was importing. Often this meant establishment of some kind of protection against international competition in order for certain industries to get started. The folly of achieving economic growth through self-sufficiency eventually was realized by observing a number of countries who had much success by exporting products in which they had a comparative advantage rather than worrying about trying to reduce or replace imports (i.e., Singapore, South Korea) (Krueger 1997).

Some countries feel very strongly that they should never be overly dependent on world markets for food and perhaps some other products critical for survival. One of the reasons that supports this position is past experience when international trade has been disrupted by wars or because of other political disagreements. It is probably difficult for many current U.S. citizens and agricultural producer groups in particular to have a full appreciation for this position because of never having experienced these types of disruptions in food supply channels.

Once a country has experienced growth of an industry based on expanding exports, often continued momentum is desired and it becomes painful if exports start to decline. This is why pressure can arise from various sources for government subsidies to pay part of the cost of exporting products to maintain or even increase exports of certain products. The United States has used an **Export Enhancement Program (EEP)** to pay some of the cost of exporting agricultural products to selected countries. EEP was initiated in the mid-1980s when exports decreased and wheat prices were especially low. At that time wheat stocks were large and European countries had become a major export competitor as a result of various subsidies. Initially, exporters of wheat and other selected commodities received payments in kind from government inventories, but later a cash payment from the U.S. Treasury was used to pay part of the cost of shipping specific products to targeted countries.

Another way that exports can be stimulated is by governments offering prospective foreign buyers some kind of financing or credit guarantees. This technique has been used by the United States to move agricultural products to developing countries that may not have the immediate means of payment.

INTERNATIONAL TRADE AGREEMENTS

As a consequence of the myriad of governmental policies related to international trade, major trading countries have had continuing interest in periodically getting together to discuss international trade policies to see if some impediments could be eliminated. One vehicle for doing this was through the establishment of a **General Agreement on Tariffs and Trade (GATT)** in 1948. This agreement resulted in a series of eight multilateral trade negotiations among a large group of nations attempting to reduce trade barriers over a period of time. The last round of GATT discussions was initiated at a meeting of representatives from 92 governments in Uruguay in 1987. This was the first time that agricultural and textile products were part of the negotiations. Philosophical differences about agricultural policies between the United States and representatives of the European countries were one of the reasons the Uruguay round of negotiations continued for more than 7 years. Finally, countries agreed to reduce tariffs on agricultural products, cut export subsidies, and not introduce any new subsidies (Sanders et al. 1996). Several other trade issues were also addressed in this agreement.

The Uruguay round of discussions also led to the establishment of the **World Trade Organization (WTO)** on January 1, 1995, which is headquartered in Switzerland. The purpose of the WTO is to increase international trade primarily in two ways. One way is to implement and monitor international trade procedures accepted by the 112 member nations at the inception of the WTO. These procedures pertain to tariffs, quotas, phytosanitary regulations, preshipment inspection regulations, subsidies, and so on. The WTO intends to encourage additional rounds of international discussions like the Uruguay round. The other primary role of the WTO is to try to settle disputes among member nations about international trade. Individual firms or corporations can not file claims. Only member nations can take that kind of initiative. Parties to a dispute select panelists from a pool of qualified individuals from each WTO country. Countries may withdraw from the WTO by providing a 6-month notice of intent.

While GATT, and now the WTO, has been a major vehicle for actions leading to more international trade, there have been various groups of countries around the world making agreements patterned after the European Common Market (now known as the European Union). These agreements are designed to encourage more trade among a set of countries that usually are geographically close together. This generally involves reducing barriers on trade among member countries and developing a set of standardized policies pertaining to trade with nonmember countries. Examples of these agreements are the recent **North American Free Trade Agreement (NAFTA),** Association of Caribbean States, Andean Common Market, Economic Community of West African States, the Association of Southeast Asian Nations (ASEAN), Council of Arab Economic Unity, and so on.

The NAFTA agreement that was implemented on January 1, 1994, consisted of actually two agreements that Mexico made with the United States and Canada. These agreements basically expanded some of the provisions that

Canada and the United States had reached as part of a separate trade agreement in 1989. The purpose of NAFTA was to promote the flow of trade among the three countries by implementing a phased reduction or elimination of import tariffs and some other trade barriers (Rosson et al. 1996). To some extent this was an extension of a path that Mexico had started a few years earlier. Currently, there is discussion of expanding the NAFTA agreements to Chile and possibly other countries in Central and South America. Where all these discussions will end up in the years ahead as countries wrestle with different political agendas, and at the same time recognize the power and benefits from the principle of comparative advantage operating at the international level is difficult to forecast. As competition on the world market increases, there appears to be increasing concern especially among some of the more developed countries about variation in labor and environmental conditions under which production occurs in other parts of the world. It will be interesting to follow the extent to which countries continue to use international trade policies to achieve various economic and political objectives.

SUMMARY

This chapter illustrates how the graphical model introduced in Chapter 10 representing economic forces affecting the geographical movement of commodities can be modified to analyze international trade. The key difference in the model is having to decide which one of the two countries' currencies will be used for measuring units on the vertical axis of the middle diagram representing the trade sector. All of the points on either the excess demand relationship for one country or the excess supply relationship for the other country must be converted into another currency before they can both be plotted on the same diagram. The intersection of excess demand and supply relationships expressed in the same currency units indicates an implied volume of international trade and a market-clearing price. The amount of production and consumption in each country can be determined after the exchange rate is used to determine market-clearing prices in each country's currency. The effects of a change in the exchange rate can be illustrated by shifting either the excess demand or supply relationship in the trade sector in a nonparallel manner. The effects of changes in domestic demand and/or supply conditions on trade can be analyzed similar to the generalization possibilities of the interregional trade model. The model can also be used to illustrate how the cost of transporting products between countries affects the amount of international trade and prices in each country. Changes in the cost of moving commodities can be generalized to include effects of tariffs and/or governmental subsidies that influence the per unit cost of moving products between countries.

Tariffs and/or quotas are frequently used to achieve particular economic and/or political objectives. Variable levies that vary inversely with price may be established by an importing country to keep domestic prices from fluctuating with world market conditions. Sometimes a high level of protection against foreign competition may be initiated presumably to provide sufficient time for a domestic industry to get started and achieve sufficient size in order to be able to compete with other countries. This type of protection is often difficult to eliminate once producers get accustomed to it. There are also many ways in which other **nontariff trade barriers** operate. For example, particular restrictions on the quality of items permitted into a country or additional inspection requirements may be established because of legitimate health and safety concerns, but economically they affect the quantity of international trade just like **tariffs** or quotas. The most extreme case of governmental intervention affecting the volume of trade is a total embargo on trading all or certain commodities with one or more countries as a sanction or retaliation for some kind of political or military action.

Over the last 50 years considerable progress has occurred in reducing trade barriers as countries realize the increasing importance and value of international trade. Some of this progress occurred through a series of trade negotiations under the General Agreement on Tariffs and Trade. The latest of these was the Uruguay round, which was the first time agricultural products were included in the discussions. This agreement outlined steps to gradually eliminate quotas and reduce the amount of tariffs and subsidization for agricultural products. This round of trade discussions also led to the establishment of the World Trade Organization. Many countries have also simultaneously entered into bilateral or multilateral trade agreements with neighboring countries following the pattern of the European Common Market (now the European Union). Implementation of the North American Free Trade Agreement was designed to reduce trade barriers between the United States, Mexico, and Canada. This agreement actually consisted of separate bilateral agreements between Mexico and Canada and between Mexico and the United States building on an earlier free trade agreement between the United States and Canada.

QUESTIONS

1. In recent weeks the French franc has been trading at approximately 5.2 francs per U.S. dollar on international currency markets. How would an appreciation in the franc (or a depreciation in the U.S. dollar) be expected to affect imports of French wine into the United States? Explain your reasoning.
2. Briefly describe how a decrease in tariffs on U.S. imports of Danish hams would affect the U.S. pork industry (consumers as well as producers).
3. Briefly explain and give an example of how a government might use a nontariff barrier to restrict or otherwise reduce the volume of international trade that might occur between two countries.
4. Briefly explain how the "infant industry" argument is often used by nations as a justification for placing tariffs on manufactured food products.
5. Briefly explain how a governmental program that subsidizes the production of agricultural commodities in one country can have detrimental effects on producers of the same commodity in other countries.

REFERENCES

Fairchild, G. F., G. A. Benson, L. D. Sanders, and J. L. Seale Jr. 1996. *Government intervention affecting agricultural trade.* Southern Rural Development Center No. 198, International Trade Leaflets 1-10, Mississippi State University. Starkeville, Miss.

Krueger, A, O. 1997. Trade policy and economic development: How we learn. *American Economic Review* 87(1): 1–22.

Rosson, C. P. III, G. A. Benson, K. S. Moulton, and L. D. Sanders. 1996. *The North American Free Trade Agreement and U.S. agriculture.* Leaflet No. 9 in Southern Rural Development Center No. 198, International Trade Leaflets 1-10, Mississippi State University. Starkeville, Miss.

Sanders, L. D., K. S. Moulton, M. Peggi, and B. Goodwin. 1996. *The GATT Uruguay round and the World Trade Organization: Opportunities and impacts for U.S. agriculture.* Leaflet No. 7. in Southern Rural Development Center No. 198, International Trade Leaflets 1-10, Mississippi State University. Starkeville, Miss.

NOTES

1. If the vertical axis of panel c of Figure 12–2 was in the currency of Country A rather than Country B, a translation of the information embodied in ED_B rather than ES_A would be required.

2. An alternative approach would be to multiply the price dependent version of the excess supply equation by 1/100; replacing 1/100 P_B by P_A and set it equal to the original version of price dependent excess demand equation. Either way produces the same set of prices and amount of trade for a given exchange rate. Alternative exchange rates influence the values that must be used to adjust both sides of the excess demand or (supply equation) before solving for the market-clearing quantity and price.

3. The final prices for the two countries illustrated in Figure 12–5 are appropriate if the two currencies had a 1 to 1 exchange rate. For other exchange rates, the horizontal line from the trade sector to County A would be correct, if the vertical axis of the trade sector is the same as Country A but the price in Country B would likely be at a different level from that indicated in the trade sector. As noted earlier, one of the two prices indicated in the trade sector diagram would need to be adjusted to different currency units using the appropriate exchange rate.

4. If the cost of transportation was sufficiently high to result in the intersections of the appropriate functions being to the left of the origin, it would imply that trade was not feasible under the existing conditions.

5. The derived excess demand and supply relationships in the trade sector may not be parallel to the primary relationships if tariffs or subsidies on international trade are not fixed amounts per unit of the product.

Part IV

Temporal Aspects of Agricultural Markets

The fourth and final part of this book concentrates on temporal dimensions of agricultural markets. The purpose of these chapters is to help students understand the important roles of storage and time in coordinating production and consumption activities. Changes in stock levels of periodically as well as continuously produced agricultural and food commodities and the way prices vary over time are important dimensions of agricultural markets. The types of market institutions and instruments available to producers and marketing firms for handling temporal variation in prices are discussed in the final three chapters of Part IV.

Chapter 13 discusses storage decisions as value-adding economic activities associated with agricultural and food products. Owning (or holding) a commodity at any stage of the marketing process involves an explicit (or implicit) decision that makes the commodity available for future consumption. Some storage occurs in conjunction with other value-adding activities that change the form or location of commodities as part of the overall coordination of production and consumption. Any type of storage activity requires the use of economic resources with implied costs. Consequently, a purposeful decision to hold a commodity means that the change in value over time is expected to be sufficient to offset the cost associated with the activity. For periodically produced commodities, changes in price provide market signals as to how rapidly an existing inventory should be consumed. The ways in which current and anticipated future market conditions influence storage decisions and price changes over time are illustrated using graphical models similar to those introduced in previous chapters of this book to analyze spatial characteristics of markets.

The final three chapters of the book are devoted to a discussion of futures and options markets that have become increasingly important to buyers and sellers of agricultural and food products. Chapters 14 and 15 are devoted to futures markets and Chapter 16 considers options markets. Chapter 14 indicates how futures markets differ from cash markets and describes some of the characteristics and procedures of futures trading. In this chapter, the trading of futures contracts is introduced primarily from the perspective of speculators who are motivated to profit from temporal changes in commodity prices. The possibility of making money when prices increase or decrease by changing the sequence of buying and selling of futures contracts market is especially attractive to speculators. The possibility of making a lot of money from small changes in prices of futures contracts because of margin requirements is an intriguing aspect of these markets. Although potential gains (or losses) from speculating in futures markets are emphasized, these markets are very dynamic and serve useful functions for individuals who wish to minimize the risk of unknown price changes.

Chapter 15 describes how futures markets can be used by producers and marketing firms to obtain some price insurance. An example of a short hedge is initially used to illustrate how opposite positions in a cash and futures market can be used to avoid much of the risk of unknown changes in price of a commodity. Alternative changes in price of a commodity are considered to illustrate how a short hedge essentially guarantees a specified return to storage under simplified assumptions. The example is then made more realistic by introducing basis uncertainty. Knowledge about spatial and temporal characteristics of price relationships discussed in previous chapters of the book is useful in understanding economic factors determining the basis for any cash market at a given time. The way in which producers can use futures markets to "lock-in" or "target" a price of a commodity they plan to produce is similar to a storage hedge. One additional element of uncertainty, however, is the effect of weather conditions on the total quantity that will be available for sale in the cash market. Another kind of hedging discussed is long hedging, which can be used to lock-in a price of a commodity that will be purchased at some time in the future. Also the way in which appropriate sequences of simultaneously purchasing and selling of different futures contracts can be used to make money and provide protection against large price movements if changes in the difference in two futures market prices are correctly anticipated is described in the final part of the chapter.

Chapter 16 indicates how options markets provide producers and marketing firms an additional element of flexibility in dealing with price uncertainty. The way in which buyers of real estate sometimes acquire additional time before making a final decision about a specific transaction is discussed before formally discussing call and put options on futures contracts. Differences in privileges and responsibilities of buyers and sellers of options are noted and compared to the obligations of traders of futures contracts. Call options that provide the privilege of purchasing a futures contract at a specified price for a specified time period are discussed before put options. The factors affecting premiums (or the prices) of options are also discussed. Put options can be used by producers to acquire some time in making decisions about whether they really want to sell a futures contract at a given price. Similarities and differences between put options and governmental programs guaranteeing producers a minimum price are noted. A comparison of alternative outcomes from using futures and options markets relative to producers accepting whatever market price exists when they decide to sell their output is graphically illustrated. The way in which net returns from buying and selling options depend on changes in the price of the underlying commodity is also graphically illustrated.

CHAPTER

13

STORAGE DECISIONS IN THE MARKETING OF AGRICULTURAL AND FOOD PRODUCTS

This chapter discusses storage activities that are an integral part of marketing agricultural and food commodities. This material helps set the stage for analyzing how prices of commodities change over time.

The major points of the chapter are:

1. The economic role of storing different kinds of agricultural and food commodities.
2. The three cost components of storage.
3. The general time pattern of inventories and prices of periodically produced commodities.
4. How a two-period graphical model can be used to show how storage decisions, the amount of current consumption, and price of commodities are influenced by current and prospective demand and supply conditions and the cost of storage.

INTRODUCTION

Holding or storing agricultural and food products for various intervals of time before using them or transferring ownership to someone else is an important component of marketing. Readers will recall that storage was identified in Chapter 1 as one of the three physical value-adding activities of marketing. The purpose of this chapter is to explore several aspects of storage decisions and to set the stage for analyzing temporal differences in prices. Of particular interest will be how *anticipated* differences in market prices influence storage decisions and in turn how *actual* changes in price are influenced among other things by how much inventory individual market participants decide to hold.

The first part of this chapter examines the role that storage plays in coordinating production and consumption activities for three different categories of food and agriculturally related products. Then three components that determine the total cost of storing a commodity are identified. Much of the rest of the chapter focuses on issues pertaining to periodically produced commodities. For example, the general pattern of inventories and prices that exist between successive harvest periods is noted and the way markets convey signals that participants use in making storage decisions is discussed. The final part of the chapter describes how a two-period graphical model can be used to illustrate how storage, consumption, and price patterns are interrelated.

WHAT IS STORAGE?

Perhaps the easiest way to begin thinking about storage is to consider the economic implications of physically holding quantities of any storable commodity for various intervals of time. From an individual consumer's or aggregate market perspective, storage is exactly the opposite of consumption. Storing or holding a commodity at some point in the marketing process essentially means that the item is made available for consumption in the future. On the other hand, a decision to consume means a certain quantity of products will not be available for future use. A decision to store essentially involves a commitment of moving products from one time period to another. Storage produces an intertemporal transfer of products in much the same way as interregional or international trade produce spatial transfers of products.

Storage is an integral component of agricultural marketing because it is part of the process that links the flow of products from where and when they are produced to where, when, and in what form they are ultimately consumed. Storage of commodities is involved in this linkage because production and consumption of agricultural and food products do not occur instantaneously. Also, the biological nature of many agricultural production processes is not always perfectly synchronized with when consumers prefer to use products. Consequently, there are numerous reasons why varying intervals of time occur between when agricultural products are initially ready to enter market channels and when food products are ultimately consumed. As commodities move through the marketing system they are stored for different intervals of time by somebody. Owning and holding agricultural and/or food products for various intervals of time at different stages of the marketing process effectively make products available for consumption at some future time. Thus, storage is the temporal component of the linkage between production of agricultural commodities and the consumption of food products.

DIFFERENT KINDS OF PRODUCTS

For some agricultural commodities, initial storage decisions originate with producers. Delaying the initial sale of items that are ready for movement into market channels is an explicit storage decision. A key question for everyone who owns something they do not necessarily expect to use themselves is how long should they hold it before transferring ownership rights to someone else. A couple of key aspects of the biological nature of agricultural production processes influence dimensions of the window of opportunity for delaying the sale of products by producers or other parties. One aspect is the degree of perishability of products when they reach maturity or are harvested and ready for possible movement into the marketing system. Another characteristic is the frequency of production. Some agricultural products tend to be produced only once or maybe twice a year.[1] Other products tend to be produced on more or less a continuing basis throughout the year, but not necessarily at a steady rate or the same rate that consumption occurs. Frequency of production and the degree of perishability can be used to classify products into three broad categories to distinguish different kinds of storage decisions. The three product categories are (1) periodically produced perishable crops, (2) periodically produced storable crops, and (3) continuously produced commodities.

Periodically Produced, Perishable Crops

All agricultural products are perishable or at least subject to losing some desirable characteristics with the passage of varying lengths of time if not properly handled. Perishability is an inherent characteristic of all biological processes, but the rate of change in desirable characteristics varies among products and depends on how products are handled once they enter or are ready to enter the marketing system. If a product tends to lose its desirable characteristics very rapidly after harvest, the window of opportunity for selling and/or use may be very short. Consequently, perishable products often need to be rushed very rapidly through the marketing system. In these circumstances, producers have little flexibility in holding the commodity or timing their sales. It is important that producers of these type of products know what avenues are going to be used to move products to consumers as rapidly as possible after harvest and perhaps establish contractual marketing arrangements prior to harvest. Preservation of desirable characteristics of fresh fruits and vegetables and other very perishable products may require cooling, refrigeration, or specialized holding facilities even for relatively short periods of time at various stages of the marketing system.

In many cases it is feasible to extend the life of perishable products by converting them into an alternative form to increase storability and expand the window of marketing opportunities. If periodically produced perishable commodities are transformed into alternative forms that are less subject to deterioration, a wider range of marketing and consumption alternatives become available through storage. For example, canning, freezing, and other forms of food preservation make it possible to consume products year-round even though they may be produced and harvested only once or twice a year. These possibilities, however, mean that decisions must be made when harvesting perishable products about how much should be converted to storable forms and how much should be sold as fresh products. Once this decision is made then future decisions about how long to hold the storable form of the product and timing of sales are required.

Periodically Produced, Storable Crops

Storage is especially important for periodically produced commodities that are not as perishable as fresh fruits and vegetables. Storage provides a source of the product for consumption during that part of the year when no new output is occurring. Deciding when to sell products is a critical element of marketing strategies of producers or owners of commodities that are storable. Agricultural producers may decide to hold some of their output in their own facilities or may contract with someone else to hold the commodity for a designated period of time while ownership is maintained. Other producers may prefer to sell all of their output immediately at harvest and let someone else worry about storing the commodity. In either case, storage of commodities requires particular kinds of facilities and a commitment of resources that could be used for other purposes. Consequently, storage must have an expected payoff to justify allocating resources to this activity.

Ownership of a commodity for any interval of time at any stage of the marketing process involves an explicit or implicit holding (or storage) decision. Some specialized firms exist simply to provide facilities for holding quantities of products for different intervals of time. Examples are grain elevators, warehouses, cold storage, and refrigeration facilities. They may rent storage space to others for a fee or operate as entrepreneurs by purchasing quantities of products at harvest or other times when prices are thought to be low with the expectation of being able to resell the product in the future when prices are higher. Unless stored products have been presold at a guaranteed fixed price, these decisions involve forecasting the expected change in market value of a commodity and comparing it to the storage costs that will be incurred. These decisions will be profitable as long as the change in price at which commodities are purchased and sold at different points in time is greater than the costs of storage. Storage is a special kind of value-adding activity in terms of facilitating the continued availability of periodically produced commodities for consumption.

Some people have difficulty accepting or understanding storage as a value-adding activity because the commodity that comes out of storage is usually identical or at least very similar to what went into storage. Storage is a value-adding activity, however, if the commodity that comes out of storage has a greater value than when it went into storage because it exists at a different point in time and market conditions changed while the commodity was being held. Even though the product may have the same physical characteristics it can have a different market value. From an economic perspective, such a change in value is no different from the change in value of products that come out of a processing plant relative to the value of basic agricultural products used as raw materials or a change in value associated with a geographical relocation of a product from one market to another. In the case of storage, a change in value is attributed to a change in time of availability instead of a change in form or location.

It is important to distinguish between explicit or conscious storage decisions from the storage of agricultural and food products that occurs as an inescapable part of other value-adding marketing activities like processing, transportation, wholesaling, or retailing. The purchase of a given quantity of an item and holding it for a particular interval of time may be the only way that other value-adding marketing activities can occur. For example, the process of transforming agricultural commodities into edible food products requires ownership (and holding) of products at least for some interval of time during the transformation process. Consequently, a certain amount of holding or inventorying of items occurs as products are being processed and transformed into edible foods. Also the geographical separation between where agricultural products originate and where they are ultimately consumed means that someone owns moving stocks as products are being transported.

Individual firms must continually evaluate how big an inventory of raw products is desirable to keep processing plants operating efficiently and how big an inventory of final products is optimal to meet potential customer demand for a specified period of time. Similarly, wholesalers and retailers make conscious decisions about how much inventory of various food products to hold in anticipation of future sales to customers. Even consumers play a role in providing storage in terms of the quantities of various food products they store in their pantries, refrigerators, or freezers. From the standpoint of aggregate market behavior, the storage of food by households is no different from storage at earlier stages of marketing. One reason households hold certain quantities of food is to eliminate having to make a trip to a retail outlet every time they want to consume certain products. Consumers also may make conscious decisions to expand or decrease household inventories of certain food products in anticipation of upward or downward price movements. In this sense, inventory or storage decisions made by consumers are similar to production and marketing firms deciding how long to own an item that they are not going to personally consume.

The decisions and behavior of all market participants regarding willingness to own and hold various quantities of periodically produced storable products for various intervals of time effectively determine the rate at which the available quantity of commodities is utilized. The periodic nature of production of many agricultural products means that the total available quantity must be allocated over time through the interaction of all those who desire to buy or sell and ultimately consume food products.

One of the functions of markets is to establish correct price signals for producers, business firms, and consumers to decide when to buy, sell, and consume products consistent with knowledge and expectations about current and future market conditions. Producers must decide not only what and how much to produce, but also when and how much to sell and/or store. Similarly, intermediate marketing firms must continually evaluate their inventories of inputs and outputs to make sure the flow of products into and out of their business is consistent with expectations about future business conditions. Ultimately, as consumers vary their purchasing and consumption rates of food products over time, marketing firms and producers alter storage decisions in response to market prices.

Governments may also get involved in influencing inventories of agricultural and food products. This can happen in a couple of ways. One approach is for governments to purchase and store agricultural and food products when prices are low with the intent of later releasing the product from storage when there is reduced production. The basic motivation for this type of governmental activity is to attempt to reduce fluctuations in price by evening out the flow of products over time. Another rationale for such governmental action is to stockpile certain quantities of food as protection against possible natural disasters. This point often resurfaces in public discussions after a sharp reduction in production of a major commodity and a subsequent steep rise in price has occurred. An important element that must be factored into these kinds of discussions is the extent to which governmental storage policies alter the incentives of businesses to store commodities.

Another way in which governments can influence inventory behavior is to pay some or all of the costs of storage. Any action that effectively reduces the private costs of storing commodities will encourage larger stocks to be held for longer periods of time than otherwise would occur.

Continuously Produced Commodities

Storage decisions are also an integral part of marketing activities for continuously produced agricultural commodities that are storable or can be converted in some form of storable product. Most types of livestock products are in this category. Much of the storage of these products is incidental to other value-adding marketing activities while these products move through market channels from production to consumption. In some cases, storage of these products at an intermediate marketing stage may be used to help synchronize seasonal changes in production and demand. For example, the ability to transform fluid milk into butter, cheese, ice cream, or other products with less perishability helps to balance fluctuations in milk production with demand.

The ability of marketing firms to stock up and hold extra inventories of certain products in anticipation of seasonal changes in demand helps to minimize some of the price fluctuations that otherwise might occur. For example, marketing firms can consciously alter storage decisions in anticipation of changes in demand knowing that the demand for certain products like turkeys and hams is especially affected by holiday eating patterns. Storage can also be used to accommodate anticipated changes in demand for food products influenced by season or temperature and thereby reduce fluctuations in production and/or prices.

STORAGE COST COMPONENTS

Any explicit or implicit decision to hold a commodity for a specific period of time involves a commitment to use a certain quantity of resources required to provide appropriate storage facilities. Consequently, any storage activity involves certain costs.[2] There are three components to the cost of storing any commodity. These are (1) direct resource costs, (2) financial costs of owning the commodity, and (3) costs associated with changes in quality of the product while it is being stored. Each of these cost elements will be briefly discussed.

Direct Costs

The direct costs of storing a commodity include the costs of owning and operating or renting holding facilities. Costs include annual fixed and operating costs of having an adequate facility to hold whatever commodity is to be stored. The facility may be a fairly simple storage shed or very special facility like controlled atmospheric facilities for storing apples. Costs associated with labor, utilities, taxes, and insurance are all part of the expenses of maintaining an adequate facility. Some of the cost elements of storage are essentially fixed costs, whereas other costs increase the longer a commodity is held. For example, commercial storage facilities often charge customers a fixed entry and exit (or loading and unloading) fee as well as a variable rate depending on how long the commodity is stored. Storage fees of commercial facilities are like prices of other goods and services in that they depend on the demand for storage space and supply of storage facilities in a given market. Storage fees generally tend to be higher at harvest time when the demand for storage is greatest. The fees will likely be less later in the year when there is less demand for storage.

Financial Costs

Another important cost of storing a commodity is the opportunity cost of the money that is tied up in owning a commodity. This cost is very obvious and explicit if someone uses borrowed money to purchase a commodity and places it in storage. Every month the commodity is held, additional interest on the borrowed money accrues. For example, suppose someone borrowed $300,000 at a 6% annual rate of interest to purchase 50,000 bushels of soybeans at $6.00 per bushel. Each month these soybeans were stored would mean an additional $1,500 of interest payments on the borrowed money. The monthly amount of interest is equivalent to a cost of $.03 per bushel. This component of storage costs definitely increases with the length of the storage period.

If someone has their own money tied up in a commodity being stored it is often easy to underestimate actual storage costs by overlooking or ignoring what might have been earned in some other kind of investment. The rate of return the financial resources could have been earning in some other kind of investment instead of having money tied up in a commodity should be considered as part of the cost of storage.

Quality Changes

A final component influencing the total cost of storing any item is what happens to the quality of the item while it is being stored. In many cases, there may be some deterioration in quality while an item is being stored due to uncontrollable circumstances. This may include a natural change in physical attributes of the merchandise or losses due to insects or other kinds of quality deterioration. Any decrease in value because of a change in quality of products while being stored is an additional cost that should be taken into account in making storage decisions.

In some cases, the quality of an item may actually be enhanced while in storage and its value increased. Examples of this are wine and cheese and other products that require a certain aging period to attain more desirable attributes.[3] In cases where storage enhances the quality of a product, the change in value is a negative cost that may compensate for some of the physical and financial costs of storage. In that case, the net cost of storage would be smaller than if only the physical and financial costs of storage were considered.

GENERAL PATTERN OF INVENTORIES AND PRICES

The key mechanisms as to how markets create incentives and communicate appropriate signals for allocating a fixed supply of a periodically produced storable commodity are rather straightforward. The forces can be most easily illustrated using some simplifying assumptions similar to what was done in introducing interregional trade concepts. After the basic principles are developed, some of the simplifying assumptions can be relaxed.

One of the simplification assumptions is that the demand for a periodically produced storable product is stable. Also for initial consideration it will be assumed that the same amount of total production occurs each year at a predictable time. Also no governmental involvement in holding explicit quantities of the commodity and no governmental subsidization of the costs of storage are assumed.

Inventory Levels

Under the previous circumstances, the inventory of a periodically produced storable commodity would be expected to be greatest at harvest and gradually decrease between successive harvests. Prior to each new harvest, inventories would be reduced to a minimum working level. In the absence of any uncertainty there would be no incentive to hold any of the commodity off the market prior to harvest in view of the forthcoming availability of a new crop. The minimum level of inventories would not necessarily be zero, however, because a certain amount of the product would be required to be moving continuously through the marketing system because new production would not be immediately available for consumption. Thus the only storage decisions to be made are associated with discretionary stocks over and above the quantities that are in the process of acquiring other value-adding marketing services.

A typical pattern of changes in stock levels of the commodity is illustrated in Figure 13–1. The total available quantity of a commodity is plotted on the vertical axis and time on the horizontal axis. The peaks in Figure 13–1, coinciding with harvest periods H_1, H_2, H_3, and so on, indicate when inventories would be replenished with new production. Stocks would begin to be reduced in varying amounts for consumption after harvest was completed.[4]

Price Levels

Changes in price corresponding to the changes in inventories are indicated in Figure 13–2. The lowest prices would occur at harvest time when the greatest quantity of the commodity exists. Higher prices would be expected during the interharvest period because of the costs associated with storing the commodity. The highest prices would be expected to occur just before the next harvest when the least amount of the commodity is available. Price changes during the interharvest period would depend on how much it cost to store the commodity for varying lengths of time.[5] The increase in price during the interharvest period creates an incentive for individuals to bear the costs of storage. If prices were not expected to increase, there would be no incentive for anyone to store any of the

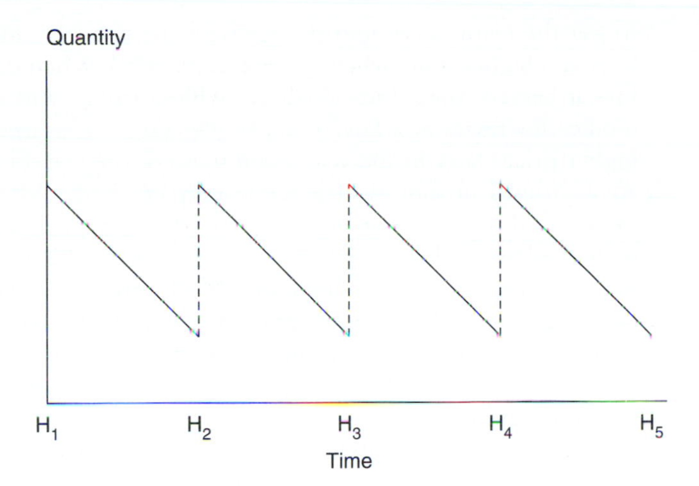

FIGURE 13–1 Stock levels of a periodically produced storable product at different points in time.

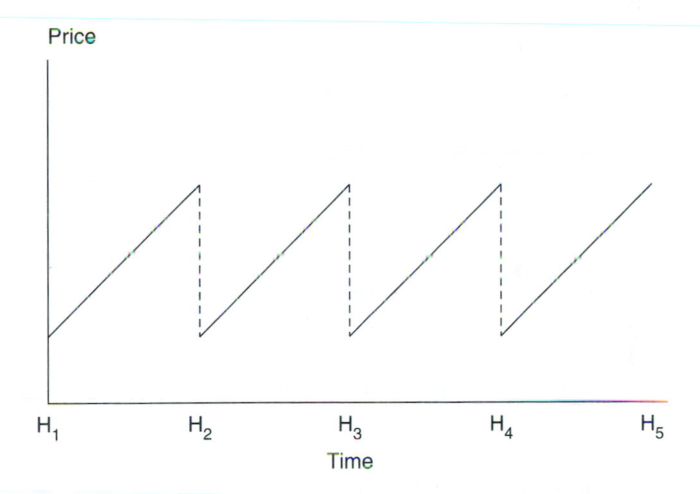

FIGURE 13–2 Price of a periodically produced storable product at different points in time.

commodity. The only reason that someone would decide to store more of a commodity than is currently required for other value-adding market activities is because of an expectation that the price at a later time will be sufficiently higher than the current price. As long as the anticipated price at some future time is sufficiently greater than the current price to compensate for the cost of storing a commodity, individuals would be willing to store commodities.

Price would be lowest at harvest time because of the availability of a new crop. A low price is an incentive to increase consumption, but also an incentive to put the commodity into storage with the expectation of being able to sell it later at a higher price when it is not as plentiful. When the price gets sufficiently low at harvest, some individuals are willing to buy and store some of the commodity for resale at a later time, anticipating consumers will be willing to pay higher prices later in the year when none of the commodity is being produced. As additional quantities of a commodity are withheld or withdrawn from the market and put into storage at harvest time, the current value or price tends to be bid up from what would exist if no one was willing to store any of the commodity. As more of the commodity is allocated to storage, the amount of expected increase in the future price or value of the commodity is likely to decrease. The reason for this is an increased amount of storage means an increase in the expected availability of the product for consumption later during the interharvest period.

Adjustments in current and future price expectations in response to storage decisions operate similar to the way spatial movement of products associated with interregional or international trade affects prices. An increased movement of a product into storage is like an increase in shipments out of a region that tends to increase current price from what it otherwise would be. Knowing that products going into storage will come back into the market at some point in the future tends to affect future price expectations. In making storage decisions, individuals will have an incentive to increase inventories as long as the difference between the expected value of a commodity in the future and its current value is greater than the cost of storing the commodity. As individuals collectively decide to increase storage stocks, there would be a tendency for current price to increase and a simultaneous decrease in expected future prices knowing that more of the commodity is being made available for use in subsequent time periods. Additional quantities would be stored until the difference between the expected future price and current price became equal to the cost of moving the product through time, in other words, the storage costs. Algebraically, this relationship can be represented as $P_i - P_1 = S_{1i}$ where P_i = the price anticipated in the ith future period, P_1 = current price, and S_{1i} = cost of moving the commodity from time 1 to time i. This is analogous to the equilibrium condition for interregional or international trade considering that S_{1i} is the cost of moving commodities through time.

Storage costs are likely to vary among individual market participants just like the costs of any other economic activity. Some individuals will have lower costs than others, because they are more efficient in providing storage services. When considering price relationships based on aggregate market behavior, however, a single cost of storage is used. The single cost can be interpreted as the cost for the marginal participant making a storage decision. Some individuals will have lower costs than the marginal participant meaning that they will have positive returns from storage whereas the marginal participant will break even. Also it is very likely that individuals will differ in their interpretations of what market conditions are likely to exist in the future. Thus, expected returns to storage can vary among market participants depending on how cheaply they can provide storage services and how much of a change they expect to occur in the value of commodities. Overall market-clearing price relationships for storage behavior are assumed to be determined by the marginal participant just like other forms of market behavior.

Although P_i could be assigned many different values in the above relationship to represent different times in the future, the most important implications can be derived considering a single relationship because storage decisions are sequential. This means that for something to be stored from time period 1 to time period 3 or 4, it must first be stored from time period 1 to time period 2. The prospective value of the commodity in time period 2 will reflect the anticipated value of having this commodity available for use or consumption in time period 2 as well as the value of continuing ownership in time period 2 because of prospective use in time periods 3, 4, and into the future. These ideas will be revisited again when a graphical approach is developed later in the chapter to illustrate the underlying market processes.

Modification of Simplifying Assumptions

Before turning to the graphical analysis, however, the effects of relaxing some of the earlier simplifying assumptions will be considered. For example, annual changes in stock levels might not be as abrupt as suggested by Figure 13–1 because harvesting may occur over an interval of time rather than occur instantaneously. Thus the replenishment process might occur more gradually as illustrated in Figure 13–3. Similarly, the readjustment of annual price levels might be expected to occur throughout the harvesting interval as illustrated in Figure 13–4.

If unexpected changes in annual production rates were to occur, inventory and price paths would not be as regular or as repetitive as illustrated in Figures 13–1 through 13–4 even with stable demand and storage costs. For example, the biological nature of agricultural production processes means that the actual amount is likely to fluctuate because of random effects even if producers planned the same amount each year. Thus, in some years the amounts of additional quantities at harvest are likely to be greater than normal and in other years the annual harvest is likely to be less than normal. The consequences of such outcomes are illustrated in Figure 13–5. As each new harvest season approaches, market participants have to make adjustments in the amount of anticipated quantities of products that are expected to exist. Smaller than normal anticipated harvests would result in an upward adjustment in price curtailing some consumption as the market developed new price signals to allocate available quantities over the next interharvest period as illustrated in Figure 13–6. For example, if it appeared that the next harvest was going to be much smaller than usual, there might be more of an incentive to hold some of the product for consumption next year. This would be reflected in bidding up price more than expected in anticipation of even greater prices in the next period. Of course, exactly the opposite tendency would be expected if market participants anticipated the next harvest to

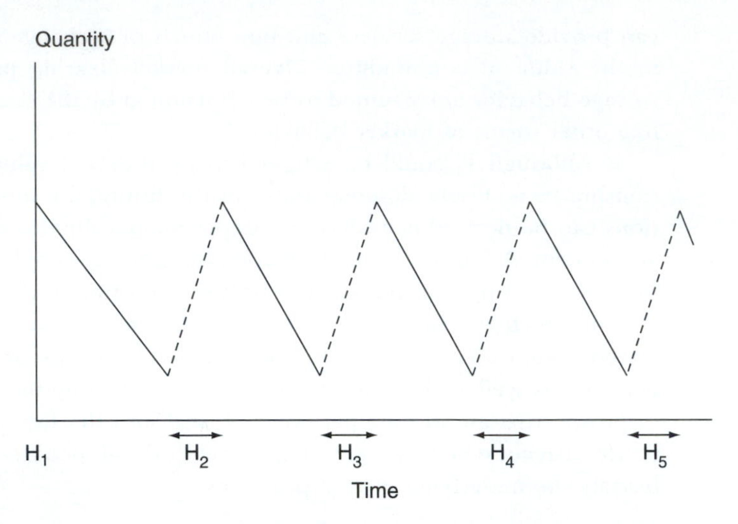

FIGURE 13–3 Stock levels of a periodically produced storable product harvested during brief intervals of time.

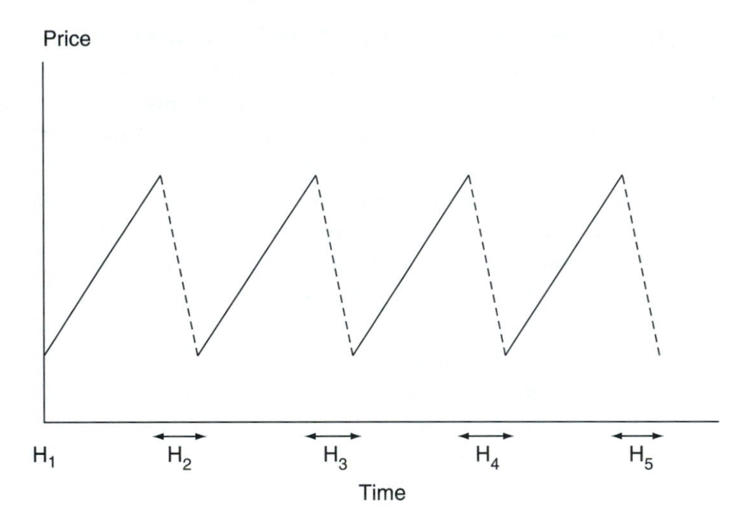

FIGURE 13–4 Price of a periodically produced storable product harvested during brief intervals of time.

be unusually big. In this case, individuals would attempt to empty out their storage facilities as much as possible and get down to the bare minimum required to keep the market pipelines operating because of an anticipated decrease in next year's average price. Even though the average annual price might be lower, there would still be reason to expect price to have an upward trend during the interharvest period to provide incentives for storing and allocating the quantity of a periodically produced storable commodity.

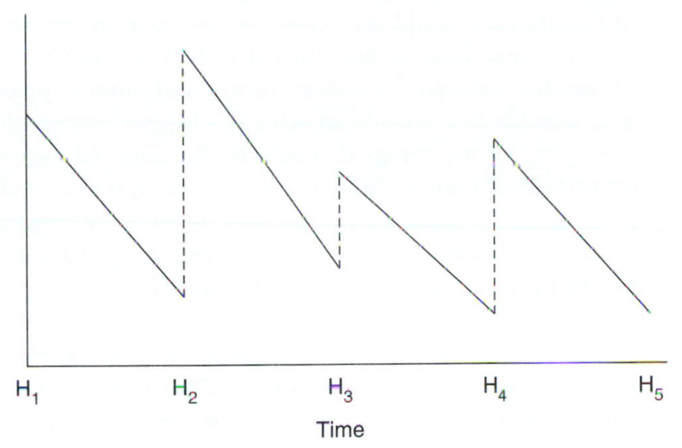

FIGURE 13–5 Stock levels of a periodically produced storable product with varying levels of production.

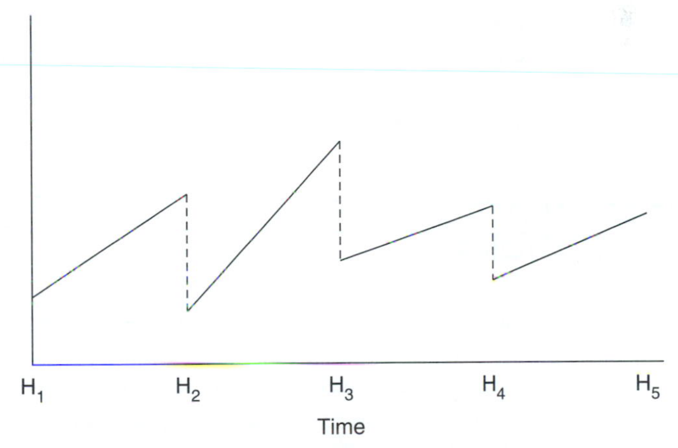

FIGURE 13–6 Price of a periodically produced storable product with varying levels of production.

Unexpected changes in demand or storage costs could also cause changes in the rates of consumption, storage, and prices even if production remained exactly the same year after year. Changes in demand or storage costs would result in the time paths of inventories and prices to be less regular than those illustrated in Figures 13–1 through 13–4. Nevertheless, the same tendency of inventories to be greatest and prices to be lowest immediately after harvest would still exist. Inventories would tend to decrease during the interharvest periods, but the rate of depletion as well as the amount carried over from year to year could vary. For example, an increase in storage costs caused by an increase in interest rates would be expected to cause people to decide to sell some of the commodity that they might have been planning to hold for longer periods of time. As this adjustment occurs, current price would decrease and a larger increase in consumption would occur as inventories decrease. A decrease in inventories would also cause an increase in the anticipated price for future time periods. Thus as current price declines and future price expectations increase, the difference between future and current prices would increase until a new equilibrium consistent with the higher storage costs occurred.

One of the things that affects the financial costs of continuing ownership or storage of commodities is a change in returns of alternative investment opportunities. As market participants continuously compare the expected changes in value of commodities with the continuing costs of storage and make adjustments in how much they wish to store (or sell), prices, inventories, and consumption rates will adjust.

In all of the above examples, storage decisions are based on how future market conditions are expected to differ from current conditions. This is similar to the kinds of decisions that are made with geographical movement of commodities in that decisions are based on current market conditions, but when the commodities actually reach their destination conditions may be different from what was expected. Thus, there is no guarantee that realizations from storage decisions will always turn out as anticipated unless the commodity is presold at a known price. In that case, the holder of the commodity is more like a provider of storage space for a fixed fee than someone trying to select an optimum selling time.

GRAPHICAL ANALYSIS OF STORAGE DECISIONS

Graphical analysis can be used to illustrate how market processes operate to determine how the total quantity of a periodically produced commodity is allocated over time. Similarities between intertemporal and interregional movement of commodities mean that some of the same graphical concepts used in earlier chapters can be used to illustrate storage decisions. The analysis uses a two-period model. The first period is the current time when choices between immediate consumption and storage decisions must be made. The second period is the next period of time into which stocks can be transferred. Supply and demand relationships are used to represent relevant market behavior for the current and future periods of time. Excess demand and supply relationships represent the storage market that links market behavior for the two periods.

Three different situations will be illustrated to demonstrate how alternative market circumstances can be analyzed for periodically produced commodities. The first situation considers period 1 to be at the end of harvesting a new crop and period 2 to be the following period when no new production occurs. The second situation considers period 1 and period 2 to be postharvest periods with no new production occurring in either period. The third situation to be presented considers period 1 to be a preharvest period and period 2 to be the time a new crop is anticipated to be available. After these three situations are analyzed, some of the ways the model can be generalized to consider storage decisions for continuously produced commodities with varying rates of consumption and/or supply are discussed.

Initially no costs of storage are introduced in the model to simplify illustrating how the demand and supply relationships for the two periods interact. This is analogous to the assumption of zero transportation costs that was initially made in analyzing interregional trade. Later an explicit cost of storage is introduced to illustrate how storage and consumption decisions are altered by this additional component.

Harvest and Postharvest Periods

The left-hand panel of Figure 13–7 contains a hypothetical demand and supply situation for a periodically produced commodity at harvest time. The demand relationship represents the schedule of quantities of this commodity desired for consumption at alternative prices in the current period. The supply schedule represents the total quantity (Q_T) of a commodity that is available for use or storage at alternative prices at this particular point in time. It includes the amount of new production that has been harvested and the inventory from the previous year being carried into the new crop year. The total supply for period 1 is illustrated in Figure 13–7 as a very inelastic supply relationship because once harvest has been completed the total quantity available for consumption or storage is a fixed amount regardless of current price. The intersection of the demand and supply relationships in the left-hand panel indicates what the market-clearing price would be if all of the available quantity had to be consumed immediately at harvest and there was no opportunity to carry any of the product forward in time. Thus the left-hand panel of Figure 13–7 is no different from the case of a periodically produced perishable or nonstorable product.[6]

The right-hand panel of Figure 13–7 represents the level of demand expected to exist in the next period. This demand represents the anticipated interest of all parties who might be willing to own varying quantities of this product at alternative prices in period 2 regardless of whether they intended to hold it for later use or consume the product in period 2. This means that the demand relationship in the right panel represents a reflection of anticipated consumption interests for several periods of time into the future. From the standpoint of allocating the available supply between consumption and storage in period 1, the only thing that matters is how much interest is anticipated in having some of the commodity carried into the second period. There is no supply relationship included in the right-hand panel of Figure 13–7 because the supply for period 2 will not be determined until the consumption and storage decisions are made for time period 1 given that no new production of the commodity is anticipated to occur in time period 2.[7]

The middle panel in Figure 13–7 represents the market for storage that links market behavior for the two periods of time. The panel for the storage market indicates the way interests in the commodity from period 1 interact with anticipated interests in the commodity for period 2. The supply of storage relationship in the middle panel is the excess supply schedule obtained from the demand and supply relationships for the current period. In other words, the excess supply represents the difference between the total quantity that is available

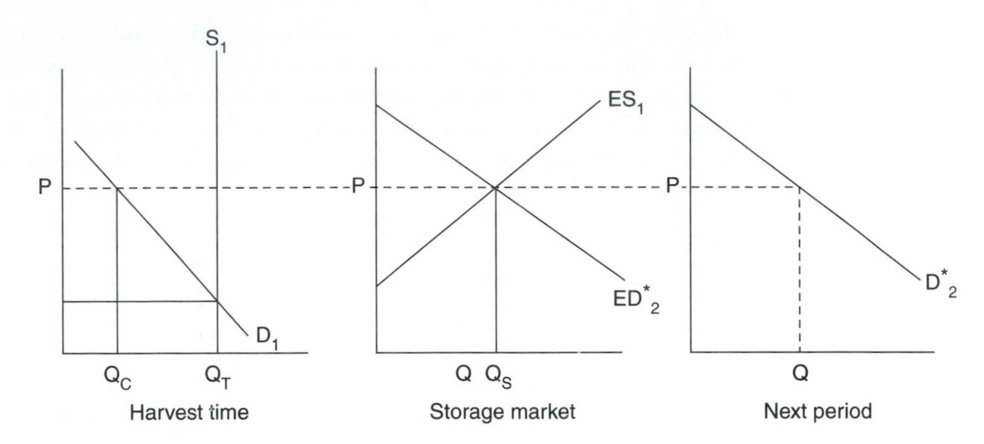

FIGURE 13–7 Two-period model of supply and demand situations for harvest and postharvest period.

and what would be consumed at alternative prices in period 1. In essence the excess supply relationship indicates the residual amounts of total supply in the current period that would not be consumed at alternative prices. Thus it is the schedule of quantities that are available for storage at alternative prices. Without any new production anticipated in period 2, storage is the only way that anticipated demand in period 2 can be satisfied.

The demand for storage in the middle panel of Figure 13–7 is identical to the anticipated demand for owning the product in period 2. It represents potential interest in having alternative quantities of the commodity stored between periods 1 and 2 at alternative prices. It also can be interpreted as the excess demand for the commodity that is anticipated to exist in period 2 because no new production is expected.

The intersection of the excess supply and demand schedule in the middle panel indicates the amount of the commodity that would be stored (Q_s) in period 1 and made available for period 2. The intersection also indicates the current price that would be consistent with this level of storage given current and anticipated demands and total availability of the product. The price is consistent with how much of the total available supply that market forces would allocate to current consumption and how much of the commodity would be stored for future use. The point at which the horizontal line representing the market price intersects the demand relationship in the left panel indicates the amount of the commodity that would be consumed (Q_c) in period 1. Given the way in which the excess supply relationship for the storage market is defined, it is clear that $Q_S = Q_T - Q_C$. Thus, the graphical solution for the amount of storage, consumption, and price is totally consistent with the demand and supply situations in period 1 and the anticipated demand for the commodity in period 2.

Given the assumption of a zero cost of storage, the solution implies that prices would not be expected to change over time. The price is exactly what would be required to induce the correct amount of consumption from current stocks of period 1 to leave the correct quantity of product available for period 2 based on anticipated demand. The middle panel of Figure 13–7 can be modified, however, to incorporate storage costs in the same manner that transportation costs were incorporated into trade models. For example, by vertically adding the cost of storage to the original excess supply schedule in the middle panel of Figure 13–7, a more realistic representation of the schedule of quantities available for the next period (at next period's prices) would be obtained. This is illustrated by the dashed line in the middle panel of Figure 13–8 assuming no change in the left- and right-hand panels from Figure 13–7. Vertically adjusting the excess supply relationship to reflect the additional per unit cost of storage is like obtaining a derived excess supply relationship in terms of period 2's prices. The derived excess supply relationship simply represents the fact that the price in period 2 has to be higher than the price in period 1 by the amount of storage cost in order to induce someone to store the commodity.

Similarly, the original excess demand relationship in the middle panel of Figure 13–7 can be shifted vertically downward to obtain a derived excess demand for the product in terms of period 1's prices (net of storage costs). This relationship is illustrated by the dashed excess demand relationship in the middle panel of Figure 13–8. Assuming that the per unit cost of storing a commodity from period 1 to period 2 is a fixed amount regardless of the amount stored results in two sets of parallel lines for the middle panel of Figure 13–8. The two intersections of the primary and derived excess demand and excess supply relationships at a quantity of Q_S indicate the amount of the commodity that would be stored. This quantity is clearly less than what would have occurred if there had been no cost of storage (or the cost was subsidized by the government). The two intersections of relevant excess demand and supply relationships in the middle panel also indicate that the prices for the two periods (P_1 and P_2) would differ by the cost of storage. The price for the current period is what actually would exist given the underlying demand and supply situations represented in the diagram. The market-clearing price for period 2 however is an expected price for the next period. Whether or not the expected price for period 2 actually occurs depends on the accuracy of the forecasted future demand conditions. If the demand for ownership of the commodity in time period 2 should turn out to be quite different from what was expected at time period 1, the market-clearing price could be quite different from what was expected when earlier consumption and storage decisions were made. This is why there is no guarantee that the returns from storing a commodity will always be sufficient to cover the costs of storage except where a guaranteed future price of the commodity has been established. In some cases the returns from holding a commodity may be greater than the costs of storage and in other cases the returns may be less depending on market conditions.

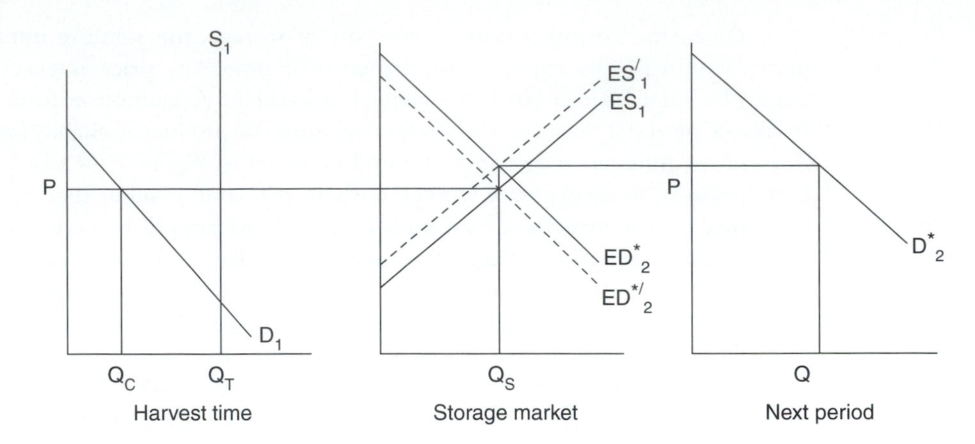

FIGURE 13–8 Incorporating cost of storage into a two-period model of storage.

The middle diagram of Figure 13–8 can also be used to demonstrate how a change in the cost of storage alters the allocation of available supply between current consumption and storage. The effect of an increase in storage costs could be analyzed by increasing the vertical distance between the original and derived excess demand and supply relationships. The original excess demand and supply relationships would remain the same, but the derived functions would be shifted.[8] Thus an increase in storage costs would result in a decreased amount of storage (and more current consumption), a decrease in current price and an increase in the expected price for period 2 relative to what would happen with lower storage costs.

Two Postharvest Periods

Analysis of storage and consumption decisions for two postharvest periods is fairly similar to the previous situation. The only difference is that in this case the total available supply for period 1 is determined entirely by the storage decisions of the previous period. Now, the first period is similar to moving forward one period from the situation previously considered. Consequently, the first period becomes the realization of what was expected to exist in the second period in the previous scenario. The demand relationship in the left-hand panel in Figure 13–9 again represents the demand for current use or consumption of the commodity. The excess supply relationship in the middle panel of Figure 13–9 represents the schedule of differences between the inventory and the quantities demanded for current consumption at alternative prices. The excess supply schedule represents the schedule of quantities that would be available for continuing storage at alternative prices. The demand schedule in the right-hand

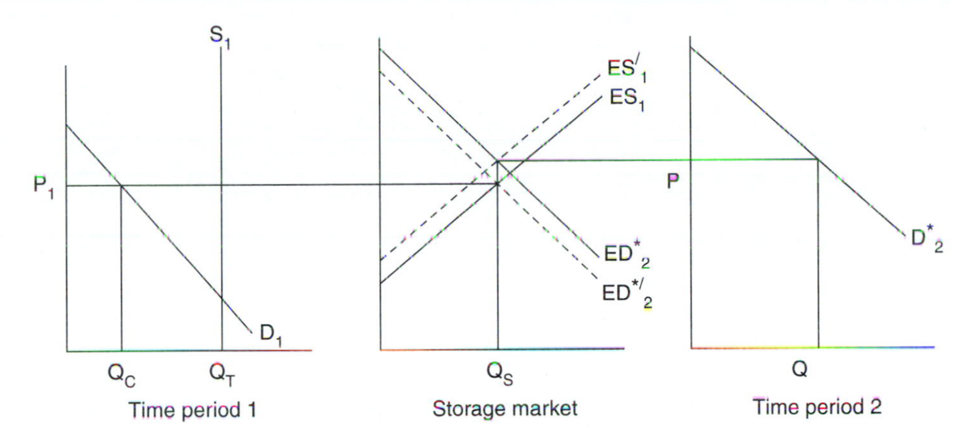

FIGURE 13–9 Two-period model for two postharvest periods of periodically produced storable commodity.

panel of Figure 13–9 would be a representation of what the future demand for owning this commodity in the next period is anticipated to be. This demand represents the next period's excess demand for storage and is reproduced in the middle panel. The intersection of the excess demand and supply relationships in the middle panel would indicate how much of the current inventory of the commodity would be held back or stored for future use. The market-clearing price and quantity available for immediate consumption are also simultaneously determined by the intersection of the excess demand and supply relationships in the middle panel.

Storage costs could be incorporated into the middle panel of Figure 13–9 in the same manner as before. This would lead to the correct price differential between the current price and anticipated price for the next period to provide the appropriate price incentive to store enough of the commodity to meet anticipated demands in the future.

Preharvest and Harvest Periods

As time continues to elapse, sooner or later the situation will arise where the first period in a two-period framework becomes the time immediately preceding the next harvest. In this situation the relationships for the left-hand panel of Figure 13–10 would be similar to those in Figure 13–9. That is the demand would be the demand for current use in that period and the supply would be the beginning inventory.

The situation for the right-hand panel in Figure 13–10 is quite different from Figure 13–9, however, because of the anticipated availability of a new crop. Consequently, Figure 13–10 has a supply relationship representing the anticipated

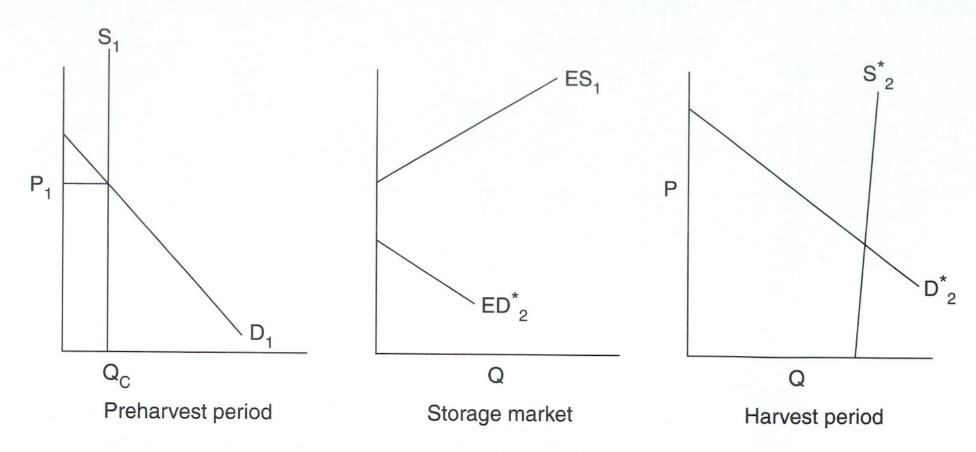

FIGURE 13–10 Two-period model for preharvest and harvest periods for periodically produced storable commodity.

new supply of the commodity that would be expected to be coming onto the market in the next period.

The introduction of a supply relationship for the next period means the demand for storage is no longer just the anticipated demand for the next period as in the two situations previously considered. Given usual anticipated supply conditions, it is likely that the excess demand relationship for period 2 would exist only at fairly low prices. In fact, in Figure 13–10 the excess demand relationship based on the relationships in the right-hand panel intersects the vertical axis of the middle panel at a price lower than where the excess supply relationship from period 1 intersects the axis. Such a situation indicates the absence of any incentive to continue to store any of the commodity from the preharvest period into the harvest period.[9] In fact if the kind of relationships depicted in Figure 13–10 were observed in regional trade models it would imply a price incentive to move some of the commodity from the lower-priced market (market 2) to the higher-priced market (market 1). This is obviously not possible with intertemporal transfers, however, because the product can move in only one direction through time.[10] There is no way that some of the commodity in period 2 can be transferred to period 1. In the case of interregional trade, a reversal in price differences may be sufficient to reverse the flow of trade between regions. In the case of storage however, it is not possible to reverse the storage of the commodity through time. This is why the price of a commodity prior to a harvest period can be considerably greater than the price that is expected to occur in the very near future. If anticipated future supply of the product is expected to be normal and have a lower price than the current value of essentially the same homogeneous product there is no incentive to store anything more than a minimum amount to keep the market pipelines operating.

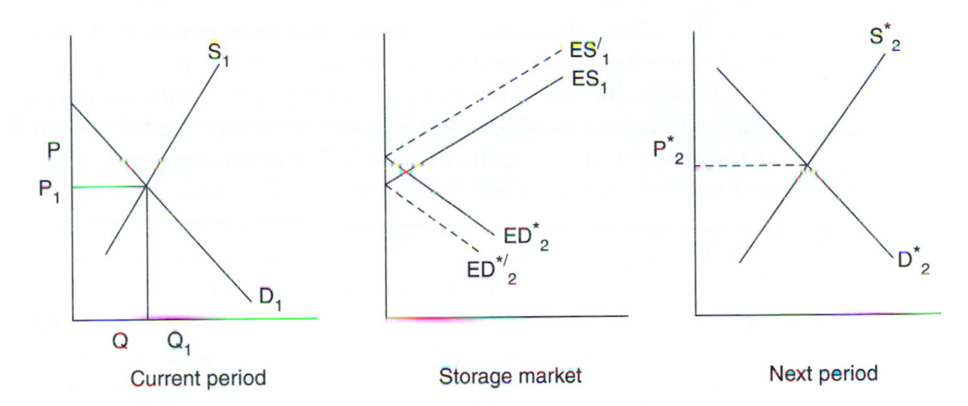

FIGURE 13–11 Two period model for continuously produced storable commodity.

Continuous Production

The introduction of a forthcoming supply of a periodically produced commodity in the previous scenario provides an easy transition to consider how the model can be used to analyze storage possibilities for continuously produced commodities. Such a situation is illustrated in Figure 13–11. The right-hand panel of Figure 13–11 is similar to the right-hand panel of Figure 13–10 in terms of representing anticipated demand and supply conditions for the next period. These relationships can be used to determine the range of prices for which there would be an excess demand for carrying some of the currently available commodity into the next period.

The left-hand panel of Figure 13–11 contains the demand for current consumption and the availability of the commodity at alternative prices from current production as well as any stocks carried into the period. The demand and supply relationships for the current period would determine the location of the excess supply relationship indicating the schedule of current quantities of the commodity that would be available for storage into the next period at alternative prices. The relative locations of the demand and supply relationships for each of the two periods would indicate whether there was any expectation of a price incentive to store any of the commodity. The market conditions could be such that essentially all of the commodity available in the current period should be consumed. On the other hand, if anticipated demand and supply relationships for the next period resulted in a high enough level of excess demand there could be an incentive to store some of the commodity even though it is produced continuously. The most likely circumstances when storage of continuously produced commodities might be expected to occur are if there were considerable seasonal variability in demand or supply. If it were impossible, or at least costly to vary production schedules to satisfy variable demand it could be advantageous to store some of the continuously produced commodity in anticipation of peak demand periods.

The above situations indicate that the basic two-period framework is very flexible to consider all kinds of changes in circumstances to illustrate the flow of commodities into and out of storage. The diagrams indicate how markets operate to influence current decisions based on current market conditions and what is anticipated to occur in the next period. Decisions about storage, consumption, and market-clearing prices in the current period are influenced by anticipated future market conditions. The incorporation of per unit storage costs into the model indicates the difference between the expected market-clearing price for the next period and current market price is equal to the **marginal cost** of storing the commodity. As new events result in new **market demand and supply** conditions, prices may differ from what was anticipated at an earlier time. Consequently, the returns to storage activities can vary from what was anticipated when initial decisions were made. Market participants have to continually reassess decisions about storage in response to new information about current and anticipated market conditions.

SUMMARY

Storage of agricultural and food commodities is one of the three physical value-adding functions of agricultural markets identified in Chapter 1. Holding or owning commodities for various intervals of time is a necessary and important element in coordinating production with consumption activities. From an economic perspective, storage is simply moving products through time or making an already existing product available for future consumption or use. Some storage occurs in conjunction with other value-adding activities that change the form or location of commodities as part of the overall coordination of production and consumption activities.

Storing products for any period of time requires the use of economic resources. Consequently, there are three distinct cost components associated with storage. The first component is the direct cost of using physical and human resources required to hold a commodity for a specific period of time. These are relevant costs whether someone owns their own facilities or hires a specialized firm to do the storage activity. A second cost component of storage is the financial cost associated with the opportunity return on the amount of money invested in owning a commodity for a period of time. The third and final cost component is the change in value associated with changes in a product's characteristics over time. Many products experience a decrease in desirable attributes as they age, although the quality of some products improve the longer they are stored and the associated change in value can offset storage cost components.

Storage is the mechanism that makes it possible to consume and use periodically produced commodities throughout the year. In some cases, highly perishable products must be converted into an alternative form before any storage is even feasible. For other products, storage helps coordinate fluctuations in production and consumption. As individual market participants make decisions to store commodities and thereby make them available for future use they have to evaluate the expected return and cost of such activities. Thus, it is not surprising that prices generally tend to increase as inventories decrease between harvest periods. As individuals decide to increase the amount of a commodity in storage, it tends to bid up the current value of a commodity and simultaneously decrease future price expectations. This is similar to the way interregional and international movement of products leads to opposite directional changes in price. Decisions that are made about the amount of a commodity to be held for future use, based on current and future price expectations, are essentially the way markets determine the rate at which products are consumed or used over time. As new information about prospective demand and supply conditions become available, market participants adjust inventory decisions.

The way in which current and prospective demand and supply conditions determine the quantity of a commodity that is stored instead of immediately consumed can be graphically illustrated using a two-period model similar to the way geographical movements of products were analyzed in earlier chapters. The model reflects the basic idea that storage is a way of moving products through time. One difference from the earlier models used to analyze geographical movements of products, however is that storage can move products only in one direction i.e., from the current time into the future and not vice versa. The amount of a commodity to be stored is determined by the intersection of the excess supply of the current period with the expected excess demand for the next time period. The cost of storing a commodity is introduced in the same way as transportation costs were handled in earlier chapters. The introduction of storage costs results in the difference between the current price and the expected price of the commodity in the next period being equal to the cost of storage consistent with current and anticipated demand and supply conditions.

QUESTIONS

1. Identify and briefly discuss the three different components of storage costs.

2. Suppose that you own 50,000 bushels of a nonperishable commodity that you currently could sell for $4.00/bu. How much of a change in price of this commodity would be required to offset the financial cost of storing this commodity for 1 month if you could be earning 6% per year in an alternative investment? Show how you arrive at your answer.

3. Briefly explain how a decrease in short-term interest rates would be expected to affect the cost of holding nonperishable commodities and individual entrepreneur's willingness to provide storage service.

4. Why is there generally little economic incentive for individual firms to hold large quantities of grains more than a year?

5. Assume that the demand and supply relationships in the left- and right-hand panels of the following figure represent the current and anticipated market conditions for a periodically purchased commodity for two adjacent time periods. Graphically illustrate how much of the commodity available in the current period would be consumed and how much would be stored. Also graphically illustrate what price you would expect to observe under these circumstances assuming all storage costs are paid by the government from general tax revenues so market participants did not have to bear or take into account any costs of storage.

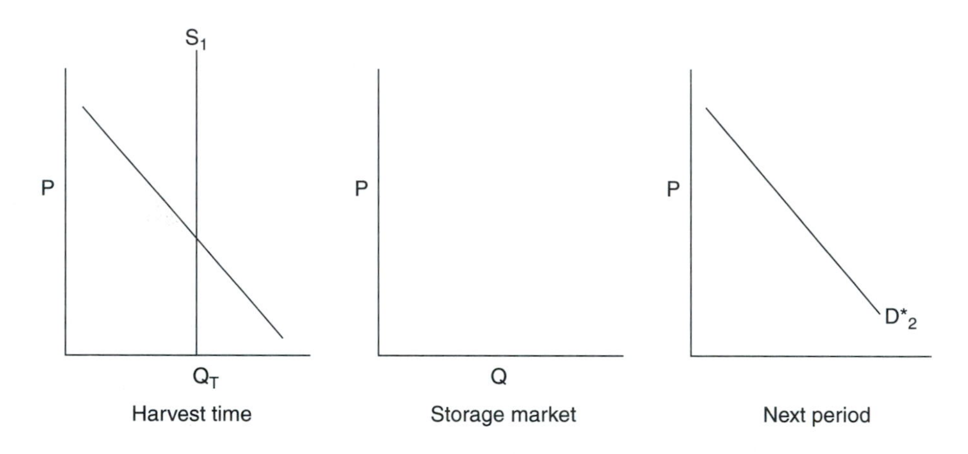

6. Assuming no change in current or future supply conditions for a storable commodity, briefly describe how a significant increase in the anticipated future demand could possibly have an immediate effect on current use and price of the commodity.

NOTES

1. A variety of growing seasons in a country like the United States and the availability of products through international trade also mean longer periods of time that fresh products are available to consumers even without additional preservation.

2. The extra costs of storage may not be as evident if the holding is an involuntary action that is part of another value-adding marketing activity then when an explicit holding decision is made in anticipation of a higher price. This is because the extra costs of storage may be absorbed as part of the costs of other activities and difficult to identify separately.

3. It is possible to argue that some aging of products is actually part of production rather than marketing activities and consequently not like the holding of other commodities that could immediately be used.

4. Even with a stable demand, the rate of consumption or use would be expected to change during the interharvest periods in response to changes in price. As the price of the commodity increased after harvest to reflect the costs of storage, the rate of consumption could be altered. Assuming no seasonal changes in demand, consumption would be greater immediately after a harvest when price was lower than other times of the year.

5. The rate of price changes illustrated in Figure 13–2 assumes that storage costs are linearly related to how long the commodity is stored.

6. In the latter case however, the supply relationship would include only the current harvest in that there would be no inventory from the previous year.

7. Once time period 1 is over with, the decision process repeats itself in terms of how much of the available supply in time period 2 should be consumed and how much should be stored for future uses.

8. This would be analogous to how changes in the marketing margin and transportation rates were analyzed in earlier chapters.

9. If the anticipated new harvest was expected to be sufficiently below normal, the excess demand relationship from the right-hand panel might be at a level providing some incentive to further reduce current consumption and hold some of the commodity for use in the next year.

10. Any excess demand relationship existing for the first period is totally irrelevant in terms of moving a commodity through time.

CHAPTER 14

FUTURES MARKETS

This chapter is the first of two chapters concerning futures markets. It indicates some of the ways in which futures markets differ from cash markets and describes some of the basic characteristics and procedures of futures trading.

The major points of the chapter are:

1. How futures markets have changed over time.
2. The nature of a futures contract and how contractual obligations are fulfilled.
3. How speculators can profit from trading in futures markets when prices increase or decrease.
4. The margin requirements and other procedures associated with taking positions in futures markets.
5. How to interpret price and other information about futures market activities.

INTRODUCTION

Futures markets are special kinds of markets designed to assist agribusinesses (including farmers) and other individuals discover prospective prices for a number of agricultural and other commodities. These kinds of markets have existed for agricultural commodities in the United States for more than 150 years. The first part of this chapter discusses some of the origin and evolution of these markets. Next some important characteristics that facilitate the trading of standardized futures contracts are described. The middle part of the chapter describes some of the procedures used for buying and selling futures contracts. Futures market trading is initially described from the perspective of speculators who can use futures markets to make money by correctly anticipating upward or downward movements in commodity prices. The remaining parts of this chapter indicate how to interpret different kinds of market information about futures market activities and the way futures markets and trading are regulated in the United States.

HISTORICAL PERSPECTIVE

Futures markets exist in the United States for many types of agricultural commodities, especially grain and livestock (e.g., wheat, corn, soybeans, hogs, cattle). Futures markets also exist for other food and fiber products such as cocoa, coffee, cotton, and orange juice. Another set of futures markets exists for lumber, plywood, and various metals (copper, gold, and silver). A number of futures markets also trade contracts consisting of financial instruments such as United States Treasury Bills or stock market indices and exchange rates such as the Japanese yen, the Swiss franc, the German mark, and so on.

The number of different types of futures contracts traded in the United States has expanded rapidly in recent years. For example, in 1998 the total number of futures contracts traded in the United States was 500,562,510 or nearly 73% more than the total number of contracts traded in 1992 (**Commodity Futures Trading Commission** 1998). Much of the recent growth has been associated with the financial futures markets. In fact, financial instruments and currencies accounted for almost 70% of all futures markets transactions in 1998 (Commodity Futures Trading Commission 1998).

The number of different kinds of futures contracts available for trading continues to change as one exchange or another initiates new contracts and the trading of other kinds of contracts is terminated. One example of this is the introduction of a lean hog carcass contract a few years ago in place of a live hog futures contract. Futures markets are very dynamic in terms of the different types of contracts traded, where they are traded, and other characteristics. All futures markets operate very similarly regardless of whether the futures contracts are for yen, lumber, or soybeans. Futures markets exist in a number of countries besides the United States.

There is some evidence that activities similar to today's futures markets existed in Japan in the seventeenth century. Something resembling a futures market may have even existed as early as medieval days. Futures markets in the United States began in the mid-1800s. At that time, the economy of the western part of the United States was just developing. Agricultural marketing and processing firms were increasing in size and becoming more specialized. These types of firms were in business to perform an economic function, such as providing transportation and storage services or converting raw commodities into a form more readily usable by consumers. The economic role of grain elevators was to facilitate the flow of storable products through market channels throughout the year between harvests. Specialized marketing firms were primarily interested in earning a return for the economic services they were providing without having to gamble or speculate on changes in the price of the commodity they were handling. A sudden drop in price could be very disastrous by more than offsetting the value added to a commodity by a marketing firm.

Increasing distances between the location of production of agricultural commodities and major processing and consumption centers meant that commodities had to be transported longer distances than in earlier periods. Longer transmittal times resulted in increased price risks associated with ownership of commodities when they were being transported. For example, market prices could change substantially between the time grain was loaded in railcars at the point of production and the time of arrival at some destination for processing or storage. Market prices could change substantially during the time the commodity was in transit. Producers or shippers did not especially want to ship commodities not knowing what they were going to receive when the product reached its destination. Similarly, buyers did not especially want to be committed to a particular price at the time the commodity was shipped without knowing what it would be worth when it reached its destination.

One reason futures markets in the United States initially began trading grains was because of substantial price variability of these commodities. Price variability of grain was especially pronounced because of the seasonality in production and variability in production from year to year due to weather conditions. A sudden change in price could occur because of changes in weather conditions influencing supply prospects. Furthermore, no public or private crop reporting systems existed to provide reliable estimates of potential crop supplies. Consequently, markets operated on fragmented pieces of information that could cause sudden and erratic fluctuations in price.

The characteristics described above provided economic incentives for people with financial interests in commodities subject to fluctuating prices desiring to renegotiate ownership positions in commodities. As market conditions changed, the value of one's commitment in a commodity would fluctuate and consequently it might become desirable to be free from a previously arranged agreement to buy or sell the commodity at a specified price. For example, in case of a commodity in transit, losses associated with a price decline could be minimized by being able to trade ownership interests in the commodity before it was received if further reductions in price were expected to occur.

As traders began to get together and renegotiate their contracts, it became efficient for all of the traders with interests in the same commodity to meet at one location to facilitate continual renegotiations of their positions. Thus, futures markets evolved because of incentives to have a meeting place allowing potential buyers and sellers of contracts to negotiate future purchase or selling commitments in particular commodities. The benefits of having uniform trading practices and ways of settling contract disputes led to organizations of businessmen to facilitate commerce in commodities. These organizations led to the development of futures markets. Futures markets are essentially places where the ownership of obligations to provide or take delivery of commodities at some *future* point in time are traded.

SOME CHARACTERISTICS OF FUTURES TRADING

The buying and selling of contracts specifying a future exchange of ownership of various commodities are the activities that occur in futures markets. A futures contract is an agreement or obligation to exchange ownership of a commodity at a fixed price at a particular point in time at a specified point of delivery.[1] The particular point in time is automatically specified in terms of when the contract expires. Futures contracts are different from forward contracts negotiated privately between two parties. Forward contracts are the result of individual negotiations and may vary in terms of contractual details such as delivery date, financial arrangements, and so on. The lack of standardization in forward contracts requires that any renegotiation of the contract involves the original two parties who made the agreement. On the other hand, all futures contracts for a given commodity are standardized and consequently are perfect substitutes for one another. This standardization means that a futures contract for May corn can be easily transferred to another party since all characteristics except price of the contract are clearly specified by the exchange where trading of that contract occurs.

Nature of Contracts

One standardization characteristic of futures contracts is the quantity of the commodity associated with each contract. If you are trading a corn contract on the Chicago Board of Trade (CBOT), it is for 5,000 bushels. You cannot sell a contract on the CBOT for 5,080 or 4,800 bushels. The quantity of corn is specified at 5,000 bushels per contract. Also, the exact quality of the commodity is specified in the contract. The price is quoted for a particular USDA grade. Futures contracts also specify single or multiple **delivery points.** For example, in the case of corn contracts traded on the CBOT, delivery can be made at either Chicago or several points along the Illinois and Mississippi Rivers with prearranged differentials in price for each location.

Anytime someone buys or sells a futures contract in a given exchange at a given price, all characteristics of the contract other than price are exactly the same as all other contracts in that commodity with the same expiration date. There can be no additional addendums or paragraphs in the contract, which change the terms of the contract. The only difference among contracts for a commodity on a given exchange will be the *price* at which the transaction occurs and the *month* the contract expires.

Futures contracts are obligations either to provide or take delivery of a specified commodity of a given quality at a particular location. An individual can sell an obligation (or a promise) to deliver a commodity at a particular price without even owning the commodity. All that is occurring is that a promise is being made to deliver the commodity at a particular time in the future. Similarly, a contract that obligates the owner to take delivery of the commodity can be acquired even if the person never wants or would not know what to do with the commodity (e.g., 40,000 lb of pork bellies). Most of the obligations undertaken through futures contracts are not fulfilled by the delivery process but are subsequently cancelled by taking an opposite position in the futures market. For example, undertaking an obligation in the futures market to buy the commodity anytime before trading in the contract expires can offset an earlier obligation made in the futures market to sell the commodity. Similarly, an obligation to buy a commodity can be offset by subsequently selling a contract. If the number of identical futures contracts purchased and sold by a given individual is equal, the person is not obligated or required to make or accept delivery of the commodity when trading in the particular contract expires.

Anybody who has bought a contract and not taken an offsetting position in the market is said to have a long position in the market. Similarly, anybody who has sold a futures contract that is not offset by a corresponding purchase is said to have a short position. The easiest way to remember this terminology is to think about what would occur if a trader did not take any further action in the futures market after establishing an initial position. If a contract has been sold, there is an obligation to deliver the commodity when the contract expires. Thus, as far as the futures market is concerned the trader would be short in the commodity if he did not do anything further. This obligation may involve a commodity the trader does not own. This is why the "short" terminology is especially appropriate. Similarly, if a contract has been purchased, the trader is obligated to accept delivery. He may not even want the commodity. Thus, he is "long" in something he may not even want, but will own if he does not take any further action in the futures market. Taking a short or long position only means that you have bought or sold a futures contract without an offsetting transaction.

Margin Requirements

Anyone can buy or sell a futures contract provided the person has sufficient money or a line of credit to cover part of the value of the contract. Usually 5%

to 20% of the value of the contract must be posted as a margin requirement prior to the purchase or sale of a futures contract. The margin requirement can be cash, a liquid financial asset, or some assurance that there is a line of credit from a financial institution. A futures market transaction cannot be executed until the margin requirement is satisfied. The margin money is placed with a broker who is either a member of a futures exchange or who can forward orders to a member of the exchange to buy and sell contracts. The size of the margin requirement depends on whether the individual wanting to take a position in the futures markets owns some of the commodity being traded or is just a speculator hoping to profit by correctly guessing the direction of price changes in a market. The size of margin requirements can change over time but the above values illustrate that a speculator may be able to put up $1,000 to $4,000 of margin money, and acquire an obligation in a contract worth $20,000. This means that a trader in futures contracts has only a fraction of the total value of a futures contract at risk when a futures contract is initially purchased or sold.

The margin requirement provides an amount of money to cover potential losses in the futures market. If the price of the futures contract moves in an adverse direction after establishing an initial position in the market, margin money is available to cover the potential financial loss. For example, if an individual has sold a corn contract and the price goes up $.01 per bushel, the owner faces a potential loss of $50 ($.01 × 5,000) in taking an offsetting position in the market since the contract is for 5,000 bushels. If the price increases by $.10 per bushel, the potential loss would be $500. If an initial margin of $.20 per bushel (or $1,000 per contract) is required to get into the futures market, a $.10 per bushel increase in price produces a 50% loss in the margin account of a person with a short position in the market. This example illustrates how substantial potential losses, as well as gains, in futures market transactions can result from relatively small price changes. An individual stands to gain or lose the effect of small changes in price on the value of the entire quantity of the contract even though the amount of money invested in a margin account is only a fraction of the total value of the contract.

The financial positions of all continuing obligations in futures markets are recalculated using the **closing price** at the end of each day's trading. The closing price is used to determine whether value is added or subtracted from each trader's margin account. The process of determining all participants' financial positions in futures markets at the end of each trading day is known as **marked-to-market.** This means that the value of all remaining or continuing obligations in the futures markets are recalculated (or marked-to-market) according to the closing market price for the particular contracts involved.

Margin money is required whether an individual is interested in buying or selling a contract. The full value of the contract is not paid or received when the initial purchase or sale is made. Thus if an individual initially sells a futures contract, he does not receive the value of the sale. In fact the person must pay an amount equal to the margin requirement in order to sell. Similarly, when a futures contract is purchased, the full purchase price of the contract is not paid. All that is paid is the initial margin requirement. If the individual makes no further transactions, the full value of the contract in terms of the price specified in the initial obligation would be required at the time the contract expired. If a person initially had bought a contract, the full value of the commodity purchased (less the initial margin) would have to be paid when the transaction expired. On the other hand, if an individual had sold a contract, the full value of the contract (plus the initial margin) would be received when the commodity was delivered to fulfill the initial obligation.[2]

If the price increases after a futures contract has been purchased, the owner (i.e., the buyer) of the contract makes money. Alternatively, if the price declines, the owner of the contract loses money but the seller of a futures contract profits. Similarly, price increases produce losses for individuals who sold contracts because something was sold at a lower price than it is currently worth. For example, if a contract was initially traded at a price of $3.00 per unit, but now is trading at a price of $3.10 per unit, the original buyer has a potential gain of $.10 per unit while the original seller would stand to lose $.10 per unit. The latter situation is the same as depleting the margin account since an obligation has been made that will now cost more money to fulfill. Thus, money can be made or lost from upward or downward changes in price depending on whether an individual initially buys or sells a futures contract. An increase in price does not necessarily mean you make money in the futures market. It depends on the trader's initial position in the market.

Once an obligation in a futures market is established, by either having bought or sold a contract, the price might behave unexpectedly resulting in that obligation having more or less value than when the initial position was taken. As prices and associated values of futures contracts change, the margin account will either increase or decrease depending on changes in value of the trader's position in the market. If the price increases after a contract has been purchased, some potential profit has been made. As long as the price remains above the price at which the contract was purchased, the trader's margin account will reflect the additional potential profit and could be withdrawn. If the contract is sold subsequently at the higher price, the trader would get the original marginal requirement back plus any increase in value of the contract based on the difference between the price at which the contract was bought and the price it was sold (assuming no profits had been withdrawn from the margin account), less a brokerage fee. Anytime the price is moving in a favorable direction for a trader, money is added to the margin account.

The margin account can shrink however if the price moves in an adverse direction. If a trader's margin account decreases to a specified minimum amount at the end of a trading day, a **margin call** will be issued. A margin call indicates that more money must be added to the margin account or the trader's position in the market will be closed out by taking an opposite position in the market eliminating any continuing obligation in the futures market. If a contract is initially sold, the obligation is offset when a similar contract is purchased at some price. If all traders knew exactly how the price was going to change, there never would be any losses. The uncertainty of prices is partly what creates a futures market because no one knows precisely how price may change. Therefore, there is always a chance of potential profits or losses depending on what happens to prices.

Exchange Membership

Any individual or firm that desires to do a large volume of futures trading may find it advantageous to become a member (or own a seat) on a futures exchange to avoid paying a brokerage fee or commission every time they wish to make a transaction in the futures market. Each exchange has a fixed number of memberships. For example, the Chicago Board of Trade has approximately 1,400 members. This means that someone has to be willing to sell their membership in order for a new trader to become a member. The price of memberships (or seats) is negotiated just like the values of other commodities and fluctuates over time depending on relative interest in memberships. Before someone can become a member of an exchange, other members of the exchange must approve the membership application. Sufficient financial resources are required to cover potential losses for transactions made by members of futures markets. Credit requirements for membership protect other parties against potential losses.

Some members of the exchange trade contracts just for themselves, rather than executing orders for the public. They make their livelihood by following price movements in the market attempting to buy at low prices and sell at high prices at appropriate times. If they sense that the market price is going up, they will buy contracts and sell them when the market starts turning around. They use their intuition and knowledge about the commodities to know when it is a good time to trade. These are the so-called **scalpers** or market makers that keep futures markets liquid by being willing to trade on a fraction of a cent difference in price. Many of these traders buy and sell exactly the same number of contracts every day so they do not carry obligations from one day to the next. They are referred to as day traders and look for profits on small price movements in the market. A confusing aspect about futures markets is that traders can sell contracts even if they do not own any of the underlying commodity. However, this is similar to a producer agreeing to sell his crop at a particular price before a crop is even planted. What must be remembered about the futures markets is that when futures contracts are bought and sold it is future obligations that are being bought and sold. It is similar to college seniors representing themselves as college graduates when interviewing for employment before all coursework for their degree is completed. Students represent themselves as university graduates in terms of promising to deliver certain qualifications in the future. In futures markets, obligations to deliver a commodity can be sold even though the commodity is not currently owned. It is understood that ownership of the necessary amount of the commodity could be arranged (at some price), if necessary, or the obligation may be offset by taking an opposite position in the futures market at anytime before the contract expires.

Nonmembers of a futures exchange who desire to buy or sell futures contracts must arrange their transactions with a representative who is a member of one of the organized markets that trade these contracts. Anyone who wants to buy or sell a futures contract must contact a broker or member of that exchange who can make the appropriate transaction. Members can buy or sell contracts for themselves or execute trades for the public for which they charge a fee. The brokerage fee charged nonmembers of the exchange depends on the commodity being traded, the number of contracts being traded, amount of previous trading, and other things. Actually the brokerage charge is usually a fee for two transactions. The charge covers getting into the futures market as well as getting out. This is called a **round turn**. Brokerage fees vary, but often range from $50 to $80 per contract. This is equivalent to 1 or 2 cents per bushel for grain contracts.

Exchange Activities

Each futures exchange establishes its own rules and regulations about how trading among its members will be conducted including specific times the market is open to conduct business. Until fairly recently, the only way trading in futures contracts took place in the United States was on a trading floor during a particular interval of time by a public outcry of willingness to buy or sell. This process was intended to provide all interested parties the best opportunity to buy and sell contracts allowing prices to reflect relative interests of all parties willing to undertake future obligations. For example, the sounding of a bell in the morning indicated the time for initiation of trading at the CBOT. Another gong sounding in the early afternoon indicated the end of all trading on the floor until the next trading day. During the trading period, members interested in trading a particular contract congregate in a designated area called a pit and shout prices at which they are willing to buy or sell in search of other traders who are willing to take the other side of the transactions. Corn futures are traded in one pit and soybean futures in a different area. Traders on the floor constantly are trying to find the most favorable price for their transactions, which means that their activity can get very hectic or almost frantic at times. Trading is facilitated by use of hand signals to indicate whether they are interested in buying or selling and at what price. As trading occurs, a price is posted and continuously adjusted according to what is happening in the pit. Bids to buy or sell are usually made in terms of differences from the last posted price. As soon as a trade is negotiated, each party records how many contracts were traded at a specified price and whom the trade was with. If there are disagreements, futures markets have committees that take care of any disputes or violations of the rules of trading within the exchange. The mutual trust existing among official members of an organized exchange is important so trades can be executed quickly and efficiently. The integrity of a trader is important for continual willingness of other traders to transact business.

For several years there have been discussions about how computer technology will eventually change the system of buying and selling futures contracts. Computers have already simplified record-keeping activities for brokers and futures market participants. Various computer innovations have also been adopted to improve the paper trail of all the activities that occur on the floor of the exchanges. In 1996, a pilot project to add an overnight electronic trading system using special computer terminals was introduced by the CBOT to extend the hours during which trading of selected futures contracts could be executed when floor trading does not occur. This experiment was designed to enable participants in futures markets to respond to new market information especially during business hours in Asia in an attempt to recapture some of the loss in market share to futures markets in other countries. Some individuals envision computer trading eventually leading to continuous 24-hour trading opportunities and perhaps the elimination of floor trading. A key element is the extent to which computers can provide the same kind of market emotion and integrity that open floor-trading procedures have provided.

The relative strength of interest in selling or buying contracts causes the price of the contracts to rise or fall. As prices change in response to market conditions, the information is communicated very quickly around the world. Thus, everyone interested in a particular commodity can tell what the market price of different futures contracts is at any given time. As orders to buy and sell come to the market from around the world, prices can fluctuate during the time trading is permitted. If there is more interest in buying than selling during the trading period, the price of contracts will rise. If this movement continues for a substantial period of time, the market is said to be **bullish**. If there is more interest in selling than buying, prices will decrease and the market is said to be **bearish**. These explanations are a little misleading because anytime transactions occur there has to be an equal number of buyers as well as sellers. If there tends to be greater interest in selling, however, people are willing to take a lower price to make a transaction and conditions would be consistent with a bearish market.

Every exchange has certain limits on how much the price of any contract can change during any trading period for most of the life of a contract. The limit indicates how much the price of a contract can change from what the price was at the close of trading on the previous day.[3] If something dramatically happens between the time the market closes one day and trading begins for the next day, the market may immediately open up or down the limit. For example, if the United States placed an embargo on all grain exports while the market was closed, the opening price on grain futures markets could very well change the limit. If the effects of the embargo were expected to be very large there might not be any trading at the new price the next trading day. Anybody willing to buy or sell futures contracts at prices outside the limit would not be able to make a transaction because the price could not move any more than the maximum amount on a given day. The maximum amount by which price can change is adjusted from time to time if the market moves the limit for a few days. There is no limit on price movements for most of the last month before a contract expires. This means that price may be quite variable during the last few weeks of trading as individuals attempt to liquidate their positions in futures markets.

TYPES OF ORDERS

There are a number of different ways an individual can place an order with a broker to execute a trade. The simplest is a **market order**. This type of order simply indicates that a trader wants to buy or sell a specified number of contracts at the best possible price. As soon as this information is communicated to the floor, the individual generally is assured that an order to buy or sell will be executed. The individual does not know what the market price will be, however, until the trade is made.

Another way of placing an order for a futures market transaction is by way of a limit order. This is a type of order where a price can be specified at which the individual would be willing to make a transaction. This requires that the order be filled at the "specified or better" price. If the individual is interested in selling, the "specified price" essentially becomes a minimum price. If the trade can be negotiated at a price higher than the level specified, obviously the seller would be happy. That is the meaning of "specified or better." If the market price is below the level specified when the limit order is communicated to the floor trader, the order will not be executed. There is no question about an individual being able to get in or out of the market at some price when using a market order, unless no trading occurs because the maximum or minimum price change for the day has been attained. The use of a limit order instead of a market order is a little more cautious strategy to protect against a sudden price change that might occur between the time the decision to enter the market is made and the trade is actually executed. In a limit order to *buy* future contracts, the specified price becomes a maximum. If the contract can be purchased at less than the price indicated, it will be done. Members of an exchange are obviously interested in making trades at the best price possible in order to please their customers since their income depends on a continuous volume of trading.

A third way of specifying interest in trading is by means of a **stop order**. These orders are used mainly by individuals who already have a position in a market. A stop order involves specifying a price at which trading should be initiated. A stop order to *sell* sets a minimum price if approached from above. An individual who had a long position in the market could issue a stop order that says in effect, "If the price starts dropping, I want to get out of the market at this price. If the market should start coming down and gets to a specified price, sell. As long as the price stays above that level, don't change my position." Similarly, a stop order to *buy* would specify a maximum price at which action should be taken if the price is approached from below. This kind of action would minimize losses for someone who had initially sold a contract at a lower price.

In addition to the various ways in which buy and sell orders can be specified, a time limit can also be placed on orders. Thus, an order can be established to be valid for a specific day. For example, a stop order or limit order could be specified to initiate certain actions on a particular day. If the conditions that have been specified are not fulfilled during trading on that particular day, then the order would no longer be in effect. On the other hand, the order can be left completely unspecified with respect to time. Such an order is called an *open order.* If the order were not completed one day, it would still be in effect and could be executed on a succeeding day if the specified circumstances were satisfied.

INTERPRETING MARKET REPORTS ON FUTURES MARKETS

There are many sources of information about futures markets on the Internet and in daily newspapers, including the *Wall Street Journal.* Most markets have an Internet address that can be used to access a variety of current and historical price information. Newspapers often report daily summary information for a selected number of futures contracts traded on several markets. The first thing that market reports usually identify is the commodity for which the contracts are applicable. Frequently, there is also an abbreviation to indicate the particular futures exchange for which prices are being reported. For example, CBOT is the acronym for the Chicago Board of Trade, CME is Chicago Mercantile Exchange, CSCE is Coffee, Sugar and Cocoa Exchange in New York, CTN is New York Cotton Exchange, and so on.

Basic Details of Market Reports

The number of units of the commodity of the contract is also usually indicated in the reports. Most grain contracts are equivalent to 5,000 bushels although some exchanges offer 1,000-bushel contracts. The prices for grain contracts are frequently expressed in cents per bushel. If you see 372 in a grain market report, it means 372 cents (or $3.72) per bushel. Livestock prices are usually expressed in cents per pound (e.g., 50.40 which is equivalent to $50.40 per cwt or hundred pounds). If the data are reported in tabular form, the left-hand column will indicate different months, such as March, May, July, September, December, and March. This refers to the different contracts being traded for the same commodity. The months indicate when trading in the various futures contracts will expire. If a May or July contract is bought or sold, the only difference is the prices at which the contract is traded and when the obligation will terminate. For example, toward the end of May, there will no longer be any information reported about the current contract because trading in the May contract will have terminated. Sometime prior to this, however, trading in another May contract that expires the following year will have begun and the price of that contract will be reported. The continuing expiration and initiation of contracts with new maturity times result in a continually changing set of contracts. At any given time, several contracts in the same commodity are traded with the only differences being the maturity dates of the contracts and the prices at which they are trading. The estimated daily volume of trading in terms of the number of contracts is usually indicated at the bottom of the columns.

Interpreting Price Movements

Typical headings for the various columns of tabular reports on futures trading are open, high, low, settle, change, lifetime low, lifetime high, and open interest. For example, for a given day the information about a May contract might be:

	Open	High	Low	Settle	Change	Lifetime Low	Lifetime High	Open Interest
May	372	375	369	373	+1/4	368	538	31,000

The number in the second column from the left indicates that when the market opened on a given day, trading began at a price of 372. The next three columns indicate information about what happened to the price of the contract during that day of trading. The high and low prices that existed sometime during the trading day are indicated. Finally, the settle (or closing) price indicates where the market was when the trading period ended for the day.

In the above example, the price varied during the daily trading period by 6 cents (from 369 to 375). That is equivalent to a total change in value of $300 on a 5,000-bushel contract. Thus, someone who bought a contract when trading was at the low point, and sold at the peak price during the day could have made a profit of $300 per contract. On the other hand, if that person had taken the opposite sequence of actions, a loss of $300 per contract would have occurred. For someone to profit from a change in price it does not matter whether the high price occurs before or after the low price provided they made the correct timing decision (i.e., sell at the high price and buy at the low price). It is important to remember that it is possible to sell before buying because only an obligation or promise is being traded.

The change column indicates what happened to the price between the end of the previous trading day to the end of trading for a given day. By looking at that column it indicates whether the market was up or down between successive days' closing prices. In the above example, a change of +1/4 indicates the market ended slightly higher than the previous day. This information is important for calculating changes in margin accounts for all traders who did not change their market positions on a given day.

The next two columns in the above example indicate what the lowest and highest prices of this contract have been since trading in the contract was initiated. The above example indicates that the price of this contract was 538 at one time. The lifetime low is just a little under the current price. The difference between the lifetime high and lifetime low prices indicates that there was a profit potential of $1.70 per bushel. If a contract had been sold at its all time high price of 538 and another contract purchased at 368, a profit of $8,500 would have been made on each 5,000-bushel contract. Considering the magnitude of the typical margins required for such contracts indicates the potential profitability (or vulnerability) associated with this type of trading.

Open Interest Information

Many market reports about futures markets also indicate the amount of **open interest** in a given market. To interpret the meaning of open interest, one must remember that anytime a contract is sold, somebody has to buy it. When an individual accepts an obligation to deliver the commodity, somebody must have been willing to accept an obligation to receive the commodity. To see how this works, let us suppose that on a given day the purchase and sale of one contract each are made by new participants as indicated by the following columns:

Buy	Sell
A	B
C	D
E	F
G	H
B	E
F	A

Initially, trader A buys a contract from trader B, C buys one from D, E from F, and G from H. After the first four trades, four people had sold a contract and four different traders had bought a contract. Suppose later in the day after the price fluctuates, trader B buys a contract from E and trader F buys one from A. Because these are standardized contracts, a contract does not have to be purchased back from the same person to which it was sold in order to fulfill the initial obligation. An opposite obligation in an identical contract offsets any initial obligation. If the above sequence of trades was a complete set for a given day, there would be two outstanding obligations to buy and sell at the end of the day. Consequently, this would be recorded as two additional open positions in this market because it was assumed that the participants had no prior positions in the market. When traders report their net positions, all traders except C, D, G, and H are free of any further obligations from this day of trading. Everyone else sold one contract and bought one contract. The traders either made money or lost money depending on the change or difference in price between their two transactions. At the end of the day, we would have traders D and H having a continuing futures market obligation from having sold a contract and traders C and G having opposite obligations. This is equivalent to two open contracts because there are two buyers and sellers who have not fulfilled their obligations at the end of this trading day. This is what is meant by two open positions.

If any of the traders had market positions in the commodity from trading on previous days it would be necessary to have a cumulative listing of all transactions to know whether the number of open positions increased or decreased on a given day. In the previous example of market information, the last column (open interest) indicates there were 31,000 open positions at the end of the trading day. That means approximately 31,000 contracts had been bought that had not been cancelled by an offsetting sale and vice versa. When the maturity time of a contract is approaching, the open position declines because anybody with an open position when the contract matures has to fulfill their obligation. When trading in a particular contract stops, the outstanding obligations must be fulfilled. For example, if a delivery of 5,000 bushels of wheat is required, it is accomplished by transferring a receipt for that amount of wheat in a particular elevator to another person who had an outstanding obligation to buy wheat because of a previous purchase of a wheat future contract. Thus, transferring a warehouse receipt for ownership of that commodity has traditionally completed settlement. Because the contracts are standardized, everybody knows that if they continue to hold positions in the futures market, they are either going to have to buy a warehouse receipt or be willing to accept a warehouse receipt for the commodity when trading in the futures contract expires.

CLEARINGHOUSE FUNCTIONS

The **clearinghouse** for each futures exchange is a separate agency that keeps track of all traders' positions and obligations. It makes sure that all outstanding obligations are eventually satisfied. When trading in the contract terminates, it notifies which traders with unfulfilled obligations need to make delivery to other traders. This operation consists of taking the open positions on the selling side and matching them with equivalent positions on the buying side. Individuals do not know who is going to deliver to whom until the clearinghouse notifies them. Most of the time before the contract matures or expires, traders with an open position either buy or sell contracts to offset their obligations to avoid going through the formal delivery process. Since all trader positions are reported to the clearinghouse, that agency essentially takes the opposite position of every trade and is able to guarantee that every obligation or commitment will be satisfied based on the creditworthiness of the individual members of the exchange. This is a key difference in the trading of futures promises or obligations on an organized exchange relative to similar kinds of transactions that occur between any two individuals who mutually depend on each other's integrity and ability to fulfill specific promises.

As trading in a futures contract draws to a close, the physical commodity specified in the contract becomes essentially a perfect substitute for the futures contract. This occurs because the futures contract will convert into the physical commodity through the delivery process if an open position is maintained. By definition perfect substitutes ought to have the same value. When the futures trading expires in a given contract, the price of the contract should be essentially the same as the cash price of the commodity because anybody holding a contract to sell has to get into the cash market to fulfill that obligation. Delivery can only be made at specific locations depending upon the specifications of the contract. The future price will be very close to the price at delivery points except for any transaction costs that might be involved in using a physical commodity to actually satisfy the obligations of a futures contract.[4]

Normal Price Patterns

A normal price pattern among different futures contracts for a given commodity that is periodically produced and storable would be an increasing price for contracts farther into the future toward the next harvest period. The reason for this is that it is going to cost something for the commodity to be stored forward into the future. For example, March corn ought to be worth more than January corn. An increasing pattern in prices indicates that the market is incorporating the added cost of storing the commodity for a longer period of time or reflecting the carrying charge. The technical term for this pattern of price relationships is **contango**. Sometimes a reverse situation where prices are lower for contracts farther away will occur. In this case the market is said to be *inverted* or reflect a pattern of **backwardation**. This pattern may initially appear to be counterintuitive because of the cost of storing commodities. When a market has an inverted price pattern, it frequently is indicating there is a near term "shortage" of the commodity. It indicates that people are willing to pay a premium to get immediate ownership of the commodity and the market is indicating an incentive not to store a commodity. Most of the time this kind of situation occurs right before a harvest when there is limited inventory of an item. In these circumstances, people may be willing to pay a higher price now rather than waiting a little while for harvest to occur.

Futures prices reflect the best judgment based on current information of what people who are interested in that commodity think is a likely equilibrium price of the commodity at the time the futures contract will expire. If anybody thinks they have a better idea about expected price, futures markets provide the opportunity to profit from superior information or intuition. Any trader who can successfully predict changes in market prices can make money. The going price in a futures market is an indication of what price people are willing to commit themselves to an obligation to buy or sell the commodity sometime in the future. As new information occurs, it may indicate that the previous expected price was too high or too low. Also as new information occurs, individuals may have an incentive to adjust their trading positions and the relative interest in buying and selling futures obligations gets reflected in the price of futures contracts.

FUNDAMENTAL VERSUS TECHNICAL ANALYSIS

Frequently, in discussions about analyzing futures market trading strategies proponents or advocates of either a fundamental or technical approach will be encountered. The fundamental approach is essentially using knowledge of economics, and potential demand and supply conditions to anticipate what is likely to happen as far as price relationships are concerned. This approach is based on the idea that market participants will arrive at prices that reflect the economic value of commodities at a point in time. The values depend, of course, on relative demand and supply situations. On the other hand, the **technical approach** essentially tries to predict what is going to happen based on past behavior. Those using a technical approach rely on patterns they see in charts and various quantitative measures of recent price behavior. For example, a trend line based on a selected number of recent low prices for a futures contract is interpreted as a *support line for an upward moving market.* Similarly, a trend line based on a selected number of recent high prices is interpreted as a *resistance line for a downward trending market.* Projections of support and resistance lines into the future are used by some traders to decide when the price appears to be breaking out of a given pattern. One difficulty with this approach is the subjectivity involved in selecting the particular set of recent high and low prices that will influence the slopes of support and resistance line. A few of the other more widely used quantitative measures of futures prices are discussed in Appendix A. The measures have been and currently are used to support various trading strategies as to when to buy and sell futures contracts. Many individuals claim to make money using particular strategies based on **technical analysis**, but the theoretical basis and scientific validification for mechanistic forecasts of future price changes are quite subjective.

EXCHANGE REGULATION

All futures markets are regulated by the **Commodity Futures Trading Commission (CFTC).** The CFTC is a federal agency much like the Securities Exchange Commission (SEC) that regulates stock and securities markets. The CFTC was established in 1974 as trading in nonagricultural futures contracts was expanding. Prior to that time, futures trading was regulated by the 1936 Commodity Exchange Act administered by the U.S. Department of Agriculture. Any type of new contract or change in futures trading practice has to be approved by the CFTC. Also, floor traders have to be licensed by this commission and each of the exchanges have internal committees to govern trading practices. In 1981, the National Futures Association (NFA) was established as a self-regulating industry organization subject to CFTC oversight. It has the responsibility of registering and monitoring trading activities of individuals and firms regulated by the CFTC. Activities that violate NFA rules of professional ethics and conduct may result in suspension of futures trading privileges.

SUMMARY

Futures markets are important institutions that perform many roles. One function of futures markets is the discovery of prospective prices of many agricultural and other commodities based on how participants interpret available information about current market conditions and the likelihood of alternative future events. Futures markets operate throughout the world and have existed in the United States for many years. Initially, most of the trading on futures commodities involved agricultural commodities, but over the last two or three decades the trading of financial instruments on futures markets has become increasingly important.

Every futures market transaction implies that two individuals have agreed to assume opposite obligations pertaining to accepting or providing a *specific* quantity of a commodity at a *specific* price at one or more *specific* locations at a *specific* time in the future. All futures market transactions involve the trading of standardized contracts that vary only in price and the time for the assumed obligations to be fulfilled. Most futures market obligations are cancelled rather than fulfilled, however, by each party taking an offsetting position in the same market (usually at a different price) before trading of the specific contract ends. The trading of contracts with later maturity times is initiated as the trading of other contracts end.

Changes in the value of a futures contract resulting from upward or downward movements in the price determines the amount of profit or loss resulting from having taken an initial long (buying) or short (selling) decision in a futures market. Each futures market has a fixed number of members who do the actual buying and selling of futures contracts. Many members, however, execute trades for a fee on behalf of anyone who establishes a margin account containing sufficient money to handle potential losses that may occur because of unexpected changes in prices. Futures markets provide opportunities for speculators to profit if they can correctly anticipate upward or downward changes in prices of agricultural commodities. Futures markets also provide a way to avoid some of the risk associated with price uncertainty. The way in which this objective can be pursued using futures markets will be discussed in Chapter 15.

Futures market prices are determined by relative interest in buying and selling contracts at alternative prices as market participants adjust expectations and make business decisions in response to changing information about current and anticipated demand and supply conditions for various commodities. There are limits on the maximum amount of change that can occur in the price of any contract from the previous day's closing price except during the final month of trading a given contract. Some traders make buying and selling decisions about futures contracts primarily on the basis of past behavior of prices attempting to profit from short run or long run adjustments in price. The prices of various futures contracts for periodically produced storable commodities often reflect the expected influence of storage costs. Sometimes, however, market circumstances can result in near-term contracts having higher prices than those maturing at a later time.

Futures market price information is readily available from multiple sources and is disseminated very quickly while trading occurs. Many newspapers and Internet sites report various types of summary information about the previous day's trading. This kind of information frequently consists of the prices that various futures contracts started trading when the market opened as well as the daily high, low, and closing prices. Daily changes in closing prices of each specific futures contract are important because this information is used to recalculate the financial position of each trader's unfulfilled (or open positions) obligations in futures market contracts at the close of each day's trading. If changes in price decrease a particular trader's margin account by a sufficient amount, it is necessary for the trader to provide additional money to maintain a continuing position in the market. Each trader's account ultimately is under the supervision of and monitored by a clearinghouse that essentially acts as a second party for all futures market transactions adding liquidity and integrity to the market. All contracts and trading practices involving futures contracts in the United States are subject to regulation by the Commodity Futures Trading Commission.

APPENDIX TO CHAPTER 14

SOME QUANTITATIVE MEASURES OF FUTURES MARKET PRICES

This appendix provides a brief introduction to some common types of quantitative measures used for various types of technical analysis of futures market prices. Exchanges and various information services regularly report many of these measures. The specific measures selected for discussion include moving averages, momentum, relative strength, stochastics, directional indicators, and volatility. Additional details about each of these measures can be found in books (e.g., Kaufman 1998; Murphy 1986) or Internet sources (e.g., http://www.tradertalk.com or http://www.barchart.com).

Moving Averages

Some of the simplest measures of changes in futures prices for a particular interval of time are moving averages of closing prices. These measures are averages of specific price information for a fixed number of trading days. Each day the measures are updated by substituting the latest days' information in place of the oldest information. For example, a 9-day moving average would always be based on information for the last 9 trading days. Each day after the market closes, a new observation would replace the oldest observation for purposes of calculating a new moving average value. An 18- or 40-day moving average would include an additional 9 or 31 previous days' information relative to a 9-day moving average.

A comparison of the most recent price data with various moving averages can be used to get an indication of how current price deviates from recent history. The more days that are used for calculating a moving average, the less sensitive the moving average will be to short-term changes. Consequently, some traders use moving averages of various lengths to identify how short run changes in price compare to long run changes.

Sometimes moving averages are calculated by applying different weights to the various observations. For example, more recent information may be weighted more heavily than earlier observations based on the rationale that all observations may not have the same informational value about market trends. One of the more popular weighting schemes for calculating moving averages is exponential smoothing that incorporates weights that decline exponentially. Different exponentially smoothed moving averages can be updated easily by combining a weighted average of the latest observation with the previous value of the moving average depending on how much weight one prefers to place on the latest observation relative to earlier values. Thus it is important to realize that all moving averages of a given set of data are not necessarily identical and can vary substantially depending on how they are calculated.

Momentum

The basic idea of a measure of momentum is to provide an indication of the total change in price for a particular interval of time. For example, one measure of momentum is the change in price expressed as a proportion of how much the market could have changed if it had consistently moved up or down the maximum amount allowed each day during an interval. Momentum measurements can be expressed in terms of actual changes in price during a particular interval of time or if expressed as a percent of the maximum change that could have occurred the value could range from zero to 100. Higher values indicate a bigger movement in one direction or the other. The closer the momentum measure is to zero indicates that the current price is essentially no different than it was x days ago.

Relative Strength

The major difference between a measure of momentum and relative strength is that the former is based on the change in closing prices between two points in time whereas relative strength is based on a comparison of the average amount of change in successive days' closing prices. Relative strength is calculated by finding the ratio (RS) between the average daily change in closing prices when the market was up during a particular interval of time to the average daily change in closing prices when the market closed down during the same interval of time. The relative strength index is computed as $100\ RS/(1 + RS)$. Different measures of relative strength will result depending on how long an interval is selected. For example, if during the last 14 trading days, the average change in closing prices on up days was the same as the average change in closing prices on down days, the relative strength index would be 50. (Actual formulas for calculating relative strength measures treat a day with no change in closing prices symmetrically as an up and down day.)

Higher values of the relative strength index indicate greater changes in upward movement of prices. Lower values indicate that average changes in downward price movements were greater than the average daily change in upward movements in price. Some traders believe that very high or low values of relative strength indicate a greater likelihood that the market will begin to move in an opposite direction because of the difficulty of sustaining movements in prices in the same direction for long periods of time.

Stochastics

Various stochastic measures have been developed to reflect how the latest closing price compares to the difference between the highest and lowest prices observed over a particular interval of time. For example, one of the simplest stochastic measures represents the difference between the closing price on a given day and the low price that occurred during the previous x days. This measure is expressed as a percent of the range over which prices fluctuated during the same period of time. For example, assume prices of a given futures contract fluctuated between $5.50 and $5.90 during the last 9 days and the latest closing price was $5.60. In this case, the basic stochastic value would be ($.10/.40) * 100 = 25. Lower values of the basic stochastic measure indicate that the most recent closing price is closer to the bottom of the range over which prices have fluctuated. Higher values of the basic stochastic indicate the most recent closing price is closer to the highest price that has occurred over a recent period of time. Different values of the stochastic could occur with differences in the length of time used to observe fluctuations in price.

The basic stochastic value described above is usually referred to as the % K measure, Murphy (1986) and Kaufman (1998). These measures can also be used to calculate different kinds of moving averages. The % D stochastic measure generally refers to a moving (or an exponentially smoothed) average of the % K stochastic. Others calculate the % D stochastic by dividing the sum of the numerators for computing the daily stochastic for x number of days by the sum of the elements in the denominators of the daily stochastics (Murphy 1986; Becker 1994).

Directional Indicators

There are several ways that directional measures are calculated to express changes in daily high and low prices relative to the maximum range in prices.

Procedures for calculating directional indicators are a little more complicated than the previously described measures so only the basic ideas for the most simplified versions will be indicated.

The first step in calculation of directional changes involves comparing the change in both high and low prices between successive days. If the change in high prices is positive and the change in low prices is negative, the largest of the two differences is selected for purposes of determining whether the directional change for a given day is up (+) or down (−). Only positive changes in high prices and negative values for changes in low prices are relevant for selecting the largest change. If the range of prices on a given day is within the previous days range of prices, the directional change is assumed to be zero for that day because the change in high prices would be negative and the change in low prices would be positive. Negative values of day-to-day changes in high prices and positive values of day-to-day changes in low prices are ignored in selecting the appropriate directional change for a given day. Separate moving averages of the appropriate + and − daily components can be calculated for a selected number of days to represent the upward and downward day-to-day market adjustments.

Another way of expressing directional indicators involves determining an appropriate value of the daily *true range.* This is defined as the largest measure resulting from three alternative calculations involving high and low prices for a given day and the previous day's closing price. One possible value of the true range is the difference between the high and low prices of the most recent day. A second possible value of the true range is the difference between the high on a given day and the previous day's closing price. The third alternative value of the true range is the difference between the closing price of the previous day and the given day's low price. One of the previous three values will provide the largest measure of how much the price of a given contract fluctuated between the close of trading of the previous day through the close of trading on the successive day. Dividing the largest change in day-to-day high or low prices by the appropriate value of the true range for a given day provides a relative measure of the strength of directional changes in high and/or low prices.

The following numerical example is intended to illustrate the above concepts. Suppose the price of a given futures contract for a given day closed at $5.70 after varying between $5.60 and $5.85 and the following day varied between $5.80 to $6.00. Comparing the high and low prices for the 2 days indicates that the greatest change would be a +$.15. (The change in the low prices is ignored or set equal to zero because the $.20 increase that occurred in the low prices is not a negative number and therefore considered to be zero for purposes of selecting the largest change in high and low prices.) The value for the true range for the above data would be $.30 (the difference between $6.00 and the previous day's close of $5.70). Dividing +$.15 by $.30 indicates that the difference in high prices for the two days was 1/2 of the maximum fluctuation in prices from the close of trading on one day through the following day. Daily values of directional measures can be averaged over varying periods of time and manipulated in other ways to produce a variety of alternative directional indicator measures.

Volatility

Volatility of futures prices is usually represented by statistical measures of the variability in the rates of return reflected by successive days closing prices. Technically, volatility is the standard deviation of the natural logarithm of the ratio of closing prices on successive days over a specified number of trading days. Different measures of volatility for the same commodity can occur depending on how long a period of time is used for the calculations. Specifically, the formula for the standard deviation is the square root of the following expression:

$$\sum_{t=1}^{n} (R_t - \text{average of } R_t)^2 / n - 1$$

where

$$\ln R_t = P_t / P_{t-1}$$

with P_t representing the closing price on day t and n representing the number of days over which volatility is measured.

QUESTIONS

1. What are the two ways in which a speculator who has taken a long position in a given futures market can satisfy his or her obligation?

2. Briefly explain the major difference between a market order and a limit order as far as establishing an initial position in a futures market.

3. Assume that the following information represents price information for a given day's trading on the May corn futures contract on the Chicago Board of Trade.

Contract	Open	High	Low	Settle	Change
May	287	287	283	284	−5 1/2

a. Based on the above information, what is the maximum gain per contract (5,000/bu per contract) a day trader could possibly have realized in this market on this day?

b. If the maximum daily price change for this contract is \$.12/bu, what is the range of prices at which futures market trading could possibly occur on the next day of trading?

c. What was the price at the close of trading on the previous day?

4. Would an unexpected increase in U.S. exports of soybeans likely be a favorable or unfavorable development for speculators who were holding short positions in March soybean futures contracts? Briefly explain the economic reasoning supporting your answer.

5. How much of a change would each \$.05/bu increase in the price of wheat futures have on the magnitude (and direction) of the margin account of a trader who had a short position in 50 December wheat futures contracts (assuming each futures contract consists of 5,000 bushels)?

6. Assume that you are a speculator in the pork bellies futures market and the margin requirement is 10%.

a. If you had taken a long position in one March pork bellies futures contract (consisting of 40,000 lb) when the price was $.60/lb, how much margin money would have been required?

b. How much and in what direction would the price of March pork bellies futures need to change in order for you to make a return of 20% on the investment you made in the margin account, ignoring brokerage fees, foregone interest, and other transaction costs?

c. Indicate how you could use a stop order to make sure you did not lose more than 50% of your margin account.

7. Briefly describe what is meant by technical analysis as far as futures markets are concerned and how it differs from **fundamental analysis.**

REFERENCES

Becker, J. 1994. Value of oscillators in determining price actions. *Futures* 23(5) (May):32–34.

Commodity Futures Trading Commission. 1998. *Annual Report.* http://www.cftc.gov/annualreport98/contents.htm

Kaufman, P. J. 1998. *Trading systems and methods,* 3rd ed. New York: John Wiley & Sons.

Murphy, J. J. 1986. *Technical analysis of the futures markets: A comprehensive guide to trading method and applications.* New York, N.Y.: Prentice Hall, New York Institute of Finance.

NOTES

1. An increasing number of futures contracts provide terms for a cash settlement of obligations to buy or sell commodities in lieu of actual delivery.

2. In addition, there would likely be some additional costs associated with making the actual transactions.

3. With the new electronic trading experiment initiated in 1996, each trading period begins with the electronic trading session and ends at the close of the floor trading the following day, approximately 15 hours later. Consequently, the appropriate price limits for trading during a subsequent 24-hour period of time are based on the closing price at the end of the traditional floor trading.

4. A maturing futures contract may not have exactly the same price as the physical commodity because of transaction costs involved in acquiring ownership or obtaining a warehouse receipt, and other costs involved in fulfilling the underlying obligation.

CHAPTER 15

FUTURE PRICES

GRAINS AND OILSEEDS

	Open	High	Low	Settle	Change	Lifetime High	Lifetime Low	Open Interest
CORN (CBT) 5,000 bu.; cents per bu.								
May	236	239½	233¾	234¼	– 1¾	261	202½	172,739
July	244	248	242½	242¾	– 1¾	278½	209	153,801
Sept	252	255	250	250¼	– 1¾	257	215¾	44,183
Nov	259	259	255½	255¾	– 1¾	260	222½	977
Dec	260	264¼	259¼	259½	– 1	279½	225¼	108,253
Mr01	268¼	271	267½	267¾	– 1	271	233¾	12,988
Dec	271	272	269½	269¾	– 1	272	246½	3,877
Est vol 68,000; vol Fri 76,212; open int 501,404, +2,820.								
OATS (CBT) 5,000 bu.; cents per bu.								
May	124	125½	123	123	+ ¾	133	112¾	4,980
July	121½	123¼	121	121	+ ¾	124¼	110½	7,414
Sept	124	124½	123	123	+ 1	130	115¾	1,584
Dec	129½	130	129	129	+ 1	135	115½	3,305
Mr01	....	....	....	135½	+ ¾	135	129	465
Est vol 2,600; vol Fri 1,654; open int 17,749, +133.								
SOYBEANS (CBT) 5,000 bu.; cents per bu.								
May	544	554	544	546½	+ 1	554	432	76,777
July	556½	566	556½	558½	+ 1	647	440	59,990
Aug	562½	569	561½	561¾	+ 1¾	569	441	9,193
Sept	568	571½	563½	563¾	+ 1¾	571½	450	4,758
Nov	573	579	571	572	+ 3½	631	453	37,555
Ja01	575¾	586	575¾	579½	+ 3½	586	504	2,477
Mar	585	591½	585	585	+ 3½	591½	502½	1,557
Est vol 60,000; vol Fri 72,880; open int 194,607, +11,245.								
SOYBEAN MEAL (CBT) 100 tons; $ per ton.								

CASH PRICES

Monday, April 3, 2000
(Closing Market Quotations)
GRAINS AND FEEDS

	Lo/Hi Range Mon	Mon	Fri	Year Ago
Barley, top-quality Mpls., bu	u2.7	sp	2.7	2.2
Bran, wheat middlings, KC ton	u41	46	43.5	53.5
Corn, No. 2 yel. Cent. Ill. bu	bpu2.115	sp	2.135	2.055
Corn Gluten Feed, Midwest, ton	42	60	51	53.5
Cottonseed Meal, Clksdle, Miss. ton	127.5	sp	125	105
Hominy Feed, Cent. Ill. ton	55	sp	57	55
Meat-Bonemeal, 50% pro. Ill. ton.	170	180	175	132.5
Oats, No. 2 milling, Mpls., bu	u1.31	1.33	1.3125	122
Sorghum, (Milo) No. 2 Gulf cwt	u4.18	4.27	4.275	411.5
Soybean Meal, Cent. Ill., rail, ton 44%	u166	174	170	132
Soybean Meal, Cent. Ill., rail, ton 48%	u176	182	179	138
Soybeans, No. 1 yel Cent.-Ill. bu	bpu5.175	sp	5.17	4.61
Wheat, Spring 14%-pro Mpls. bu	u3.575	3.725	3.73	361.5
Wheat, No. 2 sft red, St.Lou. bu	bpu2.385	sp	2.46	2.495
Wheat, hard KC, bu	2.725	sp	2.75	3.0375
Wheat, No. 1 sft wht, del Port Ore	u2.88	sp	2.87	3.25

FUNDAMENTALS OF HEDGING

This chapter describes how buying and/or selling of futures contracts can be used to reduce some risks associated with price variability of agricultural and food commodities.

The major points of the chapter are:

1. The way a **short hedge** can be used to protect the value of inventories.
2. How **basis,** the difference between cash and futures prices, affects returns to **hedging**.
3. The temporal and locational components of basis.
4. The use of futures markets for target pricing and **long hedging.**
5. The use of futures contracts for **spread and straddle strategies.**

INTRODUCTION

The first part of this chapter describes how futures markets can be used to obtain a type of price insurance for a commodity. An example of a short hedge is used to illustrate how the effects of a change in the cash price of a commodity can be offset by taking an opposite position in the futures market. Then the effect of imperfect expectations about the difference between cash and futures prices on returns from short hedging are indicated. Temporal and locational components of basis are discussed before turning attention to the use of futures markets to offset other kinds of risks associated with price variation. These uses include target pricing and long hedging. The final part of the chapter indicates how the basic idea of hedging can be extended to profit from expected changes in the difference between two futures market prices.

UNAVOIDABLE RISKS

Just because farmers' incomes may vary because crop yields are affected by weather conditions does not imply farmers like or prefer risky situations. Most farmers would prefer not to have to gamble on weather, but if they grow crops the uncertainty of weather conditions is something that must be accepted. It does not mean that they necessarily like it, but variable weather is a fact of life. The purchase of crop insurance or investing in irrigation equipment can be viewed as particular business decisions to minimize the risk of certain types of crop failure. The fact that people choose to be involved in risky situations does not necessarily mean they prefer a risky situation over a nonrisky one. It is just that the risky situation is one in which they anticipate that their satisfaction will be maximized because they view the expected payoff to be worth the cost of bearing the risk.

Hedging is one way of using the futures market to avoid some of the effects associated with uncertain prices. Hedging is like an insurance contract in the sense that you might be able to buy some protection against the unknown future price of a commodity. Hedging is not a free good, however, because there is always a cost associated with making a futures market transaction. Costs are associated with having to have a margin account or a sufficient line of credit to participate in a futures market. There are also brokerage fees that must be paid by nonmembers of futures exchanges in order to buy or sell futures contracts.

Producers, processors, or anyone else who buys and sells commodities can use futures markets to possibly reduce some risk associated with price uncertainty. The idea of hedging is related to how futures markets got started in the United States when there was a lot of uncertainty about prices. Futures markets are particular institutions that were created in response to people involved in the trading of agricultural commodities desiring some way of reducing some of the risk of changes in price because of unexpected market developments.

A SHORT HEDGE

To see how the risk of price changes can be minimized through hedging, some simplistic assumptions will be initially used to illustrate the basic concept. Later some of the initial assumptions will be relaxed to make the illustrations more realistic. To begin, suppose we consider an individual producer or an elevator operator who has sufficient storage capacity to hold a commodity from the time it is harvested until sometime later in the marketing year. The person would like to be able to earn a return for providing the services of helping to even out the flow of the commodity to consumers from when it is produced until when it is needed for consumption. Assume this individual does not want to gamble on what the price of the commodity is going to do while he owns it, but rather would be satisfied if he could earn a normal return for helping to even out the flow of the commodity into market channels. Suppose that the costs of storage from harvest until May are $.25 per bushel. If the business has a storage capacity of 500,000 bushels, it means that the person would like to know that he could make $125,000 by putting the grain in storage in the fall and taking it out of storage in the spring. One way that this could occur is if the price of the commodity happened to be $.25 per bushel higher in the spring than when it is harvested in the fall. For example, if the price of the commodity is $3.50 per bushel in the fall, the person would be happy with a price of $3.75 per bushel in the spring. If that change in price occurred, it would be just the right amount to provide returns to cover the costs of this operation.

What certainty is there, however, that if storage is undertaken in the fall that the price will be exactly $.25 per bushel higher in the spring? There is no certainty. Storing grain in the fall without knowing what it will be sold for next spring is risky. If this operation is undertaken, the owner is subjecting himself to some variability in income because of the possible variability of price while the commodity is stored. The price might be higher or lower than $3.75 in the spring because no one knows with certainty what will happen to the price between the fall and spring. There could be all sorts of developments in the world that could cause the price to rise more than $.25 or less than $.25 per bushel. In fact, there is nothing that says the price has to go up. It could drop to $3. If the price were to go up to $3.60 by the spring, the return to storage operations would be less than half of what it should be. On the other hand, if the price goes up to $4.00, then the return would be $250,000. The owner of the storage facilities might be perfectly happy to accept the risk associated with these unknown outcomes. On the other hand, he might prefer to have a little more certainty about his income even if it cost him a little money. How might he achieve this objective?

Let us suppose that a futures market exists for the commodity that is being stored. In the fall, assume that the cash market price is $3.50 per bushel when the grain is put into storage. If participants in the futures market are rational and realize that it will cost about $.25 per bushel to hold the commodity from the fall until spring, the price of a May futures contract at harvest time ought to be in the neighborhood of about $3.75. Since a buyer would incur a storage cost if grain to be used the following spring is purchased at harvest time, it is reasonable to expect to have to pay more for the commodity next spring. Thus, it would be reasonable to assume that the price of a May futures contract would be about $.25 per bushel above the cash price in the fall to reflect the carrying charge.

In order to hedge against uncertain price movements, an individual should take exactly the opposite position in the futures market that has been taken in the cash market. That is, if a grain elevator operator has purchased grain in the cash market, he would want to sell or go short in the futures market. The idea of hedging is to have offsetting positions in the cash and futures markets. When you sell in the cash market, you buy in the futures market. In the case of an elevator operator, the individual is initially buying in the cash market so he would want to sell in the futures market in the fall to establish a short hedge. Thus, if it is assumed that the hedge is established on October 15, the individual's transactions in the two markets would be the following:

TIME	CASH MARKET	FUTURES MARKET
October 15	Buy (or own) at $3.50 per bushel	Sell May contracts at $3.75 per bushel

Assuming that there were no major developments to change price expectations from October until May, when May 15 arrived the person holding the commodity ought to be able to sell it for $3.75 per bushel. Whenever the cash commodity is sold, he would buy futures contracts to offset any continuing obligation in the futures market. So, when he sold the grain, he would buy futures contracts at $3.75 per bushel. The $3.75 price per bushel in the spring assumes that nothing has happened to change people's perceptions of what the value of grain is from what traders anticipated several months earlier.

In this example it is assumed that the cash price is the same as the futures price in the spring. The reason for this is that we are considering a futures contract that is close to expiration. In May the price of a May contract should be very close to the cash price because when the contract expires the physical commodity at delivery points becomes a perfect substitute for the futures contract and vice versa. The reason that hedging provides some income certainty to participants in commodity markets is because of the relationship that exists between futures and cash prices of commodities. The reason for the relationship is because when futures contracts expire, the physical commodity is a perfect substitute for the futures contract at least at the point(s) of delivery specified in the contract. Consequently, when the future contract terminates, the cash and futures prices of goods with exactly the same quality characteristics have got to be the same. If the two prices differed, individuals would prefer to buy the lower-priced item and immediately convert it into a higher valued asset.

When a futures contract matures, locational differentials among cash prices could cause a difference between the futures prices and cash prices at nondelivery point locations. The cash price in North Carolina is not likely to be the same as the cash price in Chicago because of the cost of transporting the commodity. This will be discussed in more detail later. To illustrate the major points of a hedging strategy, it is easiest to assume for the time being that the cash transactions occur at a delivery point for the futures contract so there would be no transportation costs involved in using the physical commodity to satisfy the futures market objective.

Given the above assumptions, let us see how the individual came out by transacting in the futures and cash markets. To facilitate this calculation, the set of transactions can be updated using the following format:

TIME	CASH MARKET	FUTURES MARKET
October 15	Buy (or own) at $3.50 per bushel	Sell May contracts at $3.75 per bushel
May 15	Sell at $3.75 per bushel	Buy May contracts at $3.75 per bushel

Examining the above transactions, it is clear that $.25 per bushel was made in the cash market because prices behaved in exactly the correct manner to cover the storage costs. The futures market transaction gave the person a zero return because he bought and sold at the same price.[1]

In this example, the short hedge accomplished what was desired. That is, a return of $.25 per bushel was obtained at some cost. If the person had been sure (e.g., perfect price expectations) that the price was going to be $3.75 per bushel in the spring, there would have been no need to use the futures market.

Imperfect Price Expectations

Now let us suppose that things are more variable. Suppose that between the time the grain is put in storage and the spring, the market realizes that there is going to be an unusually big crop in Argentina or Brazil, or the world is in a recession and not as much grain will be demanded as previously thought, or other things occur that depress the price. Suppose that the cash price goes up by May but only to $3.60 per bushel. Thus, when the grain is sold in the spring it is worth $3.60, not $3.75 per bushel. Economic events that depress the cash market would also tend to depress the futures market because when the futures contract expires, the cash commodity becomes a perfect substitute for the futures contract. Thus, when the grain was sold in the cash market at $3.60 per bushel, the May contract would probably also be selling for $3.60 per bushel. Thus, the person would be able to buy futures contracts at that price. In order to see how the individual comes out in this situation, the new prices can be inserted in the tabulation of trades as follows:

TIME	CASH MARKET	FUTURES MARKET
October 15	Buy (or own) at $3.50 per bushel	Sell May contracts at $3.75 per bushel
May 15	Sell at $3.60 per bushel	Buy May contracts at $3.60 per bushel

In this situation, \$.10 per bushel is made in the cash market because the person bought at \$3.50 per bushel and sold at \$3.60 per bushel. Also, \$.15 per bushel is made in the futures market. Each contract was sold at \$3.75 per bushel and others were bought at \$3.60 per bushel. If you add the returns in the cash market to that of the futures market he still has \$.25 per bushel. If he had not participated in the futures market less than half the amount of money needed to cover the normal cost of operations would have been realized.

Another way of looking at the previous example is that the producer initially expected to be able to sell in the cash market at \$3.75 per bushel. By selling at \$3.60 rather than \$3.75 he made \$.15 less than he expected on the storage activity. On the other hand, by being able to buy back futures contracts at \$3.60 rather than \$3.75, the producer realized an unexpected gain of \$.15 per bushel in the futures market. Thus his less than expected gain in the cash market was compensated by a greater than expected gain in the futures market resulting in the same total return as if he had been able to sell in the cash market at \$3.75 per unit.

Suppose now that something happens to cause the price to go much higher than anticipated. Perhaps a new disease occurs that means not as much grain will be available this year or a war breaks out creating increased scarcity in the market. Suppose the price increases to \$4.00 per bushel by spring. In this case, a hedger would have made \$.50 a bushel by storing the grain, but he would lose \$.25 per bushel on his futures transactions. Again, however, he comes out with a net return of \$.25 per bushel or \$125,000 to cover the costs of operation. In this case, had he known with certainty that a \$4.00 per bushel price was going to occur, he would have been better off if he had never heard of futures markets. The dilemma is that there is some probability that the price could have been \$3.00 instead of \$4.00 per bushel. A \$3.00 per bushel price would have produced a loss of \$250,000 had he not participated in the futures market because of the change in value of the commodity. This example suggests that \$125,000 of income is going to occur with certainty regardless of what happens to the price because offsetting positions were taken in the two markets.

The basic principle of hedging is to take an offsetting position in cash and futures markets so that an unfavorable price movement in one market is offset by the effects of the price movement in the other market. It is a way of removing some of the uncertainty associated with whether the price of a commodity is going to increase or decrease. The reason hedging works is because of the offsetting positions in the two markets. In order to hedge, whenever an action is taken in the cash market, the opposite action must be taken in the futures market. For example, if you are long in the cash market you should be short in the futures market in order for a hedge to work.[2]

Unfortunately, when price relationships in the real world are considered, things do not always work out as nicely as the above examples. Other factors must be introduced that complicate the comparisons, but the principle is the same. If the basic principle of how offsetting positions in the two markets operate is understood, then complications are not so difficult to handle.

Locational Price Differentials

One complication that arises with actual situations is the fact that the relationship between the cash and futures prices may not always behave as nicely as illustrated in the previous examples. Suppose that we consider a cash market that is different from a delivery point specified in a futures contract. In this case, the relationship between the cash and futures prices in the month the futures contract expires that was assumed in the previous examples, may no longer hold. The cash price may differ from the futures contract price even at expiration because of locational price differences. Cash and futures prices ought to move together but when a futures contract expires, the cash price at a given location could differ from the futures price based on what it costs to move the commodity between markets. The prices for the two markets would never differ by more than the transportation charge because if the difference in prices was any larger, there would be some profit to be made by moving the commodity between locations. Traders know that the cash price at any location at the time the futures contract expires will not differ from the futures price by more than the costs of transportation because entrepreneurs will make sure there are no unusual returns to be made from moving the commodity.

Temporal Difference

In the earlier illustrations when the hedge was established it was assumed that the difference between the futures and cash prices was exactly equal to what it was going to cost to store the grain. There may be some other factors that cause the difference between futures and cash prices to differ from the full carrying charge when an initial position is established in the futures market. Local market conditions could influence the cash price in the fall to cause the difference between futures and cash prices to be different from full storage costs. For example, if there is a large harvest and producers are financially strapped because of debt payments and so on, they might be willing to sell the commodity a little faster than the marketing channels would like to take it causing the current cash price to be depressed a little more than what it otherwise would be. In these circumstances, the difference between the futures and cash prices might actually be greater than what it would normally cost to store the grain from harvest to next May.

In other cases, there may be special circumstances such as a later than usual harvest or bigger than usual local demand that cause the cash price to be bid up temporarily. A higher local cash price could result in the difference between the cash and futures prices being less than storage costs.

Changes in Basis

The thing that is critical in hedging is the relationship that exists between the futures and cash prices when the hedge is established compared to the relationship that exists between the two prices when the hedge is lifted. If we return to our initial example, suppose that the cash price had been \$3.55 per bushel instead of \$3.50 per bushel in the fall when the commodity was placed in storage. If everything else was exactly the same in terms of a \$3.75 per bushel futures price in the fall and a \$3.75 cash and futures price in the spring, the business would have grossed \$.20 rather than \$.25 per bushel. The smaller return occurs because the difference between the two prices that existed when the hedge was established was less than the full carrying charge. The operator would not be able to guarantee a full \$.25 per bushel by the hedging operation because of the initial price difference.

If market conditions resulted in the cash and futures prices increasing to \$4.00 per bushel between the fall and spring, the trader would make \$.45 per bushel from the cash market transactions, but lose \$.25 per bushel in the futures market or again end up with a return of \$.20 per bushel.

In fact, you can select any value you want for cash and futures prices in the spring and a net gain of \$.20 as opposed to \$.25 per bushel would occur provided the cash and futures market prices in the spring were identical. The reason this result occurs is because the initial relationship between the cash and the futures prices when the hedge was set up is different from the original example. Thus, the return from hedging depends heavily on changes in the relationship between the futures and cash prices or the basis.

In the above examples, the basis when the hedges were established in the fall was either \$.25 or \$.20 per bushel. In the trade, these would be referred to as negative numbers to indicate that the cash price was below the futures price. For ease of mathematical calculation (in order to subtract a smaller number from a larger one), however, basis can be algebraically defined as the difference between futures and cash prices as follows:

Eq 15-1

$$B_t = F_t - P_t$$

where

B_t = basis at time t for a specific futures contract and cash market

F_t = price of a specific futures contract at time t

P_t = cash price in a specific market at time t

To be consistent with the way basis terminology is used in the trade, the above equation would need to be multiplied by a minus sign. That operation is equivalent to reversing the terms on the right-hand side of the equation in which case the expression results in a larger value being subtracted from a smaller one. Thus, it is easier to think of subtracting a smaller value from a larger one and attaching a negative sign to the numerical value if desired to be consistent with trade terminology.

The relevant basis for a particular futures contract on a given day can differ depending on which specific cash market is being considered. Every hedger is primarily concerned about the basis for the specific cash market in which they do business. The basis for a specific location will change over time as prices in the futures market and cash market adjust to new market information. In the above hedging examples, a zero basis was consistently assumed to exist when the hedge was lifted. This may not be realistic even at the time the futures contract expires if the trader's relevant cash price is for a market that is not a delivery point for the futures contract. Location differentials in prices because of transportation costs indicate that the physical commodity at one place is not a perfect substitute for fulfilling an obligation of a maturing futures contract that requires delivery at specific sites. Thus, it is important for a hedger to consider what the relationship between the futures price and his local cash market price is likely to be when he gets ready to lift his hedge. He really is not concerned with whether commodity prices increase or decrease while the inventory is hedged. The key to his gross return is whether the relationship between the futures and cash prices change over time by a sufficient amount to cover the cost of storage.

THE ALGEBRA OF HEDGING

The outcome of short hedging can be demonstrated algebraically using general notation for any set of prices. That is, by letting F_t and P_t represent the futures and cash prices respectively at time t, a short hedge can be represented as follows:

TIME	CASH MARKET	FUTURES MARKET
1	Buy at P_1	Sell at F_1
2	Sell at P_2	Buy at F_2

The profit (or loss) per unit traded in the cash market therefore is the difference between (P_2), the price at time 2, and (P_1), the price at time 1 or $P_2 - P_1$. Similarly, the profit (or loss) per unit traded in the futures market is $F_1 - F_2$ (remembering profit is the difference between the selling and buying prices). Assuming the same quantity of product is traded in each market means the net return per unit from the hedging operation can be expressed by adding the two expressions together. That is,

Eq 15-2

$$15.2 \text{ Gross return (or loss) per unit} = (P_2 - P_1) + (F_1 - F_2).$$

Rearranging the terms and using the above algebraic definition of basis allows the gross return to be expressed (or loss) per unit in terms of the change in basis between time periods 1 and 2.

Eq 15-3

$$15.3 \text{ Gross return} = (F_1 - P_1) - (F_2 - P_2) = B_1 - B_2$$

As long as B_2 is less than B_1 (as assumed in all of the previous numerical examples) a positive return to short hedging occurs regardless of whether prices increase, decrease, or stay the same. This expression indicates why hedging can be described as replacing the variability in prices with variability in basis. Before a hedge is established, the individual can check the Internet or call his broker to determine what F_1 is and compare it to the existing local price to determine B_1 precisely. When a hedge is established, however, B_2 is not known with certainty because its magnitude will depend on future developments. However, there is usually much less uncertainty about what will happen to B_2 than what can happen to the overall price levels. This is because of the interrelationships between futures and cash prices previously discussed. To illustrate the fact that gross returns can not be predicted with certainty even with hedging, it is useful to put an asterisk on B_2 to differentiate an expected basis from an actual basis. Thus, anticipated gross returns per unit from a short hedge can be expressed as $B_1 - B_2^{\star}$ where $B_2^{\star}$ represents the best estimate at time 1 of what B_2 is expected to be. The actual return per unit to a short hedge is equal to $B_1 - B_2$, but this can not be calculated until time 2, or after the fact.

In order to make reasonable predictions of basis, it is important to study historical relationships between futures and cash prices for the cash market of interest. In this way, some evidence can be obtained on how the basis for the particular local market of interest has varied in the past.

TARGET PRICING

Once the principle of short hedging is understood, it is easy to see how producers can use futures markets to "lock-in" or target price future production. Many farmers have an interest in what price they are likely to receive for their production before they invest their time and effort and perhaps somebody else's money in some agricultural enterprise. Producers may use some type of forward contract to guarantee a price. This involves getting a commitment from someone that guarantees them a fixed price for the commodity when it is ready to be sold. There are many different ways in which forward contracts or commitments are negotiated between two parties.

Futures markets provide producers an alternative to forward contracts to obtain some price certainty. Futures markets can be used to accomplish the same objective as a forward contract but they do not lock a producer into a specific sale of the physical product as a forward contract does. The idea of target pricing is also useful to evaluate alternative uses of resources at planting time. If a producer does not obtain a forward contract or use the futures market, he accepts some risk in terms of what the price will be when the commodity is harvested and ready for sale unless the government guarantees or fixes the price. If this type of price uncertainty is too risky in terms of an individual's preferences, he might want to check what the current prices of futures contracts expiring next fall are for commodities that he is considering producing. This would simply indicate what the futures markets are suggesting about likely prices of these commodities next fall. Using this information for planting decisions is similar to the example considered previously, except the physical commodity does not exist yet. Making a planting decision in the spring, however, is equivalent to making a commitment in the cash market in terms of certain costs of production. When a decision is made to grow a commodity or go into a livestock feeding or production activity, it is similar to buying the commodity at some cost even though you are not physically purchasing so many bushels of corn or so many hundred hogs. A commitment is being made, however, as far as what the production activity is going to cost. When a production activity is undertaken that will eventually result in a long position in the cash market, an offsetting position in the futures market could be established to lock in a price by selling a futures contract. Selling in the futures market establishes an offsetting position even though you do not have the physical commodity in hand, but hope to have it barring unknown weather conditions or other unforeseen elements.

In the spring, producers may decide that a particular fall futures contract offers an attractive price. If they knew with certainty that they were going to get a given price for the commodity, they could easily calculate the profitability. After the planting decision is made and a futures contract has been sold, it is almost like locking in a fixed price for the commodity being produced. When the commodity is harvested or ready to be sold, producers can sell the physical commodity on their local cash market and simultaneously buy a futures contract to offset their position in the futures market. As long as they buy the same number of contracts in the futures market that they sold, they have no continuing obligation in the futures market. As far as the physical commodity is concerned, they can sell it in exactly the same way as if they had not participated in the futures market. The gross returns they make in the cash market are determined by whatever price exists in the local market. If the cash price exceeds their per unit costs of production, a profit has been made. This is not their only source of profit (or loss), however, if they have simultaneously traded in the futures market. If they sold a futures contract at a price of F_1 per unit and bought another one at a price of F_2 per unit, the extra profit (or loss) is $F_1 - F_2$ per unit. This information can be set up in a table as follows:

TIME PERIOD	CASH MARKET	FUTURES MARKET
At planting time	Produce at C_1 (estimated cost per unit)	Sell at F_1
Harvest or later	Sell at P_2	Buy at F_2

One difference in the above example from that considered earlier is that initially the producer makes a commitment to produce at a cost of C_1 per unit rather than physically owning or buying the commodity. Although the producer may have a fairly accurate idea of what his total production costs will be, the cost per unit of production will be uncertain to the extent that total output is variable. Output uncertainty is ignored for the time being.

The net return for the above set of transactions can be expressed as $(P_2 - C_1)$ for the cash market and $(F_1 - F_2)$ for the futures market. Adding the two components results in a net return per unit of $(P_2 - C_1) + (F_1 - F_2)$. After rearranging the terms in the preceding expression to $F_1 - (F_2 - P_2) - C_1$

and replacing $F_2 - P_2$ by $B_2^\star$, the net return per unit of production can be represented as $F_1 - B_2^\star - C_1$. By taking the initial position in the futures market, the producer essentially locks himself in or is establishing a target price of $F_1 - B_2^\star$ for the commodity. It does not really matter what happens to the price in the cash or futures markets between the time production is initiated and when the output is ready to be sold because his projected gross price per unit is $F_1 - B_2^\star$. The only thing that is uncertain is the basis that will exist when he gets ready to sell his production. When the producer is making his initial decision, he knows what F_1 is. There is no doubt about that. His only uncertainty is about B_2. He can not predict B_2 perfectly, but by studying the relationship between prices in a particular local market and the futures market in the past, he may be able to get a pretty good idea of the possible range of variation in $B_2^\star$.

An advantage of using futures markets relative to taking a chance on cash prices is that the variability of the basis is generally much less than the variability in the price. In this way, using the futures market is like buying insurance. It will involve some costs, but the reduction in uncertainty may be worth it. Securing forward contracts may also be costly in terms of the time it may take to negotiate a contract involving particular conditions that the two parties agree to. On the other hand, an individual may prefer to just take his chances on what happens to market prices. He may decide that this alternative might be optimal after evaluating his expectations and risk preferences. If producers need to finance production costs, lenders may prefer that producers have a forward contract or use the futures market to target price output in order to guarantee profitability in terms of a "three party agreement." Lenders may not want to extend credit if there is a chance the market price will go down. They may want to make sure the producer is locked in at a particular price to make sure a loan can be paid after harvest.

In order to keep the algebra simple for the above example, it was assumed that the producer would sell exactly the same quantity in the futures market as he anticipated producing. This, of course, may not be feasible because of the discrete units in which futures contracts are traded. For example, a producer anticipating 12,500 bushels of corn would not be able to sell futures contracts for exactly that quantity. Furthermore, the uncertainty associated with the actual quantity of production that will be available to sell because of weather and other uncontrollable postplanting factors often means it may be unwise for producers to make a forward commitment for all of their expected output.[3] There is no universal rule as to what fraction of anticipated output is optimal to forward sell. The results of most studies involving alternative assumptions about market conditions suggest a range of 60% to 75% of expected output as an optimal amount to forward sell on the futures market.

Now let us consider how target pricing works using some actual numbers. Suppose in the spring a producer sells a November soybean contract at a price of $6.25 per bushel. In the fall when the crop is ready for harvest, suppose the price of soybeans is $5.50 per bushel because of a bumper crop. If he had not used the futures market, he would get only $5.50 per bushel. If the cash price decreased between planting time until harvest, chances are the futures price would also have gone down perhaps to $5.75 per bushel. Assuming a harvest time basis ($F_2 - P_2$) of $.25 per bushel, how does he come out? He gets $6.25 minus $.25 so he ends up with a net price of $6.00 per bushel. His neighbors who did not know anything about the futures market would end up getting $5.50 per bushel.

Suppose now that the cash price at harvest time had increased to $6.50 per bushel. How does the producer using the futures market end up? He still gets $6.00 per bushel assuming a harvest time basis of $.25 per bushel. His neighbors who did not use the futures market end up receiving $6.50 per bushel. In evaluating these results it is important to remember what the person's original motivation was for using the futures market. The futures market was selected to avoid speculating or gambling about what the product's price was going to be at harvest time. The futures market was used to obtain a little more certainty in net returns regardless of what happens to prices.

LONG HEDGING

The type of hedging discussed in the beginning of this chapter as well as the example of how producers can use the futures market to target price their output, initially involved establishing a short position in the futures market to offset a long (or potentially long) position in the cash market. Suppose now that the initial position in the cash market is a short position. How could that occur? Suppose you are an exporter and have just negotiated a sale of grain to another country (e.g., Japan) but the buyer does not want the product delivered for another 6 months. As an exporter, you have made a commitment agreeing to sell a commodity that you may not have and furthermore, you may not want to own until you have to ship it. If you buy the commodity now, you will be tying up some money for a period of time. What you would prefer to do is wait and buy the commodity closer to the delivery time and use your capital for other purposes in the meantime.

Another example is if a firm processing an agricultural commodity makes an agreement with a retail chain to provide a specified quantity of a product at a fixed price 6 months from now. In both cases, an agreement is established at the present time that specifies the sale of a physical commodity to be delivered at a specified price sometime in the future. In essence, the agreements establish short positions in the cash market.

In such situations it would be desirable to take offsetting positions in the futures market to provide protection against unexpected changes that may occur in the cash price of the commodity. To take an offsetting position in the futures market requires the purchase of a futures contract equal to the quantity sold forward in the cash market. This would be establishing a **long hedge** because the initial position in the futures market is a long position. When it is time to deliver the physical commodity to the eventual purchaser, the cash market price might be higher or lower than what had been anticipated when the initial commitment was made. Assuming that prices in the cash and futures markets move together means that unanticipated gains (or losses) in one market would tend to be offset by unanticipated losses (or gains) in the other market.

This type of hedge can be analyzed algebraically similar to what was done before. The sequence of operations for a long hedge include the following:

TIME	CASH MARKET	FUTURES MARKET
1	Sell at C_1	Buy at F_1
2	Buy at P_2	Sell at F_2

where C_1 is the cash price per unit that will be received when the commodity is delivered. The net return from this combination of trades is $(C_1 - P_2) + (F_2 - F_1)$. Rearranging the terms in the last expression leads to $(F_2 - P_2) - (F_1 - C_1) = B_2^\star - (F_1 - C_1)$. In this formulation, the key unknown element becomes what the expected basis is going to be when the commodity is purchased for delivery and the hedge is no longer needed. The difference between the current futures price and the price specified in the forward sale (C_1) is known with certainty when the initial terms of the contract are specified. This difference is like a basis in one respect, in that it involves a difference between futures and cash prices. It is not a basis, however, because C_1 is a negotiated price at which the commodity will be sold in the future and may differ from what the current cash price is.

FUTURES MARKET EQUILIBRIUM

When considering why some traders are short hedgers and others are long hedgers, it is possible to see how a market for futures contracts could exist even without any speculators. Thus, futures trading is not necessarily just a case of risk averse individuals transferring some price risk to speculators who are willing to assume it. Of course, the potential profit from unexpected price movements that can accrue to speculators provides an incentive for speculators to participate in futures trading and adds liquidity to futures markets. Any time there are short hedgers who desire to sell futures contracts and simultaneously long hedgers who want to buy futures contracts, a market could exist. Thus futures markets are a way of mutually satisfying individuals who have interests in taking opposite directions in future ownership of a commodity. People with different interests in a particular commodity are able to negotiate contracts to limit their exposure to price uncertainty through futures markets. Futures markets can provide protection against unusual fluctuations in the cash price after a commitment to produce, buy, or sell the physical commodity has been made. Traders may want protection against unusually large losses at the cost of giving up unusually large profits in order to have more stability in their incomes. Futures markets also provide a mechanism for other individuals to profit by speculating on unknown price changes of commodities.

SPREAD AND STRADDLE STRATEGIES

Once the basic idea of hedging is understood, the world is practically unlimited as far as other trading possibilities involving transactions in more than one market. Instead of the relationship between the cash and the futures market for a given commodity being the key element as in the previous examples, the relevant price difference might be between two futures contracts in the same market or two different futures markets. For example, one could be trading May and November contracts for the same commodity at the same time in a given futures market. This is usually referred to as **spreading.** Initially, what might be done, for example, is simultaneously sell a May contract and buy a November contract. Then sometime later to fulfill the obligations in the futures markets for each contract, the opposite actions are taken. After the second set of transactions, the individual would be out of the market as far as the May and November contracts are concerned. Why might this sequence of transactions be of interest? If you think the price relationship between these two markets is going to change but are not sure whether prices in general will increase or decrease, spreading offers an opportunity to make money. Algebraically these ideas can be expressed as follows:

TIME	MARKET A	MARKET B
1	Buy at A_1	Sell at B_1
2	Sell at A_2	Buy at B_2

The return from this combination of trades is $(A_2 - A_1) + (B_1 - B_2)$ or $(B_1 - A_1) - (B_2 - A_2)$. Letting $S_1 = B_1 - A_1$ or the spread between the two markets, the return can be represented as $S_1 - S_2^\star$. An asterisk attached to S_2 indicates that in evaluating the potential profitability of such a combination of trades, A_1 and B_1 are known with certainty, but one must predict A_2 and B_2 or a value for S_2 to estimate profitability. S_1 is known with certainty because it is based on current prices. If S_1 is bigger than $S_2^\star$, then some potential profit is associated with the above sequence of trades. On the other hand, if S_1 is smaller than $S_2^\star$ (or what S_2 actually turns out to be), the above combination of transactions would be unprofitable. In the latter case, the potential or actual loss could be converted to a gain by reversing the order of transactions. That is, initially sell in Market A and buy in Market B. In this case, the sequence of operations would be:

TIME	MARKET A	MARKET B
1	Sell at A_1	Buy at B_1
2	Buy at A_2	Sell at B_2

The net outcome of this sequence of trades is $(A_1 - A_2) + (B_2 - B_1)$ or $(B_2 - A_2) - (B_1 - A_1) = S_2^\star - S_1$.

Thus, the appropriate sequence of buying and selling actions in the two markets depends on how one anticipates the price difference that currently exists between the two markets to compare to the price difference at some future date. There is no guarantee that price relationships will always change in particular ways, but people who follow market prices on a regular basis are often more successful in identifying how the relationship between two prices is likely to behave than in predicting whether prices in general will rise or fall. The existence of a profit potential through simultaneously trading in different contracts in the same commodity (spreading) or in different markets (**straddling**) helps keep price relationships from deviating too far from the values dictated by economic factors associated with time, space, or form values of physical commodities.

SUMMARY

The ownership of any commodity subject to price variability involves considerable risk. The use of futures markets can reduce some of this risk. Taking a position in a futures contract opposite from the cash market position offers some protection against unexpected changes in price of a commodity. A short hedge, for example, involves the selling of a futures contract to protect against possible decreases in the price of a commodity that is owned. The reason hedging provides some price insurance is because cash and futures prices for the same commodity generally move in the same direction. This means that if the physical commodity loses value because of a drop in price, profit is likely to result from short positions in the futures market.

The net return from hedging depends on what actually happens to the difference in cash and futures prices during the time the hedge is in place. The difference between a particular futures price and a cash price for a specific market on a given day is known as the basis. The basis for a given cash market is influenced by locational and time elements. The time component diminishes the closer a futures contract is to maturity. The locational component depends on the cost of transporting a commodity between a specific cash market and delivery points specified in the futures contract. Basis variability is generally less than price variability and is why hedging tends to offer some protection against unknown changes in price.

Futures markets can also be used to lock-in or target price agricultural commodities before they are produced. The sale of a futures contract at the time production decisions are made essentially determines the price that will be realized when the commodity is actually ready for sale. The only uncertainty is what the basis will be at the particular cash market when the commodity is sold. Analysis of historical basis behavior may be useful in establishing some guidelines about what might be expected for given markets.

Another use of futures markets to minimize price uncertainty is to simultaneously purchase futures contracts at the time a commitment is made to deliver a commodity at a fixed price at some time in the future. To acquire price protection for these situations requires the purchase of futures contracts, and this is why it is referred to as a long hedge. Long hedging is especially useful for exporting firms and agribusinesses who make commitments to provide specific quantities of commodities at specific future times at a predetermined price.

The fact that the net return from hedging depends on what happens to the difference in prices in the two markets is also applicable to consider potential profits associated with transactions involving two or more futures contracts. These can consist of simultaneously buying and selling two different futures contracts and later closing out the positions when the difference in prices has changed. The two futures contracts can be contracts with different maturity months for the same commodity or entirely different commodities.

QUESTIONS

1. What is the major factor determining whether someone buying and/or selling futures contracts is a speculator or a hedger?

2. Suppose that while you are storing 50,000 bushels of soybeans that you have hedged in the futures market, your local market's cash price declines by $.40/bu. If your local basis also has changed from $.45/bu under to $.35/bu under the futures contract at the same time the cash price has been dropping, what does this indicate about the specific amount of price change that has occurred in the futures market? Explain how you arrive at your answer.

3. Identify and briefly describe the two economic factors that determine the magnitude of the basis for any commodity for which a futures market exists.

4. Assume you sell cattle futures at a price of $69.00/cwt, later buy a futures contract at $66.00/cwt, and sell live cattle in the cash market at a price of $65.00/cwt. What is the net price you realize from this set of transactions (ignoring commissions and other transaction costs)?

5. Assume that when a long hedge is initiated, the local basis is $.75/unit under the futures price. If the basis becomes larger in absolute value (e.g., $1.00/unit under the futures price), would this be a favorable or unfavorable development as far as the hedger is concerned? Explain your reasoning.

6. If the current price of November futures contracts in soybeans is $5.95/bu and a producer expects his local basis will be $.25 under the November contract next fall when he gets ready to harvest, what price can he try to lock-in by using the futures market?

7. Assume you are employed by the XYZ exporting company and have just negotiated a 20 million bushel sales contract for grain with representatives of China at a fixed price of $5.50/unit net of shipping costs, but the commodity is not scheduled for delivery until next June.

a. If the current price of June futures of this commodity is $5.40/unit, briefly explain how your company could use the futures market to make sure a fixed cost of the commodity was locked in to "guarantee" a profit.

b. What action would your firm need to take next spring to fulfill its obligation in the futures market assuming it hedged its commitment when you made the deal with China?

NOTES

1. Note that this comparison ignores any transaction costs including brokerage fees, the opportunity cost of the margin money required to maintain a position in the futures market, and the possibility of any margin calls while the commodity was being stored.

2. These examples assume that the quantities in the futures market and cash market are identical.

3. In some cases, there may be a possibility of obtaining government or private insurance protection or other alternatives to protect against quantity uncertainty.

FUTURES OPTIONS PRICES

Monday, April 3, 2000

AGRICULTURAL

CORN (CBT)
5,000 bu.; cents per bu.

Strike	Calls-Settle			Puts-Settle		
Price	May	Jly	Sep	May	Jly	Sep
210	24½	33¾	42½	⅛	1⅝	3¼
220	15¼	26¼	35⅜	1	3¾	6
230	8	20	29⅜	4	7¼	10
240	3½	15	24¾	9½	12½	14½
250	1⅝	11⅝	20¾	17¼	18¾	20¼
260	½	9¼	17½	26¼	26¼	27

Est vol 15,000 Fr 17,762 calls 7,768 puts
Op int Fri 340,818 calls 203,139 puts

SOYBEANS (CBT)
5,000 bu.; cents per bu.

Strike	Calls-Settle			Puts-Settle		
Price	May	Jly	Aug	May	Jly	Aug
500	47¼	64¼	72½	¾	6⅛	11½
525	25⅞	46⅞	57	4⅜	13¾	21
550	11½	34	44½	15	25½	33
575	4½	24½	35	33	40¾	....
600	1⅝	18	27	55	59¼	....
625	½	13½	22¾	....	79⅜	....

Est vol 15,000 Fr 14,571 calls 9,299 puts
Op int Fri 132,817 calls 60,467 puts

SOYBEAN MEAL (CBT)
100 tons; $ per ton

Strike	Calls-Settle			Puts-Settle		
Price	May	Jly	Aug	May	Jly	Aug
160	12.70	16.10	18.00	0.60	2.75	4.50
165	8.60	12.75	....	1.40	4.25	6.50
170	4.90	10.00	12.65	2.75	6.50	9.00
175	2.75	8.00	10.50	5.50	9.35	....
180	1.50	6.25	8.75	....	....	15.00
185	0.75	4.75	7.50	....	....	....

Est vol 2,500 Fr 987 calls 2,291 puts
Op int Fri 25,357 calls 22,260 puts

SOYBEAN OIL (CBT)
60,000 lbs.; cents per lb.

Strike	Calls-Settle			Puts-Settle		
Price	May	Jly	Aug	May	Jly	Aug
175	1.190	1.870	2.150	.160	.510	.660
180	[illegible]	[illegible]	[illegible]	[illegible]	[illegible]	[illegible]

Strike	Calls-Settle			Puts-Settle		
Price	May	Jun	Jly	May	Jun	Jly
2500	0.34	0.72	0.99	0.83	1.14	1.66
2550	0.20	0.56	0.81	1.19	1.48	1.98
2600	0.10	0.44	0.65	1.59	1.86	2.32

Est vol 2,170 Fr 350 calls 0 puts
Op int Fri 20,361 calls 18,364 puts

GAS OIL (IPE)
100 metric tons; $ per ton

Strike	Calls-Settle			Puts-Settle		
Price	Apr	May	Jun	Apr	May	Jun
19500	12.35	11.05	12.35	0.10	4.30	8.35
20000	7.75	8.20	10.05	0.50	6.45	11.05
20500	4.00	5.70	8.00	1.75	8.95	14.00
21000	1.65	3.85	6.30	4.40	12.10	17.30
21500	0.55	2.60	4.90	8.30	15.85	20.90
22000	0.15	1.70	3.80	12.90	19.95	24.80

Est vol 150 Fr 100 calls 200 puts
Op int Fri 6,073 calls 1,562 puts

LIVESTOCK

CATTLE-FEEDER (CME)
50,000 lbs.; cents per lb.

Strike	Calls-Settle			Puts-Settle		
Price	Apr	May	Jun	Apr	May	Jun
8300	0.97	1.70	....	0.10	....	....
8350	....	....	....	....	....	....
8400	0.37	1.07	....	0.50	1.17	....
8450	....	....	....	....	....	....
8500	0.10	0.62	....	1.22	....	....
8550	....	....	....	....	....	....

Est vol 265 Fr 98 calls 196 puts
Op int Fri 3,341 calls 15,137 puts

CATTLE-LIVE (CME)
40,000 lbs.; cents per lb.

Strike	Calls-Settle			Puts-Settle		
Price	Apr	Jun	Aug	Apr	Jun	Aug
70	2.17	0.82	1.65	0.05	1.67	1.75
71	1.22	0.42	....	0.10	....	....
72	0.40	0.25	0.77	0.27	3.07	2.82
73	0.10	0.12	....	0.97	....	....
74	0.05	0.07	0.27	1.92	....	....
75	0.02	....	....	....	....	....

Est vol 930 Fr 847 calls 356 puts
Op int Fri 29,619 calls 47,083 puts

CHAPTER 16

OPTIONS MARKETS AND THEIR USES

This chapter discusses options on futures contracts and indicates how they can be used by producers and other firms in making marketing decisions.

The major points of the chapter are:

1. The similarities and differences between futures and options markets.
2. The privileges and responsibilities of buyers and sellers (writers) of options.
3. The intrinsic and time components of option premiums.
4. How **put options** differ from **call options** and how they can be used by agricultural producers.
5. A graphical representation of potential costs and returns to buyers and sellers of options.

INTRODUCTION

The first part of this chapter provides a brief historical review of options trading in the United States. The basic idea of an option is introduced in terms of how the concept is often used in real estate transactions to provide buyers additional time in making a final decision regarding the purchase of a particular piece of property. Similarities and differences in the trading of options and futures contracts including the privileges and responsibilities of buyers and sellers are described. The way in which prices of alternative options are related to the current price of futures contracts is explained. The chapter then focuses on several aspects of put options that provide the privilege of selling a particular futures contract at a predetermined price. The final part of the chapter indicates how the net returns from buying and selling options can be graphically illustrated.

HISTORY OF OPTIONS TRADING

Options markets for agricultural commodities are similar in many ways, yet different from futures markets. Options are purchased and sold on organized exchanges similar to futures markets. Familiarity with how futures markets operate greatly facilitates understanding of how option markets function.

In the United States, organized trading of options has been an on-again, off-again activity. As early as the Civil War, grain options were traded on the Chicago Board of Trade shortly after futures trading began. The Chicago Board of Trade tried to terminate options trading because of perceived abuses. Horner and Moriarty (1983) indicate the following about the early history of options trading:

> In 1874, Illinois banned option trading both off and on exchanges, although such trading continued despite the statutory ban. In 1887, the Chicago Board of Trade again found it necessary to restrict option trading by its members, and between that time and 1936 when agricultural options were banned by Congress they were alternately traded and banned several times. (p. 9)

The U.S. congressional ban on trading in options on domestic agricultural commodities in 1936 resulted from several factors. One key element was an attempt in 1933 to manipulate the wheat futures market using options (Kenyon 1994). The congressional ban did not prevent the trading of options tied to London commodity futures contracts. Some scandals related to these options caused the CFTC to suspend the offering and sale of commodity options originating on foreign exchanges in June 1978. At the same time, the CFTC directed its staff to develop regulations for exchange-traded options. In September 1981, the CFTC approved a 3-year pilot program allowing each futures market to begin offering one option in something other than a domestically produced agricultural commodity. Options in this pilot program were related to gold, silver, sugar, and some financial instruments. An experimental program in the trading of options on some agricultural futures contracts began in October 1984 after the Futures Trading Act of 1982 specifically lifted the 1936 ban on the trading of options on domestically produced agricultural commodities. Each exchange was permitted to begin options trading on two agricultural futures contracts. Initial offerings consisted of options on corn, wheat, cotton, live cattle, soybeans, and hog futures contracts. Based on the popularity and market success of the experimental program with agricultural options, more options have been successfully introduced over time. In 1998, there were 124,107,563 options on futures contracts traded in the United States (Commodity Futures Trading Commission 1998). This was nearly 80% more than in 1992. Agricultural products accounted for 14.3% of all options traded on futures contracts.

The renewed trading of options on agricultural commodities was preceded by the introduction of trading in stock options in the United States during the previous decade. One type of option (calls) on stocks was initiated in 1973. Four years later trading of put options in stocks was initiated in the United States.

WHAT IS AN OPTION?

Many people have some familiarity of how options can be used in conjunction with real estate transactions even if they are not aware of the trading of commodity options. Thus, it is useful to initially review some of the characteristics of an option as used for real estate transactions before examining how the trading of commodity options involve some of the same characteristics.

A potential buyer of a particular piece of property may acquire an option (or privilege) to make the purchase at a specified price before actually making a final commitment or decision to make the purchase. An option is essentially a privilege granted by the owner of an asset to a prospective buyer. The privilege gives the prospective buyer an exclusive right or opportunity to purchase the property from the owner by a specific date at an agreed-upon price. The party who obtains the option is under no obligation to actually buy the property if he or she decides not to exercise the option before it expires. On the other hand, the party who grants the option is under a continuing obligation to fulfill the terms specified in the option if the other party decides to exercise it. Usually, an amount of money is paid to compensate the grantor of the option for their willingness to assume a specified obligation for a certain period of time. If the option is not exercised within the specified period of time, the option expires and there is no continuing obligation. If the option expires, the grantor of the option keeps the money that was paid for the privilege or exclusive right to buy a specific piece of property for a specific limited period of time.

In the case of commodity options, a person can obtain the right or privilege to purchase a particular futures contract at a fixed price per unit of a commodity for a specified period of time. In this case, a particular futures contract is the asset analogous to the property in the previous example. An option that provides a right to *buy* an asset is known as a **call option.** This terminology is important in order to distinguish this type of option from other options that convey the privilege of selling a futures contract. The latter type of options will be considered later in this chapter.

Obviously, if people can buy call options on futures contracts at some price per unit of the commodity, there must be individuals willing to participate in the other side of the market. Thus, the sellers of options assume a continuing obligation to sell a futures contract at a fixed price for a specified period of time. These individuals are known as **option writers.** They simply assume an obligation to deliver a futures contract at a fixed price if called upon to do so within a limited period of time. Of course, they may never be called upon to fulfill their obligation if there is no economic advantage for the option owners to exercise or take advantage of the privilege that was purchased. This means that participants in a given call market have a choice as to whether they want to secure (buy) a call option or to write (sell) a call option. Purchasing a call option might be a desirable form of speculation if it seems likely that the privilege of being able to secure a futures contract at a fixed price will become more valuable in the future. On the other hand, if it seems likely that a call option will become less valuable in the future, or there is little likelihood that the call option will be exercised, it would be advantageous to write an option. This is because the writers get to keep the price (or premium) for granting the privilege or assuming an obligation to deliver a futures contract if called upon. In order to write (sell) a call option, it is not necessary to actually own a futures contract because all that is involved is a promise to deliver a futures contract at a specified price if called upon. Given the liquidity of futures markets there is a very high probability that a futures contract could always be obtained, if needed, at some price.

The actual price or *premium* of an option is determined by the relative strength of interest in the two sides of the options market. More interest in obtaining the privilege than assuming a continuing obligation would cause premiums of options to increase. If the relative interest in the two sides of the options market were reversed it would tend to cause market premiums of an option to decrease.

When an option is purchased, the buyer pays the total cost of the privilege at the time of the purchase. The total cost is the premium per unit of the commodity multiplied by the quantity units associated with the option. The quantity units associated with options are identical to the quantities associated with underlying futures contracts. For example, an option on a corn futures contract on the CBOT involves 5,000 bushels, or the same amount as the futures contract for the commodity. Thus, if the premium for a given call option is $.10 per bushel, the total cost of the option would be $500 ($.10 × 5,000 bushels).[1]

The total cost of an option is usually only a fraction of the total value of the asset or futures contract that might eventually be obtained if the option is exercised. In this sense, options are like futures transactions in terms of securing a potential interest in an asset by paying only a small amount of money. Option writers must post margin money to cover any potential losses that may accrue during the time they have a continuing commitment or obligation if called upon to fulfill their obligation (i.e., to sell a futures contract at a fixed price). Potential losses can be experienced by writers of call options if the price of the futures contract rises above the price specified in the option. This is because the writer may have to eventually purchase a futures contract if called upon to fulfill his obligation in the options market. Since writers of options assume continuing commitments to fulfill their obligations, they are subject to possible losses associated with adverse price changes in futures markets. Thus, writers are required to post margin accounts and are subject to margin calls similar to being in the futures market as long as they have a continuing obligation in the options market. Their accounts are adjusted every day based on what happens to the market value (premium) of the option for which they have assumed a continuing obligation.

If an option has not been exercised by the time it expires, the writer would have his margin money returned plus the initial premium less any commission charges. One way writers of options can eliminate their continuing obligation before the option expires, is to enter the options market and buy an option. Again this is similar to the way futures markets operate in that an existing obligation in a market can be eliminated by taking an offsetting position in the same market. Owners (or purchasers) of call options do not have to worry about taking offsetting positions, however, because they do not have any continuing obligation to do anything. They have a privilege to obtain a futures contract if they decide it is profitable, but no obligation to exercise the option.

The decision to exercise a call option is equivalent to converting a position in the options market into a futures market position. Once this step is taken, positions in the option market are converted into continuing obligations in the futures market. Thus, if the option is exercised, the owner of a call option assumes a long position, and the writer is assigned a short position in the futures market. The price at which these positions are established in the futures market is the price specified when the initial option was bought and sold, not the price of the futures contract that exists when the option is exercised. The futures price specified in option contracts is known as the **strike price.** Several options on the same futures contract with different strike prices are traded simultaneously. Obviously, premiums for options with different strike prices will vary depending on the relative interest in buying or writing that particular option. The premiums of call options with high strike prices will be smaller than similar options with low strike prices because of reduced interest in the privilege of buying futures contracts at higher prices. Similarly, there would be more interest in writing call options with higher strike prices than lower strike prices because of differences in the probability that the options will be exercised. The relative difference in buying and selling call options with different strike prices compared to the current futures prices is the determining factor that produces an equilibrium set of premiums that clears the call options market.

A Specific Example

In order to illustrate some of the concepts introduced in the previous discussion, it is useful to consider a specific options market. An example of information for

TABLE 16-1

PREMIUMS FOR ALTERNATIVE CALL OPTIONS ON CORN AT THE CHICAGO BOARD OF TRADE FOR A PARTICULAR DAY

Strike Price	December	March	May
150	21¾		
160	12¾		
170	5¾	14	
180	2½	9¾	14
190	1	5½	9¼
200	⅜	3¼	6½

the corn options market at the Chicago Board of Trade on a particular day is indicated in Table 16–1. The table indicates the premiums for call options with several different strike prices for three alternative futures contracts. Thus, the privilege to obtain a December, March, or May corn futures contract at the various strike prices indicated could have been purchased at the premiums indicated. The values listed under each of the three months in Table 16–1 indicate the premiums of options with different strike prices in cents per bushel. Thus, an option involving the privilege of obtaining a March futures corn contract at a price of 170 cents per bushel could have been purchased for a price of 14 cents per bushel at the close of trading on that particular day. On that same day, the closing price of a March futures corn contract was 181¼ cents per bushel. Thus, traders were willing to pay 14 cents per bushel for the privilege of being able to obtain an asset at a price of 170 cents per bushel whose current value was 181¼ cents per bushel.

The values in each of the columns indicate that the premiums (or prices) of call options decline with higher strike prices associated with a particular futures contract. For example, on the same day traders were willing to pay 5½ cents per bushel for the privilege of being able to buy a March corn contract at a price of 190 cents per bushel. The reason for their willingness to pay money for the privilege of possibly purchasing a corn futures contract at a price greater than its current value is that there is some likelihood that the price of March futures contracts might exceed 190 cents per bushel before the option expires. If the March futures price should increase to any level above 195½ the owner of a 190 option could profit by exercising his option before it expires. By exercising the option, a futures contract could be obtained at a price of 190 cents per bushel and could be immediately sold for more than enough to cover the cost of the option. Any price of the futures contract above 195½ bushel would result in a net return (ignoring brokerage and transaction costs) to the option based on the difference between the futures price when the option is exercised and the strike price less the cost of the option. On the other hand, any price above 195½ cents per bushel would lead to a net loss for writers of such options in that they would be required to sell a futures contract whose current value is greater than the fixed price specified in the option. Part of the loss associated with the difference in prices, however, would be offset by the premium that was received when the option was written.

If the futures price of the March contract decreased below its current value of 181¼ cents per bushel, a 190 March option would lose some of its value and might never be exercised. For example, if the March futures price decreased to 175 cents per bushel, there would be no incentive to exercise the option to acquire an asset at a price of 190 cents per bushel if its current value was 175 cents per bushel. In this case, the owner of the option would likely let it expire. If the option expired, the option writer would not be required to do anything and be able to keep the premium of 5¼ cents per bushel.

COMPONENTS OF OPTION PREMIUMS

The premium or price of any particular option can be decomposed into two components: an **intrinsic value** and **time value.** This relationship can be expressed by the following relationship:

$$\text{Options premium} = \text{Intrinsic value} + \text{Time value} \qquad \textbf{(Eq. 16-1)}$$

The intrinsic value of a call option is the difference between the current futures price (F) and the strike price (S) specified in the option. This means that the intrinsic value for a given option can be expressed as $F - S_i$ where S_i represents the strike price for the *i*th option. For example, if the current price of March futures is 181¼ cents, a March option with a strike price of 170 has an intrinsic value of 11¼ cents. This option provides the right to buy an asset at a price of 170 cents whose current value is 181¼ cents. Thus, 11¼ cents of the premium for this option reflects the amount by which the current futures price exceeds the strike price. Call options with lower strike prices on a given futures contract obviously have larger intrinsic values. Intrinsic values for call options exist only for those with a strike price less than the current futures price. The intrinsic value can never be negative even though the above formula could lead to negative values if F were less than S_i.

A graphical representation of how the intrinsic values for a given option with a specific strike price varies with the price of a given futures contract is illustrated in Figure 16–1. Call options for which the intrinsic values are positive are said to be **in-the-money**. If the current futures price is less than the strike price of a call option, it is said to be **out-of-the-money**. Finally, if the futures price coincides with the strike price, the option is said to be **at-the-money.** As changes in economic conditions cause price adjustments in futures markets, intrinsic values of various options change and consequently is one reason option premiums change.

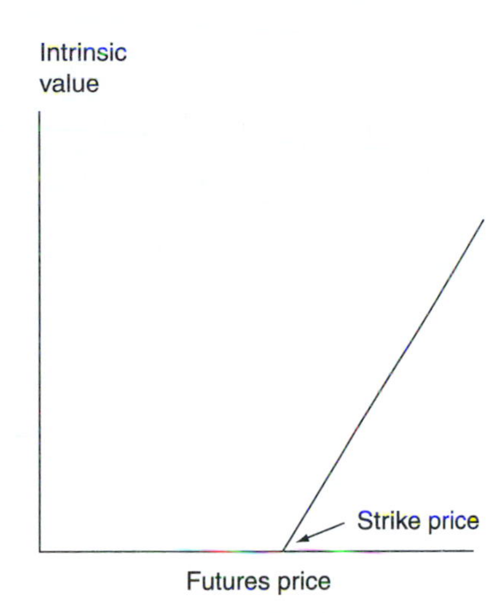

FIGURE 16–1 Relationship between intrinsic values of a call option with a specific strike price and alternative futures price.

The difference between the premium and intrinsic value of any option represents the **time value** of the option. In the previous example, the 170 March option has a time value of 2¾ cents which is the difference between 14 (the premium) and 11¼ (the intrinsic value when the futures price is 181¼). The time value represents the amount of money per unit of the commodity individuals are willing to pay over and above the intrinsic value of an option in order to have an opportunity (or privilege) to possibly profit from prospective changes in the price of the futures contract. Time values for options on a given future contract usually vary inversely with the absolute difference between the current futures price and strike price. This means that the largest time values are usually observed for options with strike prices closest to the current future price. Options with higher and lower strike prices generally have smaller time values reflecting the fact that there is a lower probability that futures market prices will change by a sufficient amount to make the option profitable. A representation of this relationship is illustrated in Figure 16–2.

The time values of options also vary with the amount of time before options will expire. Trading in a given option stops usually about a month before trading in the futures contract ends. This provides a few weeks for anyone who desires to exercise an option before it expires to have some time to offset the obligation in the futures market that is established when an option is exercised.

A clearinghouse acts as an intermediary between buyers and sellers of options just like futures markets. Option writers really are obligated to the clearinghouse instead of particular individuals who buy options. Since all option contracts with the same strike price are standardized, it doesn't matter who sold or

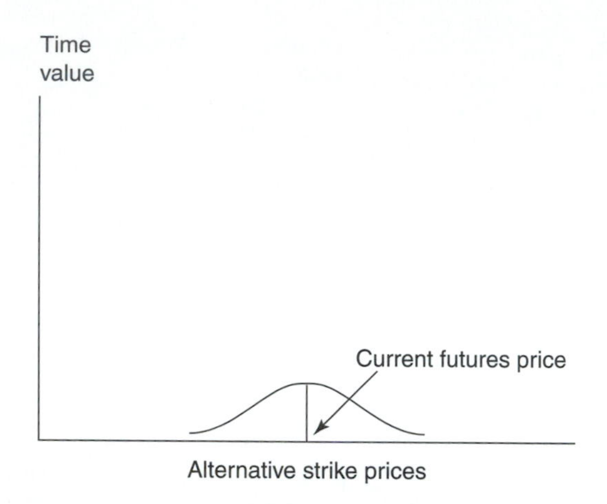

FIGURE 6–2 Typical relationship for time values of options with different strike prices at a particular point in time.

bought a particular option. The clearinghouse can match up buyers and sellers of options with the same strike price regardless of which pair of individuals made the initial transaction.

An individual has essentially three alternatives after acquiring a call option. One alternative is to do nothing and let the option expire. In this case, all that is lost is what was paid for the option. A second alternative is to exercise the option before it expires as described above. A third alternative is to transfer the privilege (or option) to someone else. The way a transfer occurs is by the initial purchaser of an option selling (or writing) an option with the same identical strike price on the same underlying futures contract. Once this is done, the two transactions cancel out as far as the options market is concerned in exactly the same way futures markets operate. Since all the specifications of a given option on a given futures contract with a specified strike price are identical, it is not necessary for the original buyer and seller of an option to get together to cancel earlier transactions. As long as an individual has bought and sold the same number of identical options there is no continuing obligation as far as the clearinghouse is concerned. Since every option transaction involves one buyer and one seller, there will always be an equal number of outstanding obligations to sell a futures contract at a set price as there are privileges to buy a futures contract at the same set price regardless of how many option market transactions have occurred.

Once a decision to exercise an option is made, the clearinghouse will select one of the parties who has an outstanding obligation to fulfill that option. Once an owner of an option makes a decision to exercise the option, it is immaterial who provides a future contract at the specified strike price. Anyone who has written an option at the specified strike price (that has not been offset by the purchase of an option) is committed to fulfill the obligation assumed if called upon.

Call options can be especially attractive alternatives instead of futures contracts in situations where a firm has made a bid to sell a commodity at a specified price for some time in the future but will not know whether the offer will be accepted for a period of time. Between the time the initial bid is submitted and the final decision is made, market conditions could change drastically and the firm's profits from such a transaction could be vulnerable. Once a firm knew its bid was accepted, it could protect itself by establishing a long hedge in a futures market, but it might not want to take such action if there was uncertainty about whether the firm's bid was going to be accepted. A call option provides the flexibility to be able to get into the futures market at a specified price if it became clear at a later time that such action was desirable.

PUT OPTIONS

Previously, the discussion in this chapter has been related to options that provide the owner with the privilege of acquiring an asset (a futures contract or property) at a specified price within a designated period of time. These are not the only kinds of options on futures contracts that are traded. Another kind of option that can be acquired provides the owner the privilege of selling a futures contract at a specified price within a designated period of time. These are referred to as put options, which operate similarly to call options but are entirely separate as far as market transactions are concerned. Even though March corn call and put options may have identical strike prices for the same futures contract, they involve two entirely different privileges. Trading in corn put options occurs in one location and trading in corn call options occurs at a different location of an exchange. A transaction in the call market cannot be used to cancel an obligation taken in the put market and vice versa. Understanding put and call options is simplified by considering them as if they were two separate entities like wheat and sugar.

At first glance, the put and call terminology can be confusing. Both kinds of options involve buying and selling of privileges and it may be difficult to remember which option conveys the privilege of buying something and which one conveys the privilege of selling. One simple way to help keep the distinction straight until sufficient familiarity is gained with the concepts is to think about the association of the first letters in *c*all and *p*ut with *b*uy and *s*ell. Call options provide the privilege to *b*uy, whereas *p*ut options provide the privilege to *s*ell. Remembering the closeness of *c* and *b* as well as the closeness of *p* and *s* helps to keep the terminology straight.

Someone who obtains a put option has the privilege or opportunity of selling a futures contract at a specified price if they choose to do so anytime before the put option expires. As far as agricultural producers are concerned, put options are somewhat similar to government programs that guarantee a minimum price for a commodity at harvest time. If the market price is above the minimum price guaranteed by the government, producers do not sell their output to the government.

On the other hand, if market price is less than the price guaranteed by the government, producers have the opportunity of selling the commodity to the government at the guaranteed price. One difference between put options and government price support programs, however, is that producers have to pay for put options. The existence of a well-organized put options market has the potential of providing an alternative way of guaranteeing a minimum price to producers.

Almost everything about the way put options and markets operate is similar to the previous discussion about call options. The only difference is the kind of privilege or right that is being purchased and sold. Anytime someone buys a put option, some other party is willing to take the other side of the transaction. Writers of put options obligate themselves to buy a futures contract at a specified price if the owner of the put option should decide to exercise it. In cases where options are not exercised, writers may never be called upon to fulfill their obligations and get to keep the premium. Buyers of put options (just like call options) have limited risk, since the most they can lose on a transaction is the premium they pay. Potential profits to buyers of put options are tied directly to how far the price of the futures contract falls relative to the strike price. Exercising a put option is equivalent to the owner of the option establishing a short position in the futures market at the same time a writer is assigned a long position. Thus, the exercise of a put option is equivalent to transforming the initial option transaction into opposite positions in the futures market at the strike price specified by the option. Of course, it is not necessary for an owner of a put option to exercise it to profit. If market developments cause a particular kind of put option to become more valuable, that right could be transferred to someone else by selling a put option.

Put Option Premiums

The way in which premiums of put options vary with the strike prices is quite different from that of call options. This can be seen by looking at the additional array of premiums reported for a given set of options on a particular futures contract. For example, premiums for various put options on a particular day in Table 16–2 clearly indicate that the premiums increase with the strike price instead of decreasing as in the case of call options. This is consistent with individuals' willingness to pay more for a right to sell a futures contract at a higher price.

Premiums for put options can be decomposed into an intrinsic value and time value just like the premiums for call options. The intrinsic value for a put option is the difference between the strike price and the current futures price. For example in the case of a March put option with a strike price of 190 in Table 16–2, 8¾ cents of the 14¼-cent premium is the intrinsic value associated with the privilege of being able to sell a March futures contract at a price of 190 cents per bushel when the current price of the March futures contract is 181¼ cents per bushel. In this particular example, an intrinsic value of 8¾ cents implies that market participants have decided that the time value of this privi-

TABLE 16-2

PREMIUMS FOR CORN OPTIONS AT THE CHICAGO BOARD OF TRADE ON A PARTICULAR DAY

	Calls			Puts		
Strike Price	Dec	March	May	Dec	March	May
150	21¾			5/8	3/4	
160	12¾			1½	1½	2
170	5¾	14		4¾	4½	5
180	2½	9	14	10¾	8	8
190	1	5½	9¼	19¼	14¼	13¾
200	⅜	3¼	6½	29	23	19

lege is 5½ cents (i.e., the difference between the premium and intrinsic value). Any change in the current price of the future contract underlying this particular option will cause a change in the intrinsic value and a likely change in the overall premium.

The intrinsic value of any put option that has a strike price less than the current price of the futures contract is zero. Thus, the premiums in Table 16–2 for all of the March put options with strike prices equal to 180 or less reflect only the time values of the options because the current price of March futures was 181¼ cents. Time values of put options have two characteristics similar to those of call options. First, time values tend to be largest for the options with strike prices closest to the current futures prices. Second, time values tend to be larger for more distant maturing futures contracts.

Producer Uses of Options

Options provide producers with an alternative to using forward contracts or the futures market to reduce some of the uncertainty about the price that will be received when they are ready to sell their products. A forward contract may provide a firm commitment about price to be received or sometimes the contract will specify a differential from whatever the price of a given futures contract turns out to be on the day the commodity is delivered.[2]

As discussed in the previous chapter, taking a short position in the futures market when production begins can help producers lock-in or target price their output. This market strategy does not guarantee an exact price, however, because of basis uncertainty. Using the futures market can help a producer protect himself in case prices turn out to be much lower than anticipated when production decisions are made. Selling short in the futures market, however, prevents producers from enjoying as much revenue as they might have received if the market price turns out better than expected. Of course, once it becomes clear that prices definitely are going to be better than expected, producers can decide to close out their position in the futures market and assume the risk of taking whatever cash price occurs when they are ready to sell their output.

The existence of put option markets provide producers with the opportunity to buy a little time before they have to decide whether they really want to assume a short position in a futures contract. Put options provide producers with the opportunity to get some price protection if at a later time they decide it is desirable. On the other hand, producers are under no particular obligation to exercise a put option. Later, if it appears advantageous to take a chance on whatever happens to the cash price, the maximum loss from having purchased a put option is the premium paid for the option. Alternatively, the option could be exercised if subsequently it appears advantageous to get locked into a short position in the futures market at the strike price specified in a put option.

Acquiring additional flexibility in decision making through put options should be evaluated in terms of potential benefits and costs since options are not free. Differences in the price realized by producers based on alternative marketing decisions is graphically depicted in Figure 16–3. The 45-degree line in the diagram illustrates the relationship between the market price and the price someone would receive if they took no action to establish a price until a product is ready to be sold. This type of individual receives whatever price exists in the cash market at the time of sale. If market price is low, a low price is realized. Similarly, if the market price is high, a high price is received. If market prices fluctuate significantly over time, this kind of strategy can produce substantial variation in gross and/or net returns.

The horizontal line in Figure 16–3 that intersects the vertical axis at point A represents the price that could be realized by an individual who used some type of forward or futures contract to lock-in a price at a given level prior to the time the product is sold. In this case, the realized price will be basically the price that was locked-in regardless of what the cash price is when the product is sold. This type of strategy may not always produce the targeted price exactly because of basis uncertainty, but it is likely to lead to more certain prices than taking a chance on the cash market. The actual price at which producers can lock-in will vary with market conditions, but once the decision is made it does not depend on subsequent developments in the cash market. If the futures price starts rising after a short position has been established, potential losses in the futures market may cause margin calls before a producer can recoup any potential gain from sale of the physical commodity at higher than anticipated prices.

The remaining parts of Figure 16–3 illustrate the possible outcomes of using a put option. The option is assumed to have a strike price of S and a premium equal to the vertical distance represented by distance between points O and R. The latter amount is a fixed cost per unit of the commodity that will reduce returns from what they otherwise would be if the resulting cash price (and futures price at maturity) turn out to be higher than S. If the cash and futures price turn out to be higher than the strike price, there would be no interest in

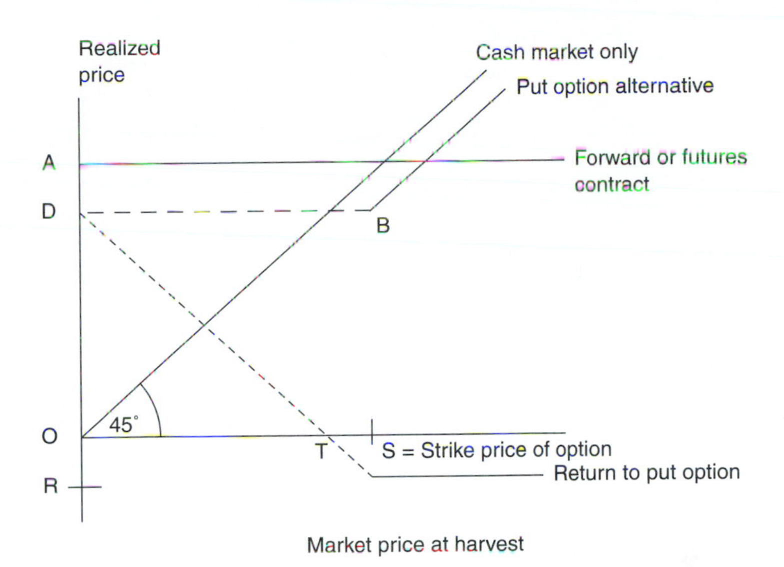

FIGURE 16–3 Comparison of realized price versus market price for alternative market strategies.

exercising the option and the producers who had selected this strategy would not fare as well as those who depended totally on the cash market determining their eventual income. This is illustrated by the returns involving an option being below the cash market alternative by the amount of the premium for prices greater than S. In the case of high prices the options strategy, however, is better than a forward or futures contract.

If the market price turns out to be below the strike price of the put option, there will be some value to exercising the option. In fact at some price T (less than S), the return on the option would be enough to offset the initial premium and producers would break even as far as the option is concerned. For all market prices less than T, the option becomes increasingly valuable because of its increasing intrinsic value. The potential return to the option is illustrated by the dashed line DT.

The actual return realized by a producer who had purchased a put option is the money that is earned from the option as well as what is realized from the sale of the commodity. Thus, if the cash market price was zero, he would still realize a return of OD per unit because of the value of the option. At higher prices, a smaller return to the option is offset by some return from the cash market. Thus the total realized return per unit of commodity is obtained by vertically adding the return from the put option to the return from the cash market. For all market prices below S, the return per unit of the commodity does not fluctuate. At high prices, however, producers have the potential of getting greater returns than what is possible through forward contracting or taking a short position in the futures market.[3]

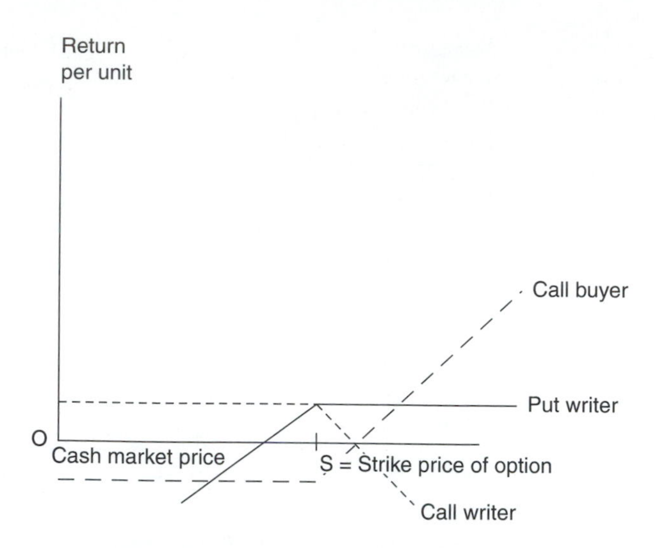

FIGURE 16–4 Alternative returns to writers of put option and buyers and writers of call option.

RETURNS TO CALL OPTIONS AND WRITERS

Having illustrated the relationship between the return to a put option and cash price in Figure 16–3, it is easy to use the same concepts to show the net returns to purchasers of call options and writers of either put or call options. First, the return to the writer of a put option with a strike price S would be just the opposite of the return to the person who purchased the option, Figure 16-4. For example, the maximum potential loss per unit to a purchaser of a put option in Figure 16–3 was the premium. This becomes the maximum return per unit to the writer of the put option and the maximum potential loss per unit is the strike price less the initial premium. At very low prices it may be very costly for writers of put options to fulfill their obligations.

The costs and returns to buyers and writers of call options with a strike price of S are illustrated by the dashed lines in Figure 16–4. For a buyer of a call option, the premium is the maximum net loss and losses decrease until the cash price exceeds the strike price. The returns from exercising a call option increase at prices above the strike price. The cost and returns to writers of call options vary in exactly the opposite manner as the relationships for purchases of call options.

SUMMARY

Call and put options on futures contracts provide some additional alternative strategies that producers and other firms need to evaluate in making marketing decisions. The trading of agricultural options in the United States has expanded rapidly after being reinstated in the 1980s following a 1936 congressional ban on organized trading of options.

A call option provides the privilege of purchasing a particular futures contract at a set price during a specified period of time. The trading of options is very similar to the trading of futures contracts, but there are some differences in buyer and seller privileges and responsibilities. For example, option buyers pay a particular amount of money (a premium) for the privilege of being able to purchase a futures contract at a fixed price, but are under no obligation to take any further action. On the other hand, option sellers or (writers) receive and get to keep the premium for granting the privilege. Writers, however must post margin money to meet any potential loss if called upon to deliver a futures contract at the price specified by the option. Buyers can exercise options they purchase, sell them to someone else, or let them expire. Option writers can eliminate continuing responsibilities to the options market by purchasing an equivalent number of options with the same strike price as what they sold.

The price (or premium) of call options is determined by the willingness to purchase and offer the privilege of purchasing a futures contract at a fixed (or strike) price. If a call option has a strike price less than the current futures price, the option is said to be in-the-money. Conversely, if the strike price is greater than the current futures price, the option would be out-of-the-money. An option with a strike price equal to the futures price would be referred to as at-the-money.

Option premiums consist of two components: the time value and the intrinsic value. The intrinsic value is the difference between the futures price and the strike price and exists only for in-the-money options. Premiums for out-of-the-money and at-the-money options reflect only time values associated with the possibilities that futures prices may change by a sufficient amount to make the options profitable before they expire. Options generally expire 1 month prior to expiration of trading the underlying futures contract.

Put options are similar to call options except they provide the privilege of being able to sell, rather than purchase, a specific futures contract at a fixed price during a specified period of time. Writers of put options assume the obligation of being willing to purchase a futures contract at a fixed price if called upon. Put options are similar to government programs guaranteeing a minimum price to producers except they are not free. Put options provide producers with additional time before they have to decide if they really want to assume a continuing obligation in the futures market. This additional flexibility comes at a cost. If the eventual cash market price exceeds a producer's initial price expectation, a higher return would be realized with a put option than using forward contracts or the futures market to lock-in a price.

The net cost or profit resulting from buying (or selling) call (or put) options depends on what happens to the prices of commodities after the option is acquired. An increase in cash (and futures) prices tends to benefit owners of call options and writers of put options. On the other hand, decreasing prices would tend to benefit writers of call options and buyers of put options. The way in which the net return from buying and selling options varies with changes in commodity prices can be graphically illustrated.

QUESTIONS

1. Briefly describe the alternative courses of action that an owner of a call option has at his or her disposal.

2. What kind of price movements would be detrimental to a writer of a call option on crude oil futures? Explain why.

3. How would a decrease in the price of a given futures contract be expected to affect the premiums of call options on the same futures contract? Briefly explain the rationale behind your answer.

4. How would an unexpected increase in exports of pork to Japan in December be expected to affect the premiums of call options on April hog futures? Briefly explain the rationale behind your answer.

5. Assume the premium of a call option with a strike price of $5.00/bu is $.35/unit when the current futures price is $5.25. Is this particular option in-the-money, out-of-the-money, or at-the-money? Explain your answer.

6. Identify the various alternatives that an owner of a $4.50 put option on March wheat futures has once the premium necessary to acquire the option has been paid. What alternatives does the writer of such an option have?

7. Assume that a soybean producer has acquired a put option with a strike price of $6.50/bu on September bean futures for a premium of $.35/bu. Briefly describe what kind of changes in the soybean market between now and next fall would result in the producer being better off having purchased the put option than using the September futures market or a forward cash contract to lock in a price of $6.50/bu?

REFERENCES

Commodity Futures Trading Commission. *1998. Annual report.* http://www.cftc.gov/annualreport98/contents.htm

Horner, D. L., and E. Moriarty. 1983. The CFTC options pilot program: A progress report. *Education Quarterly* 3(3): 9–14. Division of Economics and Education, Commodity Futures Trading Commission: Washington, D.C.

Kenyon, D. 1994. *Farmers guide to trading agricultural commodity options.* Agric. Info. Bull. 463. National Economics Division, Economic Research Service, USDA. Washington D.C.

NOTES

1. If a buyer desired to have the option of purchasing more than one futures contract at a specific price, multiple options would need to be purchased.

2. This kind of contract can guarantee a market for commodities, but does not guarantee a specific price because it depends on what happens in the futures market.

3. The relationships illustrated in Figure 16–3 consider only the cost and returns of a put option assuming it either is exercised or expires. No consideration is given to costs and returns that might occur if the initial transaction is offset by a subsequent transaction in the same option before expiration.

GLOSSARY

Agribusiness A term that refers to an individual firm or set of activities associated with the production and marketing of agricultural products.

Agricultural industrialization The coordination of subsequent production and marketing activities associated with agricultural products by a firm through ownership or forward contracting with other parties.

At-the-money An option with a strike price equal to the current futures price. See *in-the-money* and *out-of-the-money.*

Backwardation A term used to describe situations in which near-term futures contracts have a higher price than more distant contracts. See *Contango.*

Basis The difference between a cash price for a particular market at a specific time and the price of a futures contract (often the one closest to maturity) for the same commodity. Also used to refer to the difference between cash and futures prices in the opposite order.

Bearish market A term used to characterize a market with downward price movements resulting from relatively stronger interest in selling than buying.

Bullish market A term used to characterize a market with upward price movements resulting from relatively stronger interest in buying than selling.

Call option An agreement that provides the owner the privilege of purchasing a specific futures contract at a fixed price during a specified period of time if so desired. See *Put option.*

Cartel A group of firms using formal or informal arrangements to act as a single entity in controlling price and/or total volume of business.

Ceteris paribus A phrase frequently used in economic analysis to indicate all other things are assumed to remain unchanged.

Clearinghouse A separate agency associated with each futures exchange that keeps track of all traders' obligations and makes sure they are fulfilled when trading in a given contract ends.

Closing price The price at which the last transactions of futures contracts occurs on a given day. See *Settle price.*

Coefficient of price flexibility The percentage change in price associated with each 1% change in quantity demanded along a demand relationship. Similar to the reciprocal of the price elasticity of demand.

Commodity Futures Trading Commission (CFTC) A governmental agency that regulates all trading practices on futures markets in the United States.

Competitive imports Commodities purchased from other countries that are similar to those available from domestic production.

Complementary imports Commodities that are generally not produced in a given country, but obtained through international trade.

Consumer sovereignty An economic concept indicating that individual and collective consumer desires and willingness to purchase alternative goods and services ultimately determines the way available resources in an economy are used if markets are permitted to function.

Contango A term used to describe when near-term futures contracts have a lower price than more distant contracts. See *Backwardation.*

Cross-price elasticity of demand The percentage change in quantity demanded of one good associated with each 1% change in the price of another good.

Delivery point The explicit location where ownership of a physical commodity is transferred to settle remaining obligations of futures contracts after trading in a given contract expires.

Demand A schedule of alternative quantities of an item that would be purchased at alternative prices by one or more individuals. See *Supply.*

Double auction A process in which participants enter bids to buy or sell and multiple exchanges of ownership can occur simultaneously.

Dutch auction A bidding process involving successively decreasing prices. See *English auction.*

Eight-firm concentration ratio The share of a total market accounted for by the eight largest firms in an industry. See *Four-firm concentration ratio.*

Embargo A government decision to prohibit the exporting of one or more products to one or more destinations.

Engel's law The observation that expenditures for many food products increase less rapidly than household income (income elasticity < 1) implying an inverse relationship between the proportion of a household's income spent on food and the amount of income.

English auction A bidding process involving successively higher bids. See *Dutch auction.*

Exchange rate The rate at which the units of one currency can be converted into units of another currency.

Export Enhancement Program (EEP) The use of government funds to pay part of the cost of exporting specific goods to particular countries.

First-price auction An auction in which the winning bidder pays the price bid. See *Second-price auction.*

Formula pricing The specification of an explicit procedure for determining the price or remuneration that will be paid for the transfer of ownership rights in lieu of identifying a specific price.

Forward contract An agreement between two parties that specifies terms and conditions related to a transfer of ownership or completion of certain activities at a specified time under specific conditions.

Four-firm concentration ratio The share of a total market accounted for by the four largest firms in an industry. See *Eight-firm concentration ratio.*

Free Alongside Ship (FAS) A term that refers to a sellers obligation to deliver specified items to a carrier designated by the buyer.

Fundamental analysis (as applied to futures and options markets) The use of market information about factors influencing current and prospective demand and supply conditions about a commodity to predict likely prices. See *Technical analysis* (as applied to futures and options markets).

Futures contract An agreement to accept or provide a specific quantity of a specific commodity at a specific location at a specified time for a specified price.

General Agreement on Tariffs and Trade (GATT) A framework initiated in 1948 and continued for nearly 50 years as a way for nations to reach agreements on particular issues such as tariffs that affect international trade. Replaced by World Trade Organization in 1995.

Group bargaining Negotiations between one party and another entity representing the combined interests of a set of buyers or sellers.

Income elasticity of demand The percentage change in demand of an item associated with each 1% increase in income.

Incoterms A set of terms developed and periodically revised by the International Chamber of Commerce to facilitate understanding and reduce uncertainty about alternative ways buyers and sellers assume responsibilities for transportation, insurance, and other aspects of international transactions.

Infant industry A term often used as a justification for a trade barrier to prevent imports of particular goods in order to allow a "new" industry to become established and be able to compete on the world market.

Isoprice line The collection of all points in a geographical market area at which producers receive (or consumers pay) the same price for a commodity.

In-the-money An option with some intrinsic value, that is, a call option with a strike price below the current futures price or a put option with a strike price above the current futures price. See *At-the-money and out-of-the-money.*

Intrinsic value of options The positive difference between current price of a futures contract and a strike price of a call option on the same futures contract. In the case of a put option it is the positive difference between the strike price of the option and the current price of the underlying futures contract. See *Time value of options.*

Limit order The instructions issued to a broker to buy or sell a certain number of futures contracts at a specified or better price.

Long hedge The purchase of futures contracts at the same time a commitment has been made to deliver a physical commodity to a specific buyer at a fixed price at some point in the future. See *Short hedge.*

Margin call The requirement that additional money or assets must be added to a trader's account in order to maintain a particular position in the futures market to replace some of the value that was lost because of adverse price changes after the initial position was established.

Margin requirement The amount of money or assets required by a broker from a trader before buying or selling a specific quantity of futures contracts.

Marginal cost The change in total costs associated with each additional unit of product that is produced, handled, sold, or purchased (i.e., marginal costs exist for each activity in the marketing chain between production and consumption).

Marginal revenue The change in total revenue associated with each additional unit sold.

Marked-to-market The practice of using prices of futures contracts at the close of trading each day to recalculate each trader's financial position with respect to all continuing commitments in futures markets. See *Settlement price.*

Market order The instructions issued to a broker to buy or sell a certain number of futures contracts at whatever price exists as soon as the order can be executed.

Monopolistic competition A term used to describe a market structure in which there are many sellers of differentiated products.

Monopoly A term used to describe a market structure in which there is only one seller of a given product.

Monopsony A term used to describe a market structure in which there is only one buyer of a given product.

Nontariff trade barriers Governmental rules or regulations other than tariffs that restrict the amount of international trade that otherwise would occur.

North American Free Trade Agreement (NAFTA) The set of arrangements that began a phased reduction or elimination of import tariffs and other trade barriers in 1994 between Canada, Mexico, and the United States that extended and expanded provisions that were established in 1989 to expand international trade between Canada and the United States.

Oligopoly A term used to describe a market structure in which there are only a few sellers.

Oligopsony A term used to describe a market structure in which there are only a few buyers.

Open interest The number of futures contracts of a given commodity representing continuing obligations to buy or sell at the close of trading each day that have not been cancelled by taking an opposite position in the market.

Option writer One who grants a privilege in exchange for money for selling an option and assumes responsibility for fulfilling the terms of the option if called upon.

Out-of-the-money A call option with a strike price above the current futures price or a put option with a strike price below the current futures price. See *At-the-money and in-the-money.*

Predatory pricing A term used to characterize deliberate attempts by one firm to sell products at less than cost in hopes of increasing its market share and gaining a longer-term market advantage.

Price determination The use of economic concepts of demand and supply to indicate how the terms of trade (or price) associated with transactions involving an exchange of ownership become established.

Price discovery A mechanism by which buyers and sellers find and agree on a mutually satisfactory price for a transfer of ownership of something.

Price discrimination A deliberate decision to set different prices for similar products purchased by different groups of buyers who are perceived to have different price elasticities of demand.

Price elasticity of supply The percentage change in quantity supplied for each 1% change in price along a supply relationship.

Price elasticity of demand The percentage change in quantity demanded for each 1% change in price along a demand relationship.

Production contract A formal agreement between a producer and a buyer indicating the amount of merchandise that will be delivered and final terms of payment often agreed to before production is initiated.

Put option An agreement that provides the owner the privilege of selling a specific futures contract at a fixed price during a specified period of time if so desired. Also see *Call option.*

Round turn A sequence of two transactions involving futures transactions consisting of initially getting into the market and subsequently closing out the position.

Scalper A trader who is willing to buy and/or sell futures contracts (or other items) to profit from small changes in prices.

Sealed bids An auction in which offers to buy or sell are submitted secretly and all bids are evaluated simultaneously.

Second-price auction An auction in which the winning bidder pays the price of the next highest bid. See *First-price auction.*

Self-sufficiency The extent to which a household or country produces all products used by its citizens and does not take advantage of the principle of comparative advantage and international trade to enhance the country's economic welfare.

Settle price A price determined by officials of futures markets based on the range of prices of transactions at close of trading used for determining gains and losses on continuing open positions in market. See *Closing price* and *Marked-to-market.*

Short hedge The sale of futures contracts at the same time one has purchased (or owns) a certain amount of the same commodity in order to guard against unexpected changes in price of the physical commodity. See *Long hedge.*

Speculation Deciding to buy or sell an item anticipating to profit from an anticipated change in price. Often used in conjunction with trading in futures and options markets, but also applicable with respect to producing, owning, or committing to the purchase of physical commodities whose future value is uncertain.

Spreading in futures contracts Taking an opposite position in different contracts for the same commodity in anticipation of being able to profit from a change in the price difference of the two contracts. See *Straddling in futures contracts.*

Stop order The instructions issued to a broker to buy or sell a certain number of futures contracts if the price reaches a particular level. Often used to establish the conditions under which a trader desires to liquidate a position in the market after achieving a given level of profit or to make sure losses do not exceed a specified amount.

Straddling in futures contracts Taking an opposite position in two futures markets in anticipation of being able to profit from a change in the price difference in the two markets. See *Spreading futures contracts.*

Strike price The price specified in a call or put option at which futures contracts can be purchased or sold.

Supply A schedule of alternative quantities of an item that will be produced and sold at alternative price by one or more individuals. See *Demand.*

Tariff A tax placed on particular products being imported (or exported) by a given country to generate governmental revenue or as a means of trying to influence the volume of trade.

Tariffication The use of tariffs instead of quotas and other restrictions on shipments of commodities between countries.

Tariff-rate quota (TRQ) A system of variable tariffs that includes a lower rate until a certain quantity of imports occurs and then a higher tariff rate becomes applicable.

Technical analysis (as applied to futures and options markets) The measurement and identification of various characteristics of price movements to forecast the kind of future price movements. See *Fundamental analysis (as applied to futures and options markets).*

Time value of options The amount of a premium that does not reflect intrinsic value of an option. See *Intrinsic value of options.*

Transaction costs The value of time and other expenses incurred by individuals in arranging and finalizing transfers of ownership.

Variable levy A tax on imported items that varies inversely with world prices that attempts to stabilize domestic prices.

Vertical integration See *Agricultural industrialization.*

World Trade Organization (WTO) An organization established in 1995 as a successor to GATT to implement and monitor international trade procedures, settle disputes, and encourage additional trade negotiations among nations.

INDEX